16th Annual International Astrophysics Conference (AIAC 2017)

Turbulence, Structures, and Particle Acceleration Throughout the Heliosphere and Beyond

Journal of Physics: Conference Series Volume 900

Sante Fe, New Mexico, USA
6 – 10 March 2017

ISBN: 978-1-5108-4898-6
ISSN: 1742-6588

Printed from e-media with permission by:

Curran Associates, Inc.
57 Morehouse Lane
Red Hook, NY 12571

Some format issues inherent in the e-media version may also appear in this print version.

Printed by Curran Associates, Inc. (2017)

For permission requests, please contact the Institute of Physics at the address below.

Institute of Physics
Dirac House, Temple Back
Bristol BS1 6BE UK

Phone: 44 1 17 929 7481
Fax: 44 1 17 920 0979

techtracking@iop.org

Additional copies of this publication are available from:

Curran Associates, Inc.
57 Morehouse Lane
Red Hook, NY 12571 USA
Phone: 845-758-0400
Fax: 845-758-2633
Email: curran@proceedings.com
Web: www.proceedings.com

Table of contents

Volume 900

16th Annual International Astrophysics Conference: "Turbulence, Structures, and Particle Acceleration Throughout the Heliosphere and Beyond"

6–10 March 2017, Santa Fe, New Mexico, USA

(previous editions: AIAC 2015, AIAC 2016)

Accepted papers received: 13 September 2017
Published online: 26 September 2017

Preface

Papers

IOP Conf. Series: Journal of Physics: Conf. Series **900** (2017) 011001 doi:10.1088/1742-6596/900/1/011001

Preface

G P Zank
Editor, *Proceedings of the 16th Annual International Astrophysics Conference,*
Center for Space Plasma and Aeronomic Research (CSPAR) and Department of Space Physics, University of Alabama in Huntsville, Huntsville AL 35805, USA

E-mail: garyp.zank@gmail.com

The 16th Annual International Astrophysics Conference was held at the La Posada de Santa Fe hotel, Santa Fe, New Mexico, USA from March 6 to 10, 2017. The meeting entitled, "Turbulence, Structures, and Particle Acceleration Throughout the Heliosphere and Beyond", addressed the remarkable ability of nature to accelerate charged particles to extraordinarily high energies. This problem remains, after 100 years, one of the outstanding puzzles of solar and astrophysical plasmas. Diffusive shock acceleration (DSA) is thought to energize charged particles at shock waves. Steady-state DSA predictions include the particle intensity peaking at the shock, after which it is constant, and that the accelerated particle distribution is a power law with an index depending only on the shock compression ratio. DSA predictions are often but not always met. The anomalous cosmic ray spectrum was observed to peak behind the heliospheric termination shock and to possess a spectrum far harder than predicted by classical DSA theory. This is frequently true of shocks in the inner heliosphere and in astrophysical settings. Shocks are effective in generating magnetic turbulence and structures downstream and amplifying pre-existing turbulence, all factors in the further energization of charged particles. Furthermore, certain regions such as the heliospheric current sheet naturally produce complex turbulent environments in which numerous structures are present. Not surprisingly, in these regions observed energetic particle events, sometimes called anomalous solar energetic particle events, have characteristics quite unlike those predicted of typical impulsive or gradual solar energetic particle events. The purpose of this meeting was to explore the role of turbulence and structures (including magnetic reconnection-related processes, shock waves, etc.), in the acceleration of particles throughout the heliosphere and beyond. The meeting addressed current and past observations from spacecraft in the inner heliosphere, the distant heliosphere and very local interstellar medium, explored expectations and predictions for missions such as Solar Orbiter and Solar Probe Plus, now renamed as the Parker Solar Probe, and of course remote observations.

Finally, we thank Adele Corona and ICNS for her continued excellent organization of the AIAC meetings and her help in providing the logistical support for this volume of papers.

IOP Conf. Series: Journal of Physics: Conf. Series **900** (2017) 011002 doi:10.1088/1742-6596/900/1/011002

Peer review statement

All papers published in this volume of *Journal of Physics: Conference Series* have been peer reviewed through processes administered by the proceedings Editors. Reviews were conducted by expert referees to the professional and scientific standards expected of a proceedings journal published by IOP Publishing.

Nearly incompressible turbulence for different 2D and slab energy ratios

L Adhikari[1], G P Zank[1,2], P Hunana[1,2], R Bruno[3], D Telloni [4], Q Hu[1,2] and D Shiota[5]

[1]Center for Space Plasma and Aeronomic Research (CSPAR), University of Alabama in Huntsville, Huntsville,
AL 35899, USA
[2]Department of Space Science, University of Alabama in Huntsville, Huntsville, AL 35899, USA
[3]INAF-IAPS Instituto di Astrofisica e Planetologia Spaziali, Via del Fosso del Cavaliere 100, 00133 Roma, Italy
[4]INAF - Astrophysical Observatory of Torino, Via Osservatorio 20, 10025 Pino Torinese, Italy
[5]Institute for Space-Earth Environmental Research, Nagoya University, Nagoya, Aichi 464-8601, Japan

E-mail: la0004@uah.edu

Abstract. Zank et al 2017 [1] developed equations describing the transport of 2D and slab turbulence based on nearly incompressible magnetohydrodynamic (NI MHD) theory. Zank et al [1] discussed preliminary solutions of the model equations, and Adhikari et al [2] provided a detailed analysis of the model equations. Their analyses of the 2D and slab turbulence transport model equations were based on the ratio of the inner boundary values of the 2D and slab energies being of the order 80:20. Here, we investigate the effect of choosing the 2D and slab energy ratios to be 80:20, 70:30, 60:40, and 55:45 for the boundary energy values, and solve the 2D and slab turbulence transport model equations. We compare our theoretical results with observations made by Voyager 2 and Ulysses. We find that the 2D and slab turbulent quantities such as the energy in forward and backward propagating modes, the fluctuating kinetic and magnetic energy, for example, are different for different boundary values, but show similar trends with increasing heliocentric distance. In addition, we also investigate the model equations by including a shear source of turbulence in the slab turbulence transport equations. This was not considered by [1] and [2]. We find that the slab turbulent quantities behave more reasonably in the presence of both shear and pickup ion sources of turbulence.

IOP Conf. Series: Journal of Physics: Conf. Series **900** (2017) 012001 doi:10.1088/1742-6596/900/1/012001

1. Introduction

The plasma beta β is defined as the ratio of the plasma pressure and the magnetic pressure. The plasma beta of the solar wind and the solar corona is of the order of $\beta \sim 1$ and $\beta \ll 1$ when away from the heliospheric current sheet region. Therefore, a turbulence transport model that reflects plasma beta values of order $\beta \sim 1$ or $\beta \ll 1$ is required to describe the evolution of fluctuations in the solar wind, in contrast to previous 3D incompressible turbulence transport models [3–17], which are appropriate to a plasma beta $\beta \gg 1$ [18]. The nearly incompressible magnetohydrodynamic (NI MHD) theory allows for the formulation of turbulence transport model equations in the $\beta \sim 1$ or $\beta \ll 1$ regime [18, 19]. In such a description, the turbulence can be decomposed into a dominant 2D and minority slab component so that the total turbulence field is a superposition of 2D and slab turbulence. Zank & Mattaheus [19] and Bieber et al [20] showed theoretically and observationally that the ratio of the 2D and slab turbulence energy is 80:20. This ratio of the energies can have different values in the slow and fast solar wind [21]. Smith et al [22] showed that the ratio of the energy between the two populations of slab (1D) and 2D wavevectors is roughly equally distributed at high latitudes during solar minimum.

Initially, a form of NI MHD was developed to describe the observed $k^{-5/3}$, (where, k is a wave number), spectrum of the interstellar electron density fluctuations [23, 24]. The NI theory was mostly developed in the early 1990s for homogeneous flows [18, 19, 25–29] before being extended to inhomegeneous flows [30–33]. Zank et al 2017 [1] derived the turbulence transport model equations using an Elsässer variables formulation for both homogeneous and inhomogeneous flows. Zank et al 2017 [1] introduced a further higher-order coupling of the 2D and slab turbulence model equations for inhomogeneous flows, which is discussed further by Zank et al 2017 (this issue) for homogeneous flows. Zank et al 2017 [1] show that the Alfvén velocity does not appear in the leading order 2D turbulence transport model equations in contrast to models that start from 3D incompressible MHD (for which $\beta \gg 1$) [34–36], that have the Alfvén velocity in their 2D turbulence transport equations. The 2D turbulence transport equations of Zank et al 2017 [1] based on NI MHD [18, 19, 30], shows that the fluctuations lie in a plane orthogonal to a direction of the large scale magnetic field, and are convected by the large scale solar wind speed. Further, the 2D transport equations for an inhomogeneous flow do not introduce any type of Alfvén critical point as do other turbulence transport model equations.

In this manuscript, we solve the Zank et al [1] 2D and slab turbulence transport model equations for a variety of energy ratios 80:20, 70:30, 60:40, and 55:45, and compare the numerical solutions with Voyager 2 and Ulysses measurements. The Voyager 2 observed values are obtained from Adhikari et al [2], and the Ulysses observed values are obtained from Bavassano et al [37] and Yang et al [38]. Zank et al [1], and Adhikari et al [2] presented numerical solutions of the 2D and slab model equations. Both papers used a boundary condition in which the 2D and slab turbulence energy has the ratio 80:20. In Zank et al [1] (and also [2]), the coupled turbulence transport equations included a shear source of turbulence in the 2D turbulence transport equations only, and a pickup ion source of turbulence in the slab turbulence transport equations. Here, along with solving the [1] turbulence transport equations for different energy ratios, we also solve the

IOP Conf. Series: Journal of Physics: Conf. Series **900** (2017) 012001 doi:10.1088/1742-6596/900/1/012001

coupled equations when both sources of turbulence are included in the slab turbulence transport equations. Section 2 describes the 2D and slab turbulence transport model equations. Section 3 presents a comparison of the numerical solutions of the 2D and slab turbulence transport equations with Voyager 2 and Ulysses observations, where a shear source is included in the 2D turbulence equations and a pickup ion source in the slab turbulence equations. Section 4 presents a comparison of the numerical solutions with observations, when a shear source is included in the 2D turbulence model, and shear and pickup ion sources are included in the slab turbulence model. Finally, Section 5 presents some discussion and conclusions.

2. 2D and Slab Model Equations

In this section, we present the 1D steady-state coupled 2D and slab turbulence transport model equations. The details of the derivation of the 2D and slab model equations are given in [1] (see also [2]). The 1D steady state 2D turbulence transport model equations in a spherical coordinate system r are [1]

$$U\frac{d}{dr}\langle z^{\infty\pm 2}\rangle + \frac{U}{r}\langle z^{\infty\pm 2}\rangle + \frac{U}{r}E_D^{\infty} + \frac{1}{r}\langle z^{\infty\pm 2}\rangle^{1/2}(\langle z^{\infty\pm 2}\rangle - E_D^{\infty}) = -2\frac{\langle z^{\infty\pm 2}\rangle\langle z^{\infty\mp 2}\rangle^{1/2}}{\lambda_{\infty}^{\pm}} + 2\frac{\langle z^{*\pm 2}\rangle\langle z^{*\mp 2}\rangle^{1/2}}{\lambda^{*\pm}} + 2C_{sh}^{\pm}\frac{r_0|\Delta U|V_{A0}^2}{r^2}; \tag{1}$$

$$U\frac{dE_D^{\infty}}{dr} + \frac{U}{r}E_D^{\infty} + \frac{U}{r}E_T^{\infty} + \frac{1}{2r}\left(E_D^{\infty} - \langle z^{\infty+2}\rangle^{1/2}\langle z^{\infty-2}\rangle^{1/2}\right)\left(\langle z^{\infty+2}\rangle^{1/2} + \langle z^{\infty-2}\rangle^{1/2}\right) = -E_D^{\infty}\left(\frac{\langle z^{\infty-2}\rangle^{1/2}}{\lambda_{\infty}^{+}} + \frac{\langle z^{\infty+2}\rangle^{1/2}}{\lambda_{\infty}^{-}}\right) + E_D^{*}\left(\frac{\langle z^{*-2}\rangle^{1/2}}{\lambda^{*+}} + \frac{\langle z^{*+2}\rangle^{1/2}}{\lambda^{*-}}\right) + 2C_{sh}^{E_D}\frac{r_0|\Delta U|V_{A0}^2}{r^2}; \tag{2}$$

$$U\frac{dL_{\infty}^{\pm}}{dr} + \frac{U}{r}L_{\infty}^{\pm} + \frac{U}{2r}L_D^{\infty} - \frac{1}{2r}\langle z^{\infty\pm 2}\rangle^{1/2}(L_D^{\infty} - 2L_{\infty}^{\pm}) = 0; \tag{3}$$

$$U\frac{dL_D^{\infty}}{dr} + \frac{U}{r}L_D^{\infty} + \frac{U}{r}(L_{\infty}^{+} + L_{\infty}^{-}) + \frac{1}{2r}\left(L_D^{\infty}\left(\langle z^{\infty+2}\rangle^{1/2} + \langle z^{\infty-2}\rangle^{1/2}\right) - 2L_{\infty}^{+}\langle z^{\infty-2}\rangle^{1/2} - 2L_{\infty}^{-}\langle z^{\infty+2}\rangle^{1/2}\right) = 0, \tag{4}$$

where $\langle z^{\infty+2}\rangle$ is the energy in forward propagating modes, $\langle z^{\infty-2}\rangle$ is the energy in backward propagating modes, E_D^{∞} is the residual energy, $L_{\infty}^{\pm}$ are correlation functions corresponding to $\langle z^{\infty\pm 2}\rangle$, and L_D^{∞} a correlation function for E_D^{∞}. The first terms on the right-hand side (RHS) of Equations (1) and (2) are the non-linear dissipation terms, the second terms are the higher order correction terms associated with the dissipation of slab turbulence and the corresponding injection of 2D energy through the 3-wave resonance condition [39]. The third terms are stream shear sources of turbulence [1, 2]. The parameters $C_{sh}^{\pm}$ and $C_{sh}^{E_D}$ are parametrized strengths of the shear source of energy in forward (+) and backward modes (-) and the residual energy (E_D), respectively. Here, $r_0(= 1\text{AU})$ is a reference location, ΔU is the difference between fast and slow solar wind speed, and V_{A0} is the Alfvén velocity at 1 AU. The normalized cross-helicity is defined as $\sigma_c^\infty(\equiv E_C^\infty/E_T^\infty)\neq 0$, where $E_C^\infty=(\langle z^{\infty+2}\rangle-\langle z^{\infty-2}\rangle)/2$ is the cross-helicity and $E_T^\infty=(\langle z^{\infty+2}\rangle+\langle z^{\infty-2}\rangle)/2$ is the total turbulent energy.

Similarly, the 1D steady-state slab turbulence transport model equations are [1]

$$\begin{aligned}(U\mp V_{A0}\frac{r_0}{r})\frac{d\langle z^{*\pm 2}\rangle}{dr}-(2b-1)\frac{U}{r}\langle z^{*\pm 2}\rangle+2(3b-\frac{1}{2})\frac{U}{r}E_D^*\pm(4b-1)\frac{V_{A0}}{r_0}\left(\frac{r_0}{r}\right)^2E_D^*\\ \pm\frac{V_{A0}}{r_0}(\frac{r_0}{r})^2\langle z^{*\pm 2}\rangle+\frac{1}{r}(\langle z^{*\pm 2}\rangle-E_D^*)\langle z^{\infty\pm 2}\rangle^{1/2}=-2\frac{\langle z^{*\pm 2}\rangle\langle z^{\infty\mp 2}\rangle^{1/2}}{\lambda_\infty^\pm}\\ -2\frac{\langle z^{*\pm 2}\rangle\langle z^{*\mp 2}\rangle^{1/2}}{\lambda_*^\pm}+2C^{*\pm}_{sh}\frac{r_0|\Delta U|V_{A0}^2}{r^2}+\frac{f_Dn_H^\infty UV_{A0}}{n_{\rm sw}^0\tau_{\rm ion}^0}\exp\left(\frac{-L}{r}\right);\end{aligned} \tag{5}$$

$$\begin{aligned}U\frac{dE_D^*}{dr}-(2b-1)\frac{U}{r}E_D^*+2(3b-\frac{1}{2})\frac{U}{r}E_T^*-(4b-1)E_C^*\frac{V_{A0}}{r_0}(\frac{r_0}{r})^2\\ +\frac{1}{2r}\left[E_D^*(\langle z^{\infty+2}\rangle^{1/2}+\langle z^{\infty-2}\rangle^{1/2})-\langle z^{*-2}\rangle\langle z^{\infty+2}\rangle^{1/2}-\langle z^{*+2}\rangle\langle z^{\infty-2}\rangle^{1/2}\right]\\ =-E_D^*\left(\frac{\langle z^{\infty-2}\rangle^{1/2}}{\lambda_\infty^+}+\frac{\langle z^{\infty+2}\rangle^{1/2}}{\lambda_\infty^-}\right)-E_D^*\left(\frac{\langle z^{*-2}\rangle^{1/2}}{\lambda_*^+}+\frac{\langle z^{*+2}\rangle^{1/2}}{\lambda_*^-}\right)+2C^{*E_D}_{sh}\frac{r_0|\Delta U|V_{A0}^2}{r^2};\end{aligned} \tag{6}$$

$$\begin{aligned}(U\mp V_{A0}\frac{r_0}{r})\frac{dL_*^\pm}{dr}-(2b-1)\frac{U}{r}L_*^\pm+2\left(\frac{3}{2}b-\frac{1}{4}\right)\frac{U}{r}L_D^*\pm\left(2b-\frac{1}{2}\right)\frac{V_{A0}}{r_0}\left(\frac{r_0}{r}\right)^2L_D^*\\ \pm\frac{V_{A0}}{r_0}\left(\frac{r_0}{r}\right)^2L_*^\pm+\frac{1}{r}(L_*^\pm-\frac{L_D^*}{2})\langle z^{\infty\pm 2}\rangle^{1/2}=0;\end{aligned} \tag{7}$$

$$\begin{aligned}U\frac{dL_D^*}{dr}-2\left(b-\frac{1}{2}\right)\frac{U}{r}L_D^*+2\left(3b-\frac{1}{2}\right)\frac{U}{r}(L_*^++L_*^-)-(4b-1)(L_*^+-L_*^-)\frac{V_{A0}}{r_0}\left(\frac{r_0}{r}\right)^2\\ -\frac{1}{r}\left[(L_*{}^--\frac{L_D^*}{2})\langle z^{\infty+2}\rangle^{1/2}+(L_*{}^+-\frac{L_D^*}{2})\langle z^{\infty-2}\rangle^{1/2}\right]=0,\end{aligned} \tag{8}$$

where $C^{*\pm}_{sh}$ and $C^{*E_D}_{sh}$ are the strengths of the shear source of turbulence. Equations (5) and (6) are different from the Zank et al [1] and Adhikari et al [2] in that these equations have an additional shear source of turbulence which were not included in [1] and [2]. We solve the coupled equations for two cases; i) $C^{*\pm}_{sh}=0$ and $C^{*E_D}_{sh}=0$, and ii) $C^{*\pm}_{sh}\neq 0$ and $C^{*E_D}_{sh}\neq 0$. The parameter $\langle z^{*+2}\rangle$ is the slab energy in forward propagating modes, $\langle z^{*-2}\rangle$ is the slab energy in backward propagating modes, E_D^* is the slab residual energy, $L_*^\pm$ are the correlation functions corresponding to $\langle z^{*\pm 2}\rangle$, and L_D^* the correlation function for E_D^*. E_T^* is the total slab turbulent energy, and E_C^* is the slab cross-helicity. The parameter $L(\sim 8$ AU) is the ionization cavity length scale, and f_D is the fraction of pickup ion energy transferred into excited waves [40]. The parameter

$n_H^\infty = 0.1\ \mathrm{cm}^{-3}$ is the number density of interstellar neutrals entering the heliosphere, $\tau_{ion}^0 = 10^6$ s is the neutral ionization time at 1 AU, $n_{sw}^0 = 5\ \mathrm{cm}^{-3}$ is the solar wind density at 1 AU, and $0 < b < 1$ is a constant. The third term on the RHS of Equation (5) is a pickup ion source of slab turbulence [8].

The 1D steady-state variance of density fluctuations is [1]

$$U\frac{d}{dr}\langle\rho^{\infty 2}\rangle + \frac{4U}{r}\langle\rho^{\infty 2}\rangle + \frac{4}{r}\langle u^{\infty 2}\rangle^{1/2}\langle\rho^{\infty 2}\rangle = -\frac{\langle u^{\infty 2}\rangle^{1/2}\langle\rho^{\infty 2}\rangle}{\lambda_u^\infty} + \eta_1\langle\rho^{\infty 2}\rangle_0\frac{r_0^2|\Delta U|}{r^3} + \eta_2\langle\rho^{\infty 2}\rangle_0\frac{U}{V_{A0}}\frac{n_H^\infty}{\tau_{ion}^0 n_{SW}^0}\exp\left(\frac{-L}{r}\right), \tag{9}$$

where η_1 and η_2 are constant, and $\langle u^{\infty 2}\rangle$ and λ_u^∞ are the variance of the velocity fluctuations and the corresponding correlation length, i.e.,

$$\langle u^{\infty 2}\rangle = \frac{\langle z^{\infty+2}\rangle + \langle z^{\infty-2}\rangle + 2E_D^\infty}{4}, \text{ and}$$
$$\lambda_u^\infty = \frac{(E_T^\infty + E_C^\infty)\lambda_\infty^+ + (E_T^\infty - E_C^\infty)\lambda_\infty^- + E_D^\infty\lambda_D^\infty}{2(E_T^\infty + E_D^\infty)}. \tag{10}$$

In Equation (9), the first term on the right-hand side is the non-linear dissipation term. It shows that the 2D incompressible velocity fluctuations $\langle u^{\infty 2}\rangle$ are responsible for the dissipation through mixing of the density fluctuations. In Equation (9), if $\langle u^{\infty 2}\rangle \sim 0$, the non-linear dissipation term (first term on the rhs of (9)) and mixing term (third term on the left-hand side) become zero. In such a case, along with the absence of a source of turbulence (second and third terms on the rhs), the variance of density fluctuations decays as r^{-4}, which is due to the expansion of the solar wind. However, $\langle u^{\infty 2}\rangle \neq 0$ ensures that the non-linear dissipation and mixing terms are non-zero, in which case the rate of dissipation of $\langle\rho^{\infty 2}\rangle$ is faster than r^{-4} [1, 2]. The 1D steady-state solar wind temperature equation is given by

$$U\frac{dT}{dr} + (\gamma - 1)\frac{2UT}{r} = \frac{1}{3}\frac{m_p}{k_B}\alpha\left[\frac{2\langle z^{*+2}\rangle\langle z^{\infty-2}\rangle^{1/2}}{\lambda_\infty^+} + \frac{2\langle z^{*-2}\rangle\langle z^{\infty+2}\rangle^{1/2}}{\lambda_\infty^-} + E_D^*\left(\frac{\langle z^{\infty-2}\rangle^{1/2}}{\lambda_\infty^+} + \frac{\langle z^{\infty+2}\rangle^{1/2}}{\lambda_\infty^-}\right) + \frac{2\langle z^{\infty+2}\rangle\langle z^{\infty-2}\rangle^{1/2}}{\lambda_\infty^+} + \frac{2\langle z^{\infty-2}\rangle\langle z^{\infty+2}\rangle^{1/2}}{\lambda_\infty^-} + E_D^\infty\left(\frac{\langle z^{\infty-2}\rangle^{1/2}}{\lambda_\infty^+} + \frac{\langle z^{\infty+2}\rangle^{1/2}}{\lambda_\infty^-}\right)\right], \tag{11}$$

where $\gamma = 5/3$ is the adiabatic index, m_p is the proton mass, k_B is the Boltzmann constant, and α is the von Kármán-Taylor constant. The RHS of (11) describes the heating of the thermal background plasma by the dissipation of 2D turbulence, i.e., the loss terms on the RHS of the 2D equations (1)–(4). Only the correlation lengths of 2D turbulence are associated with the solar wind heating. We solve the coupled turbulence transport equations (1)–(11) using a 4th-order Runge-kutta method from 1 to 75 au.

IOP Conf. Series: Journal of Physics: Conf. Series **900** (2017) 012001 doi:10.1088/1742-6596/900/1/012001

3. Results: $C^{*\pm}_{sh} = 0$ & $C^{*E_D}_{sh} = 0$

In this section, we present numerical solutions of the coupled turbulence transport equations (1)–(11), setting $C^{*\pm}_{sh} = 0$ & $C^{*E_D}_{sh} = 0$, and compare with Voyager 2 observations from 1 – 75 au. We also compare the energy ratio of the inward and outward propagating modes with Ulysses measurements from 1 – 5 au. We use the boundary conditions shown in Tables 1 and 3 to solve the coupled equations (1)–(11) and then compute several turbulent quantities such as the energy in forward and backward propagating modes, the total turbulent energy, the normalized residual energy and cross-helicity, the correlation functions and correlation lengths corresponding to forward and backward propagating modes and the residual energy, the fluctuating kinetic and magnetic energy, the Alfvén ratio, the variance of density fluctuations and the solar wind temperature. In Tables 1 and 3, the first column identifies the turbulent quantities. The second, third, fourth, and fifth columns represent the boundary conditions corresponding to the 80:20, 70:30, 60:40 and 55:45 energy ratios between 2D and slab turbulence, respectively. Table 2 shows the parameter values that are used in the model equations [2]. Figures 1–4 are solutions of the turbulence transport model equations corresponding to the inner boundary values shown in Table 1, and Figures 5–8 are solutions corresponding to the boundary conditions shown in Table 3.

Parameters	80:20	70:30	60:40	55:45
$\langle z^{\infty+2} \rangle$ (km^{2}s^{-2})	4000	3500	3000	2750
$\langle z^{\infty-2} \rangle$ (km^{2}s^{-2})	800	700	600	550
E_D^∞ (km^{2}s^{-2})	-100	-87.5	-75	-68.75
L_∞^+ (km^{3}s^{-2})	8.25 $\times 10^9$	7.22 $\times 10^9$	6.19 $\times 10^9$	5.67 $\times 10^9$
L_∞^- (km^{3}s^{-2})	1.33 $\times 10^9$	1.16 $\times 10^9$	9.93 $\times 10^8$	9.09 $\times 10^8$
L_D^∞ (km^{3}s^{-2})	-2.11 $\times 10^8$	-1.85 $\times 10^8$	-1.58 $\times 10^8$	-1.45 $\times 10^8$
$\langle z^{*+2} \rangle$ (km^{2}s^{-2})	1000	1500	2000	2250
$\langle z^{*-2} \rangle$ (km^{2}s^{-2})	200	300	400	450
E_D^* (km^{2}s^{-2})	-25	-37.5	-50	-56.25
L_*^+ (km^{3}s^{-2})	7.4 $\times 10^8$	1.1 $\times 10^9$	1.47 $\times 10^9$	1.66 $\times 10^9$
L_*^- (km^{3}s^{-2})	5.96 $\times 10^8$	8.93 $\times 10^8$	1.19 $\times 10^9$	1.33 $\times 10^9$
L_D^* (km^{3}s^{-2})	-5.27 $\times 10^7$	-7.92 $\times 10^7$	-1.06 $\times 10^8$	-1.19 $\times 10^8$
$\langle \rho^{\infty 2} \rangle$ (cm^{-6})	1.4	1.4	1.4	1.4
T (K)	8×10^4	8×10^4	8×10^4	8×10^4

Table 1. Boundary values for 80:20, 70:30, 60:40, and 55:45 energy ratios between 2D and slab turbulence.

Figure 1 shows the comparison between the theoretical 2D and slab turbulent energies with observed values as a function of heliocentric distance. The observed turbulence energy corresponding to Voyager 2 observations is considered as being a superposition of 2D and slab energy [2]. The top-left plot of Figure 1 shows the energy in forward propagating modes $\langle z^{+2} \rangle$, the top-right plot the energy in backward propagating modes $\langle z^{-2} \rangle$, the middle-left the total turbulent energy E_T, the middle-right the fluctuating magnetic energy $\langle B^2 \rangle$, and the bottom-left the fluctuating kinetic

Parameters	Values	Parameters	Values
b	0.26	f_D	0.25
C_{sh}^{+}	0.9	τ_{ion}^{0}	10^6 s
C_{sh}^{-}	0.9	n_{sw}^{0}	5 cm^{-3}
$C_{sh}^{E_D}$	-0.5	n_H^{∞}	0.1 cm^{-3}
U	400 km s^{-1}	η_1	1.8×10^{-2}
ΔU	200 km s^{-1}	η_2	5×10^{-5}
V_{A0}	40 km s^{-1}	α	0.2
r_0	1 AU		

Table 2. Model parameters [2].

energy $\langle u^2 \rangle$. The bottom-right plot shows a comparison of the theoretical energy ratio between the backward and forward propagating modes, i.e., $\langle z^{-2} \rangle / \langle z^{+2} \rangle$ (also represented as R_{vA}^2(cal)) and the corresponding Ulysses observations R_{vA}^2(obs) [38]. R_{vA}^2(cal) is obtained from the observed Walén slope through an analytical expression, and the observed R_{vA}^2 is based on a direct decomposition of original Alfvénic fluctuations into outward- and inward-propagating Alfvén waves [38, 41]. Walén slope describes to what extent the fluctuations are purely Alfvénic and can be related to the amplitude ratio of inward (sunward) to outward (antisunward) Alfvn waves. In Figure 1, the solid curves illustrate 2D quantities, the dashed curves the slab quantities, the dashed-dotted-dashed the superposition of 2D and slab quantities, and the scatter plots the observed values. Furthermore, the red curves are the numerical solutions corresponding to the 80:20 energy ratio, the blue curves the 70:30, the green curves the 60:40, and the black curves the 55:45. Figure 1 shows that the 2D and slab energies are slightly different from each other for different boundary values, but nonetheless show similar trends with increasing heliocentric distance. As we change the energy ratios from 80:20 to 55:45, the 2D energies decrease slightly (solid red to black curves in Figure 1) and the slab energies increase slightly (dashed red to black curves in Figure 1) in the heliosphere. However, the total energy, i.e., the sum of 2D and slab energy (dashed-dotted-dashed curves), remains approximately the same in each case with increasing heliocentric distance.

Figure 1 also shows that the slab energies $\langle z^{*\pm 2} \rangle$, E_T^*, $\langle b^{*2} \rangle$, and $\langle u^{*2} \rangle$ decay faster than the 2D energies $\langle z^{\infty \pm 2} \rangle$, E_T^{∞}, $\langle B^{\infty 2} \rangle$, and $\langle u^{\infty 2} \rangle$ below $\sim 2-4$ au, similar to [1] and [2]. The slab energies start to increase beyond $\sim 4-5$ au, due to the presence of pickup ions driving turbulence in the outer heliosphere [1, 2]. The 2D and slab turbulence exhibit different levels of energy throughout the heliosphere, indicating that an anisotropy is present in the 2D and slab turbulence [1]. Furthermore, it shows that the backward propagating modes $\langle z^{*-2} \rangle$ increases initially unlike the forward propagating modes $\langle z^{*+2} \rangle$. This result indicates that backward propagating modes are created by the reflection of forward modes in the presence of magnetic field, solar wind velocity, and solar wind density gradients.

In the bottom-right plot of Figure 1, the black triangles are taken from [37], the blue squares and red circles are taken from [38]. The comparison between the theoretical energy ratio and the observed energy ratio is quite reasonable. The energy ratio increases towards ~ 1 with increasing heliocentric distance, most rapidly within ~ 10 au, and more slowly beyond ~ 10 au. Here, $\langle z^{-2}\rangle/\langle z^{+2}\rangle < 1$ below ~ 10 au, which indicates that the turbulence is imbalanced, whereas $\langle z^{-2}\rangle/\langle z^{+2}\rangle \sim 1$ beyond ~ 10 au, indicating that the turbulence is approximately balanced. We re-emphasize the point made in [1], which is that the balanced turbulence state is a consequence of turbulence driving by the creation of PUIs and not an intrinsic property of MHD turbulence.

The top-left, top-right, and middle-left plots of Figure 2 show the comparison between the theoretical and observed normalized residual energy σ_D, the normalized cross-helicity σ_c, and the Alfvén ratio r_A as a function of heliocentric distance. Similarly, in these plots the solid red, blue, green, and black curves illustrate 2D quantities, and the corresponding dashed curves the slab quantities. As in Figure 1, the 2D and slab σ_D, σ_c and r_A for each 2D:slab ratio evolve in a way that is almost similar with increasing heliocentric distance.

The middle-right plot of Figure 2 shows the comparison between the theoretical and observed variance of density fluctuations $\langle \rho^{\infty 2}\rangle$ with increasing heliocentric distance. The solid curves are the theoretical $\langle \rho^{\infty 2}\rangle$, the scatter "+" symbols are the observed $\langle \rho^{\infty 2}\rangle$, and the dashed red line illustrates the r^{-4} curve, where r is the heliocentric distance. Similarly, it also shows that the theoretical $\langle \rho^{\infty 2}\rangle$ evolves similarly throughout the heliosphere for each choice of 2D:slab ratio.

The bottom plot of Figure 2 shows the comparison between the theoretical and observed solar wind temperature T as a function of heliocentric distance. In the plot, the dashed red line is an $r^{-4/3}$ curve, i.e, depicting an adiabatic profile of the solar wind temperature. Again, the temperature profiles in each case are approximately similar throughout the heliosphere, and are in good agreement with the observed solar wind temperature.

The top-left, top-right, and bottom plots of Figure 3 show the normalized 2D and slab correlation functions corresponding to forward and backward propagating modes and the residual energy as a function of heliocentric distance. In Figure 3, the solid and dashed curves illustrate the 2D and slab correlation functions, respectively. Similarly, the red, blue, green, and black curves are the numerical solutions corresponding to the 80:20, 70:30, 60:40, and 55:45 energy ratios, respectively.

Figure 4 shows the comparison between the theoretical and observed correlation length with increasing heliocentric distance. The top left and right plots of Figure 4 illustrate the correlation length corresponding to forward and backward propagating modes. The middle left and right plots show the correlation length for the residual energy and the variance of the velocity fluctuations. The bottom left and right plots depict the correlation length for the variance of the magnetic field fluctuations and the cross-correlation of the co-variance of the velocity and magnetic field fluctuations. The blue triangles are the observed correlation lengths, and are scattered widely.

In this manuscript, we also solve the coupled equations (1)–(11) for a different set of boundary conditions as shown in Table 3. Here, the inner boundary values of the 2D and slab energies in the Table 1 and Table 3 are same, but the boundary values of the correlation functions are different. Use of different boundary values allows us to explore their effect on the evolution of turbulence in the heliosphere.

Figure 5 illustrates the comparison between the theoretical and observed turbulent

Parameters	80:20	70:30	60:40	55:45
$\langle z^{\infty+2}\rangle$ (km^{2}s^{-2})	4000	3500	3000	2750
$\langle z^{\infty-2}\rangle$ (km^{2}s^{-2})	800	700	600	550
E_D^∞ (km^{2}s^{-2})	-100	-87.5	-75	-68.75
L_∞^+ (km^{3}s^{-2})	2.95 $\times 10^9$	2.59 $\times 10^9$	2.22 $\times 10^9$	2.03 $\times 10^9$
L_∞^- (km^{3}s^{-2})	1.33 $\times 10^9$	1.16 $\times 10^9$	9.99 $\times 10^8$	9.15 $\times 10^8$
L_D^∞ (km^{3}s^{-2})	-2.11 $\times 10^8$	-1.85 $\times 10^8$	-1.59 $\times 10^8$	-1.46 $\times 10^8$
$\langle z^{*+2}\rangle$ (km^{2}s^{-2})	1000	1500	2000	2250
$\langle z^{*-2}\rangle$ (km^{2}s^{-2})	200	300	400	450
E_D^* (km^{2}s^{-2})	-25	-37.5	-50	-56.25
L_*^+ (km^{3}s^{-2})	1.03 $\times 10^9$	1.54 $\times 10^9$	2.07 $\times 10^9$	2.34 $\times 10^9$
L_*^- (km^{3}s^{-2})	9.93 $\times 10^8$	1.49 $\times 10^9$	1.99 $\times 10^9$	2.25 $\times 10^9$
L_D^* (km^{3}s^{-2})	-2.64 $\times 10^8$	-3.96 $\times 10^8$	-5.31 $\times 10^8$	-5.98 $\times 10^8$
$\langle \rho^{\infty 2}\rangle$ (cm^{-6})	1.4	1.4	1.4	1.4
T (K)	8×10^4	8×10^4	8×10^4	8×10^4

Table 3. Boundary values for 80:20, 70:30, 60:40, and 55:45 energy ratios between 2D and slab turbulence.

energies corresponding to the boundary values shown in Table 3. Figure 5 shows that the trends of the turbulence energies are similar to the observed values, and also these energies are approximately similar to the results shown in Figure 1. Figure 5 shows that the theoretical 2D energies are larger than the theoretical slab energies within $\sim$ 10 au. However, beyond $\sim$ 10 au, the 2D energy is larger than the slab energy only for the 80:20 ratio, while in the other cases the slab energies are larger than the 2D energies. In contrast, the theoretical 2D energies are larger than the theoretical slab energies throughout the heliosphere for the boundary conditions shown in Table 1. Moreover, in Figure 5 (bottom-right) the ratio of the backward and forward slab energies is closer to the R_{vA}^2(cal) (red circles), whereas the ratio of the 2D backward and forward propagating modes energies (and also ratio of total backward and forward propagating energy (dashed-dotted-dashed curves)) is larger than the observed values. In Figure 1 (bottom-right), the ratio of 2D backward and forward propagating modes energies (and also the ratio of total backward and forward propagating energy) is closer to the [37] and [38] (R_{vA}^2(obs)–blue circles) observed values.

Figure 6 shows the comparison between the theoretical and observed normalized residual energy and cross-helicity, the Alfvén ratio, the variance of the density fluctuations, and the solar wind temperature as a function of heliocentric distance. The theoretical results are approximately similar to the results shown in Figure 2. Figure 7 illustrates the normalized correlation functions as a function of heliocentric distance. Figure 8 shows the comparison between the theoretical and observed correlation lengths as a function of heliocentric distance. The correlation lengths shown in Figure 8 and Figure 4 have similar trends, although with different values.

4. Results: $C^{*\pm}_{sh} \neq 0$ & $C^{*E_D}_{sh} \neq 0$

In this section, we show numerical solutions of the coupled turbulence transport equations (1)–(11) setting $C^{*\pm}_{sh} \neq 0$ & $C^{*E_D}_{sh} \neq 0$. Here, we use the boundary conditions shown in Table 4. We use the same parameter values shown in Table 2 except $\eta_1 = 4.5 \times 10^{-2}$ and $\eta_2 = 10^{-4}$. Furthermore, we use $C^{*\pm}_{sh} = C^{\pm}_{sh}/2.9$ and $C^{*E_D}_{sh} = C^{E_D}_{sh}/2.9$, where the values of $C^{\pm}_{sh}$ and $C^{E_D}_{sh}$ are given in Table 2 i.e., we assume that the slab shear source terms are weaker than those for 2D turbulence.

The solutions of the coupled turbulence transport model equations are shown in Figures (9) to (12). The descriptions of Figures (9), (10), (11), and (12) are similar to Figures 1, 2, 3, and 4, respectively. Figures (9)–(12) show that by including the shear source of turbulence in the slab turbulence transport equations, the slab quantities decay much more slowly with increasing heliocentric distance, compared to the results in Section 3. The 2D turbulent quantities show similar trends to those of Section 3. With a shear source, the slab forward and backward propagating energy $\langle z^{*\pm^2} \rangle$ (top left & right of Figure 9), the total turbulent energy E^*_T (middle left of Figure 9), the fluctuating kinetic energy $\langle u^{*2} \rangle$ (bottom left of Figure 9), and the fluctuating magnetic energy $\langle B^{*2} \rangle$ (middle right of Figure 9) decay gradually with increasing heliocentric distance.

Parameters	80:20	70:30	60:40	55:45
$\langle z^{\infty+^2} \rangle$ (km^{2}s^{-2})	1600	1400	1200	1100
$\langle z^{\infty-^2} \rangle$ (km^{2}s^{-2})	160	140	120	110
E^{∞}_D (km^{2}s^{-2})	-80	-70	-60	-55
L^+_{∞} (km^{3}s^{-2})	2.95 $\times 10^9$	2.58 $\times 10^9$	2.21 $\times 10^9$	2.03 $\times 10^9$
L^-_{∞} (km^{3}s^{-2})	2.65 $\times 10^8$	2.32 $\times 10^8$	1.99 $\times 10^8$	1.82 $\times 10^8$
L^{∞}_D (km^{3}s^{-2})	-1.7 $\times 10^8$	-1.49 $\times 10^8$	-1.27 $\times 10^8$	-1.16 $\times 10^8$
$\langle z^{*+^2} \rangle$ (km^{2}s^{-2})	400	600	800	900
$\langle z^{*-^2} \rangle$ (km^{2}s^{-2})	40	60	80	90
E^*_D (km^{2}s^{-2})	-20	-30	-40	-45
L^+_* (km^{3}s^{-2})	2.96 $\times 10^8$	4.44 $\times 10^8$	5.93 $\times 10^8$	6.67 $\times 10^8$
L^-_* (km^{3}s^{-2})	1.66 $\times 10^8$	2.49 $\times 10^8$	3.33 $\times 10^8$	3.75 $\times 10^8$
L^*_D (km^{3}s^{-2})	-4.25 $\times 10^7$	-6.38 $\times 10^7$	-8.5 $\times 10^7$	-9.57 $\times 10^7$
$\langle \rho^{\infty 2} \rangle$ (cm^{-6})	1.4	1.4	1.4	1.4
T (K)	8×10^4	8×10^4	8×10^4	8×10^4

Table 4. Boundary values for 80:20, 70:30, 60:40, and 55:45 energy ratios between 2D and slab turbulence.

The slab normalized residual energy σ^*_D plot (top left of Figure 10) shows the slab fluctuating magnetic energy dominates within the ionization cavity and beyond (< 25 au) due to the presence of a shear source. Also, a shear source of slab turbulence is responsible for the normalized cross-helicity σ^*_c (top right of Figure 10) decreasing from $\sim$ 1 au, unlike in Section 3 where it decreases from beyond 2 au. The Alfvén ratio r^*_a (middle left of Figure 10) is also different from Section 3 (middle left of figures 2 & 6) in that r^*_a now decreases towards smaller values initially i.e., the variance of the slab magnetic field fluctuations dominates the variance of the slab solar wind velocity fluctuations.

The 2D correlation functions and the 2D correlation lengths (Figures 11 & 12, respectively) show similar trends to that of the previous results. Also, the slab correlation functions shown in Figure 11 look approximately similar to the previous results. However, the slab correlation lengths (dashed curves in Figure 12) are different from the previous results. A reason for this is that the shear sources appear only in the transport equations of the forward and backward propagating modes, and the residual energy, and not in the correlation functions. Figure 12 shows that the slab correlation lengths corresponding to different energy modes increase gradually as a function of heliocentric distance. In Figure 12, there are no peaks in the slab correlation lengths as in the previous results (Figures 4 & 8).

5. Discussion and Conclusions

We solved the coupled 1D steady-state turbulence transport equations (1)–(11) from 1 to 75 au for two cases; i) ${C^*}^{\pm}_{sh} = 0$ & ${C^*}^{E_D}_{sh} = 0$, and ii) ${C^*}^{\pm}_{sh} \neq 0$ & ${C^*}^{E_D}_{sh} \neq 0$, and compared the solutions to Voyager 2 and Ulysses measurements. In the first case, we solve the model equations (1)–(11) by including a shear source of turbulence in the 2D turbulence transport equations, and a pickup ion source of turbulence in the slab turbulence transport equations. In second case, we solve the equations (1)–(11) by including a shear source of 2D turbulence, and a shear and pickup ion source of slab turbulence. Here, the first case is similar to that solved by Zank et al [1] and Adhikari et al [2]. We computed several 2D and slab turbulent quantities: the energy in forward and backward propagating modes, the total turbulent energy, the normalized residual energy and cross-helicity, the correlation lengths and normalized correlation functions corresponding to forward and backward propagating modes and the residual energy, the fluctuating kinetic and magnetic energies, the solar wind temperature, and the variance of the density fluctuations for the boundary conditions shown in Table 1 and Table 3 for the first case, and Table 4 for the second case.

Zank et al 2017 [1] developed the coupled 2D and slab turbulence transport model equations, and presented preliminary results. Later, Adhikari et al [2] presented a detailed analysis of these model equations. Both papers used boundary values for the 2D and slab energies of the order of 80:20 [19, 20]. However, these ratios of the 2D and slab energies can be different in the fast and slow solar wind, at higher and lower latitudes, and in the ecliptic plane. Solar maximum and minimum conditions can also affect the value of the ratio. In this manuscript, we solved the model equations for four 2D to slab energy ratios, 80:20, 70:30, 60:40, and 55:45, and compared the solutions with Voyager 2 and Ulysses measurements. We found that the turbulence parameters in each case are different, however, the trends with heliocentric distance are similar. We found that the boundary values can change the turbulence energies in the outer heliosphere - in certain cases, the 2D energy is larger than the slab energy, and in others the slab energy can be larger than the 2D energy. In addition, the slab turbulent quantities in the presence of both sources of turbulence behave more reasonably than when only a pickup ion source of slab turbulence is included.

In summary, we find that the choice of energies in the 2D and slab components, together with the correlation functions, at the inner boundary and the sources of turbulence are important in determining the evolution of turbulence throughout the heliosphere.

Acknowledgments

We acknowledge the partial support of NASA grants NNX14AC08G, NNX15AI65G and a Parker Solar Probe Plus SWEAP subcontract SV4-84017. This work is partly supported by the International Space Science Institute (ISSI) in the framework of International Team 504 entitled "Current Sheets, Turbulence, Structures and Particle Acceleration in the Heliosphere".

References

[1] Zank G P, Adhikari L, Hunana P, et al 2017 *ApJ* 835 147
[2] Adhikari L, Zank G P, Hunana P, et al 2017 *ApJ* 841 85
[3] Marsch E, & Tu C Y, 1989 *JPlPh* 41 479
[4] Zhou Y, & Matthaeus W H, 1990a *JGR* 95 14881
[5] Zhou Y, & Matthaeus W H, 1990b *JGR* 95 14863
[6] Matthaeus W H, Oughton S, Pontius D H Jr, & Zhou Y, 1994 *JGR* 99 19267
[7] Matthaeus W H, Zank G P, Smith C W, & Oughton S, 1999 *PhRvL* 82 3444
[8] Zank G P, Matthaeus W H, & Smith C W, 1996 *JGR* 101 17093
[9] Zank G P, Dosch A, Hunana P, 2012 *ApJ* 745 35
[10] Smith C W, Matthaeus W H, Zank G P, 2001 *JGR* 106 8253
[11] Breech B, Matthaeus W H, Minnie J, 2008 *JGRA* 113 8105
[12] Usmanov A V, Matthaeus W H, Breech B A, & Goldstein M L, 2011 *ApJ* 727 84
[13] Adhikari L, Zank G P, Hu Q, & Dosch A, 2014 *ApJ* 793 52
[14] Adhikari L, Zank G P, Bruno R, 2015a *ApJ* 805 63
[15] Adhikari L, Zank G P, Bruno R, 2015b *JPhCS* 642 012001
[16] Wiengarten T, Fichtner H, Kleimann J, & Kissmann R, 2015 *ApJ* 805 155
[17] Shiota D, Zank G P, Adhikari L, 2017 *ApJ* 837 75
[18] Zank G P, & Matthaeus W H, 1993 *PhFl* 5 257
[19] Zank G P, & Matthaeus W H, 1992 *JGR* 97 17189
[20] Bieber J W, Wanner W, & Matthaeus W H, 1996 *JGR* 101 2511
[21] Dasso S, Milano L J, Matthaeus W H, & Smith C W, 2005 *ApJL* 635 L181
[22] Smith C W, 2003 *in AIP Conf Ser 679 Solar Wind Ten ed M Velli (Melville, NY: AIP)* 413
[23] Armstrong J W, & Woo R, 1980 *Rep IOM* 3331-80-070
[24] Armstrong J W, Cordes J M, & Rickett B J, 1981 *Nature* 291 561
[25] Klainerman S, & Majda A, 1981 *CPAM* 34 481
[26] Klainerman S, & Majda A, 1982 *CPAM* 35 629
[27] Montgomery D, Brown M R, & Matthaeus W H, 1987 *JGR* 92 282
[28] Zank G P, & Matthaeus W H, 1991 *PhFlA* 3 69
[29] Zank G P, & Matthaeus W H, 1992 *JPlPh* 48 85
[30] Hunana P, & Zank G P, 2010 *ApJ* 718 148
[31] Hunana P, Zank G P, Heerikhuisen J, & Shaikh D, 2008 *JGR* 113 A11105
[32] Hunana P, Zank G P, & Shaikh D, 2006 *PhRvE* 74 026302
[33] Bhattacharjee A, Ng C S & Spangler S R, 1998 *ApJ* 494 409
[34] Oughton S, Dmitruk P, & Matthaeus W H, 2006 *PhPl* 13 042306
[35] Oughton S, Matthaeus W H, Smith C W, Breech B, & Isenberg P A, 2011 *JGRA* 116 8105
[36] Wiengarten T, Oughton S, Engelbrecht N E, 2016 *ApJ* 833 17
[37] Bavassano B, Pietropaolo E, & Bruno R, 2000 *J Geophys. Res.* 105 15959
[38] Yang L, Lee L C, Li J P, Luo Q Y, Kuo C L, Shi J K, and Wu D J *(submitted to ApJ 2017)*
[39] Shebalin J V, Matthaeus W H, & Montgomery D, 1983 *JPlPh* 29 525

IOP Conf. Series: Journal of Physics: Conf. Series **900** (2017) 012001 doi:10.1088/1742-6596/900/1/012001

[40] Isenberg P A, 2005 ApJ 623 502
[41] Yang L, Lee L C, Chao J K, 2016 *The Astrophysical Journal* 817 178

IOP Conf. Series: Journal of Physics: Conf. Series **900** (2017) 012001 doi:10.1088/1742-6596/900/1/012001

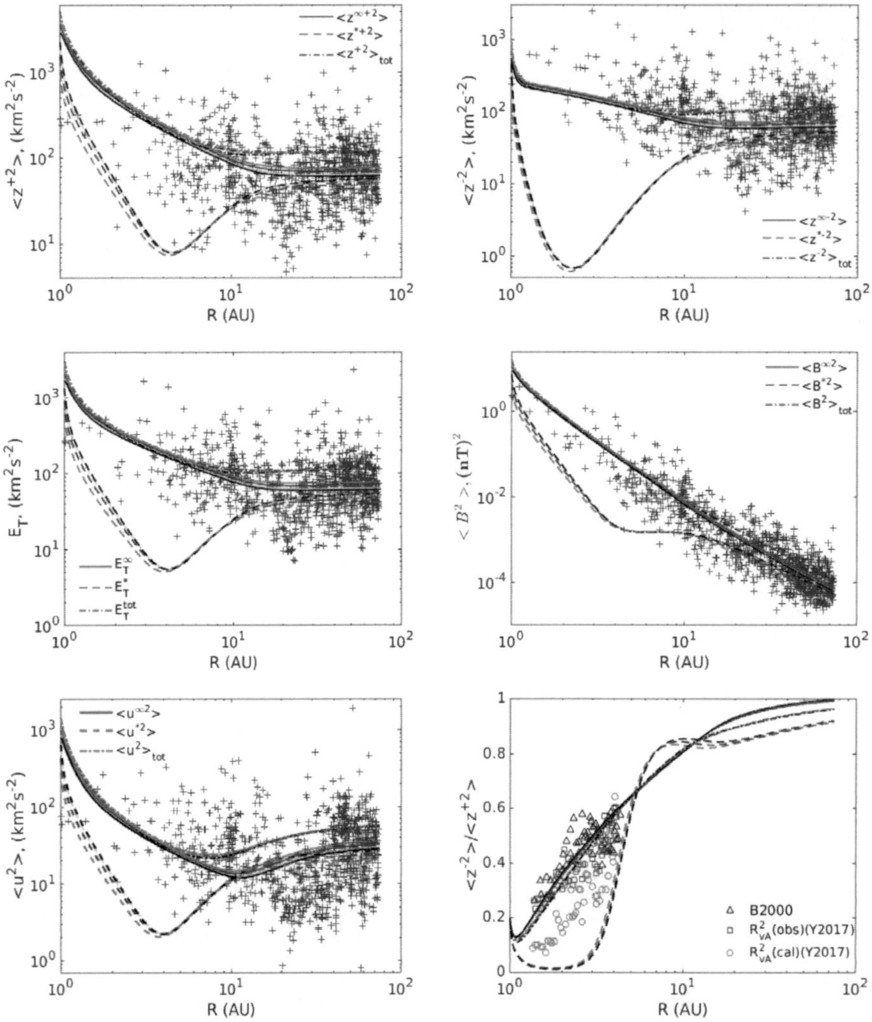

Figure 1. Comparison between the theoretical results and the values observed by Voyager 2 (V2) and Ulysses (bottom-right) as a function of heliocentric distance. Top left and right: the energy in forward and backward propagating modes. Middle left and right: the total turbulent energy and the fluctuating magnetic energy. Bottom left and right: the fluctuating kinetic energy and the energy ratio between inward and outward propagating modes. The solid curves describe 2D quantities, the dashed curves the slab quantities, and the dashed-dotted-dashed the sum of the 2D and slab quantities. Red curves correspond to boundary 2D and slab energy ratios of 80:20, the blue curves 70:30, the green curves 60:40, and the black curves 55:45. The blue scatter "+" symbols are observed values taken from [2]. The black triangles are obtained from [37] and the blue squares and red circles are obtained from [38]. R_{vA}^2 is the energy ratio of the backward and forward Alfvén waves.

IOP Conf. Series: Journal of Physics: Conf. Series **900** (2017) 012001 doi:10.1088/1742-6596/900/1/012001

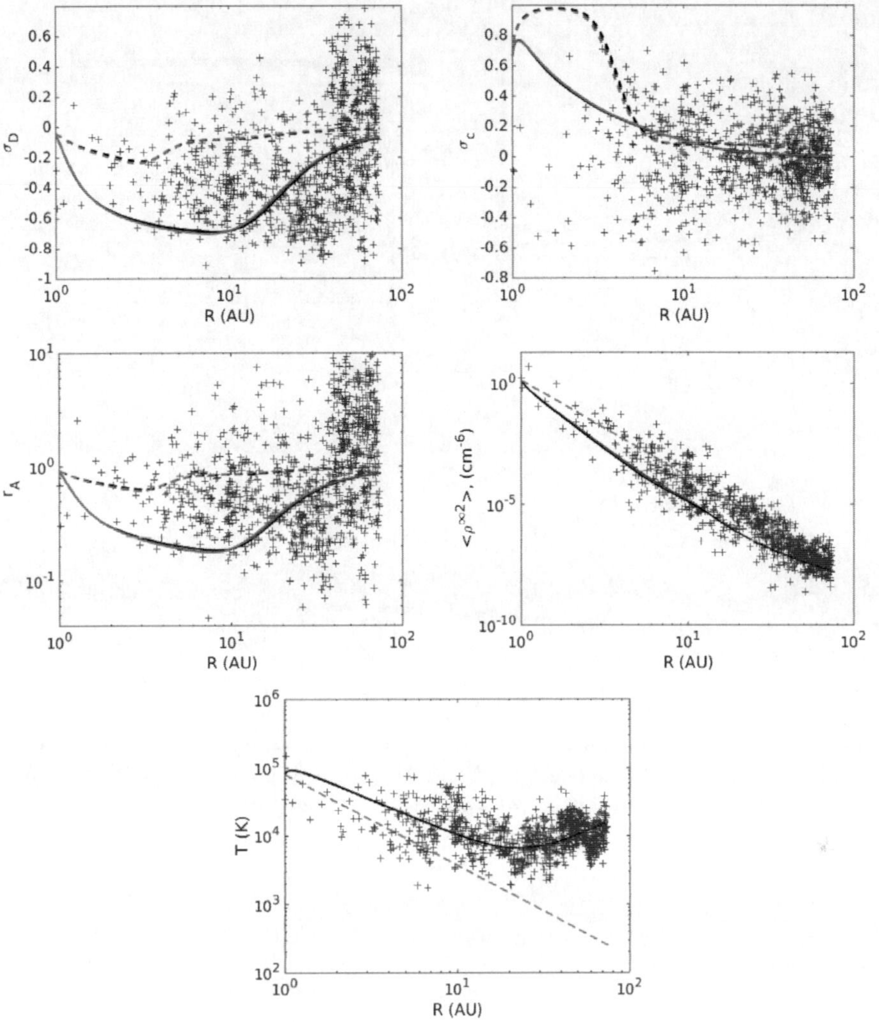

Figure 2. Comparison between the theoretical results and observations as a function of heliocentric distance. Top left and right: the normalized residual energy and cross-helicity. Middle left and right: the Alfvén ratio and the variance of the density fluctuations. Bottom: the solar wind temperature. The red dashed curve of the middle right panel indicates a r^{-4} curve, and the bottom plot $r^{-4/3}$, where r is the heliocentric distance. The description of the curves in the figure is similar to Figure 1.

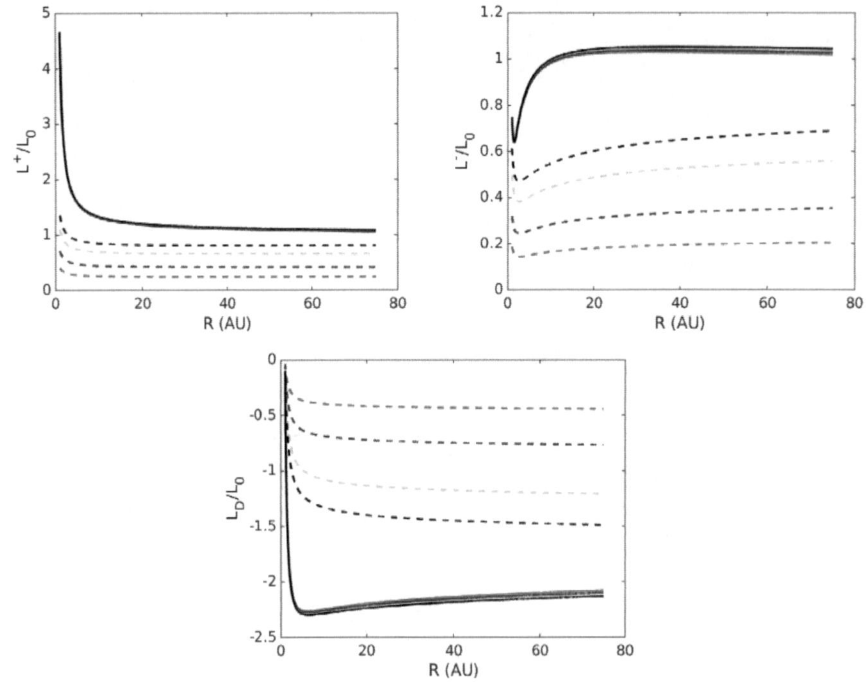

Figure 3. Normalized correlation functions as a function of heliocentric distance. Top-left, top-right and bottom plots are the correlation functions corresponding to forward and backward propagating modes, and the residual energy. L_0 is the normalization constant. Description of the curves in the figure is similar to Figure 1.

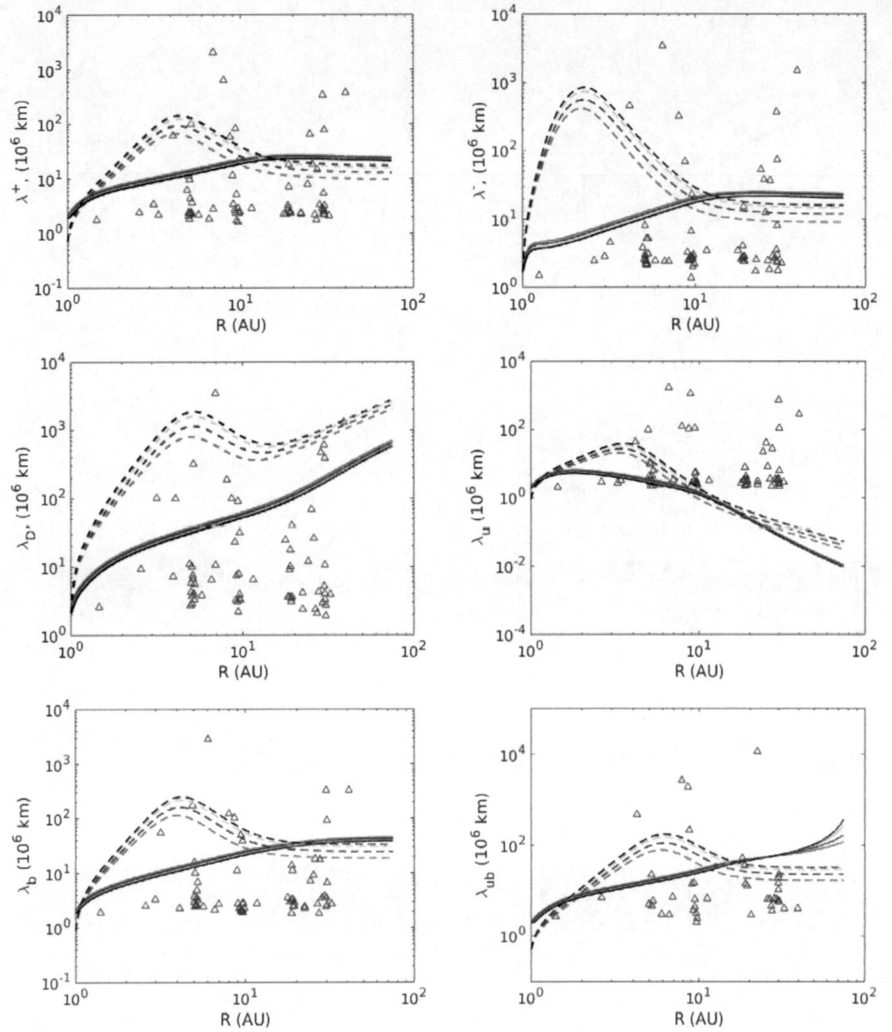

Figure 4. Comparison of the theoretical and observed correlation lengths as a function of heliocentric distance. Top left and right: correlation length corresponding to forward and backward propagating modes. Middle left and right: correlation length corresponding to the residual energy and the variance of velocity fluctuations. Bottom left and right: correlation length corresponding to the variance of magnetic field fluctuations and the cross-correlation between the co-variance between the velocity and the magnetic field fluctuations. The blue triangles are the observed correlation lengths. Description of curves in the figure is similar to Figure 1.

IOP Conf. Series: Journal of Physics: Conf. Series **900** (2017) 012001 doi:10.1088/1742-6596/900/1/012001

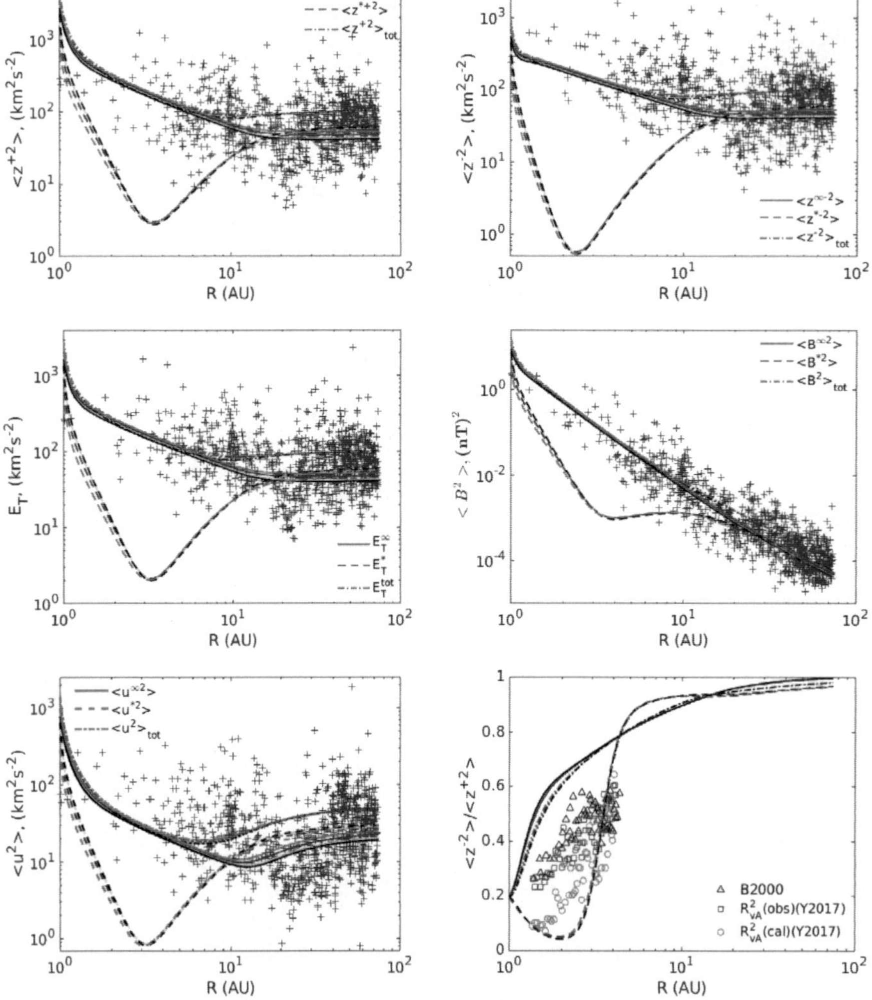

Figure 5. Comparison between the theoretical results derived from Table 3 and the observed values as a function of heliocentric distance. Description of the curves of the figure is similar to Figure 1.

IOP Conf. Series: Journal of Physics: Conf. Series **900** (2017) 012001 doi:10.1088/1742-6596/900/1/012001

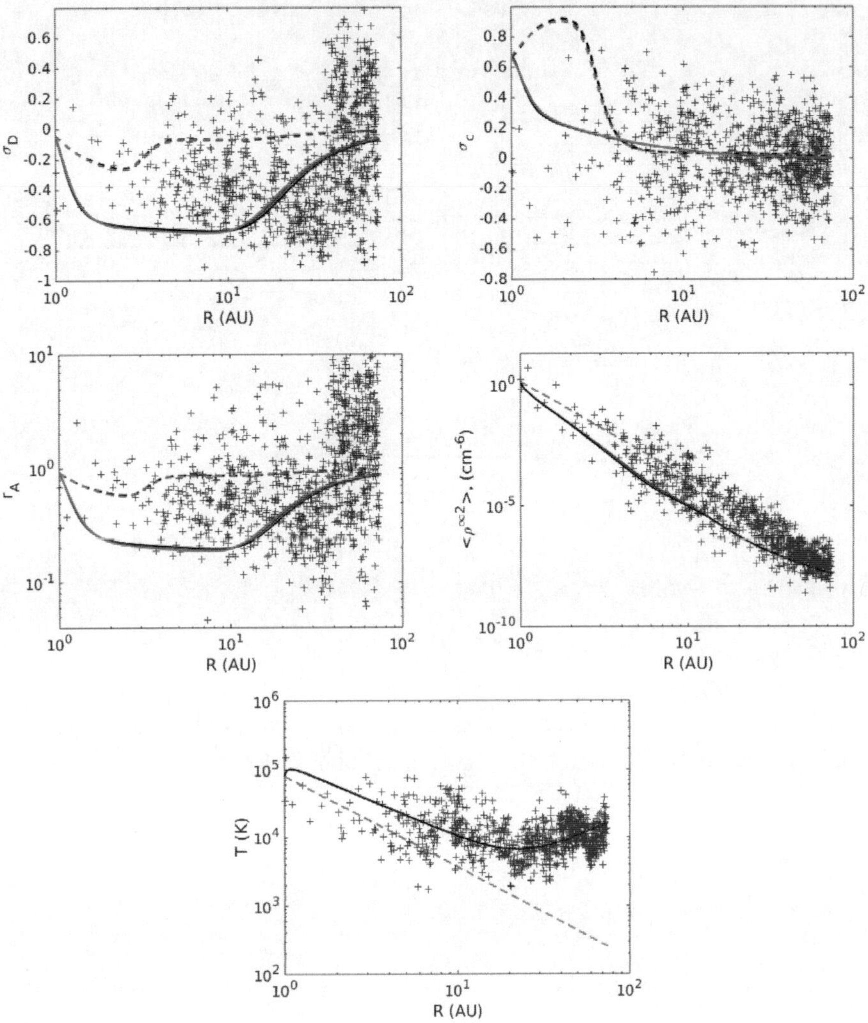

Figure 6. Comparison between the theoretical results derived from Table 3 and the observed values as a function of heliocentric distance. Description of the curves of the figure is similar to Figure 2.

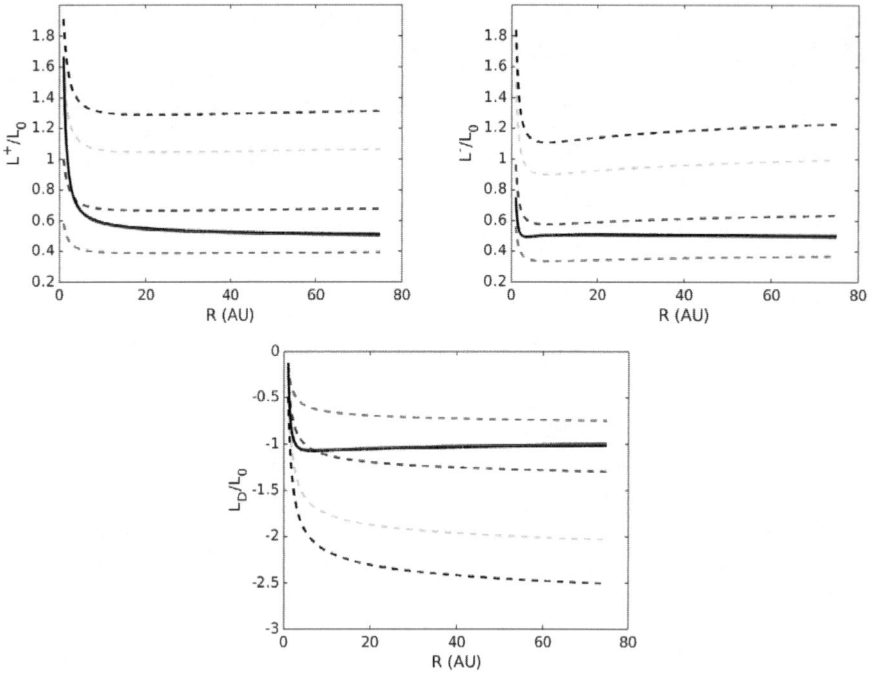

Figure 7. Normalized correlation functions derived from Table 3 as a function of heliocentric distance. Description of the figure is similar to Figure 3.

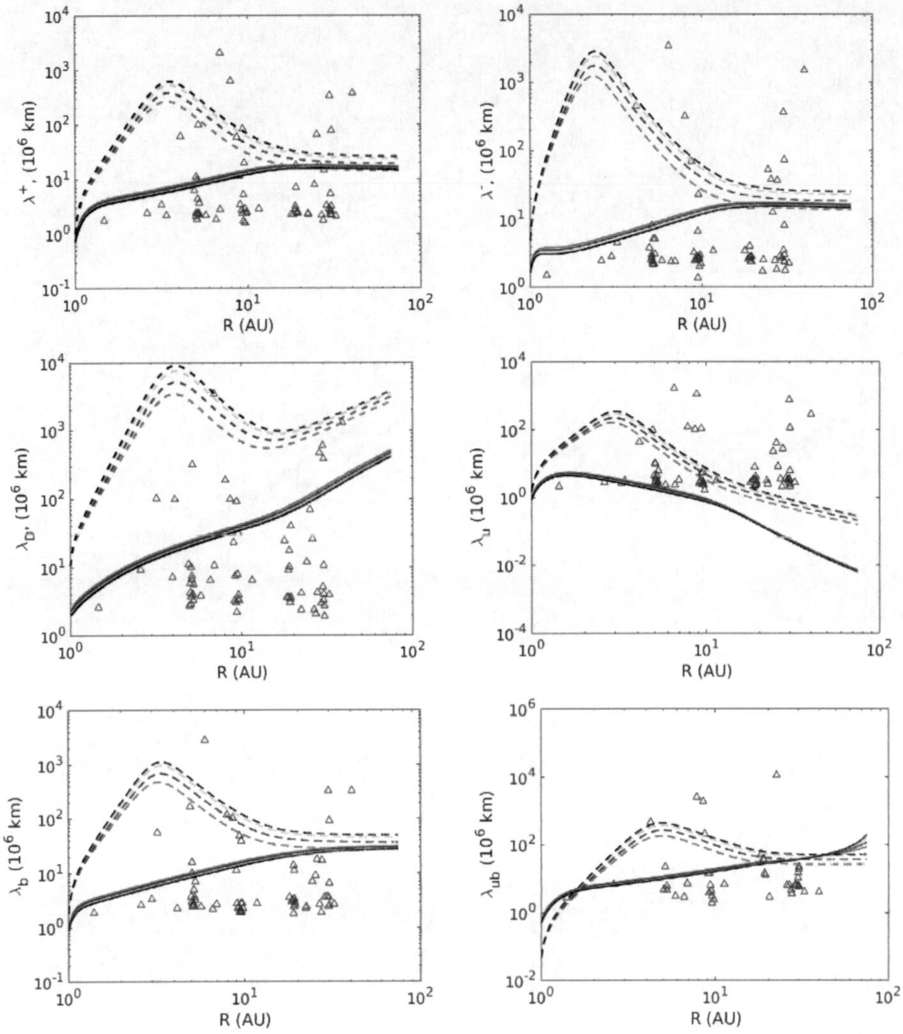

Figure 8. Comparison of the theoretical and observed correlation lengths derived from Table 3 with increasing heliocentric distance. Description of the figure is similar to Figure 4.

IOP Conf. Series: Journal of Physics: Conf. Series **900** (2017) 012001 doi:10.1088/1742-6596/900/1/012001

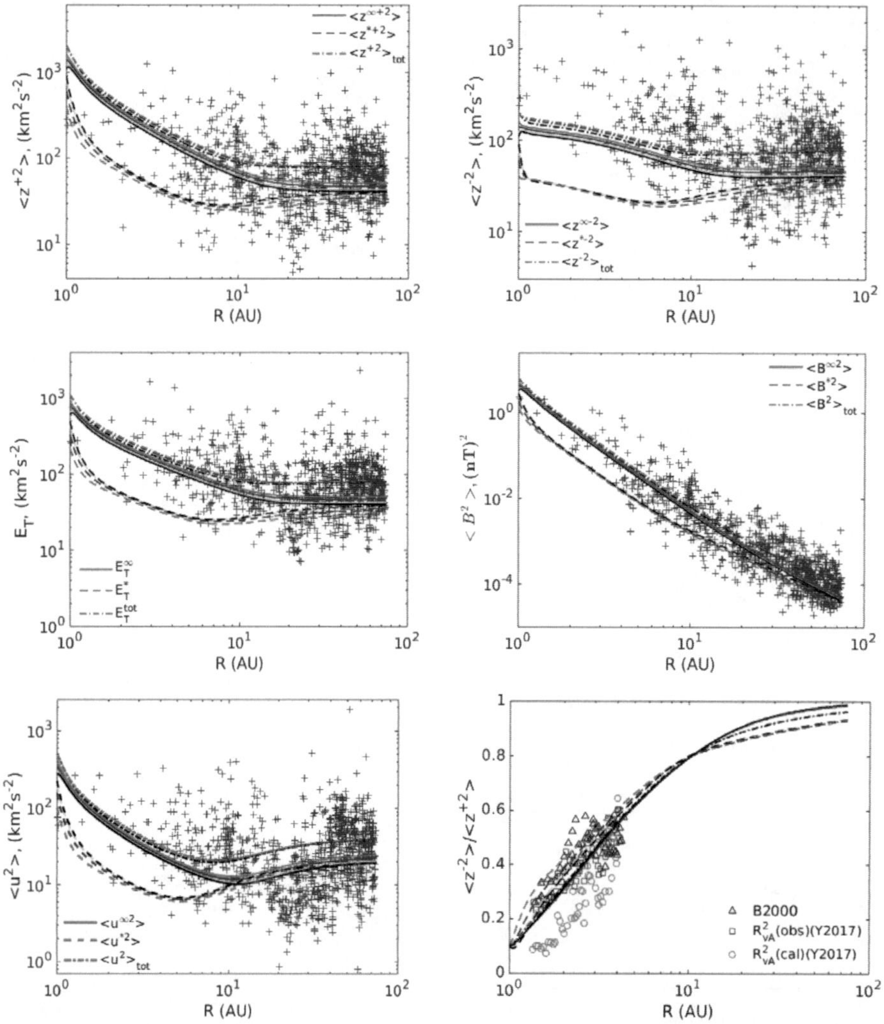

Figure 9. Comparison between the theoretical results derived from Table 4 and the observed values as a function of heliocentric distance. Description of the curves of the figure is similar to Figure 1.

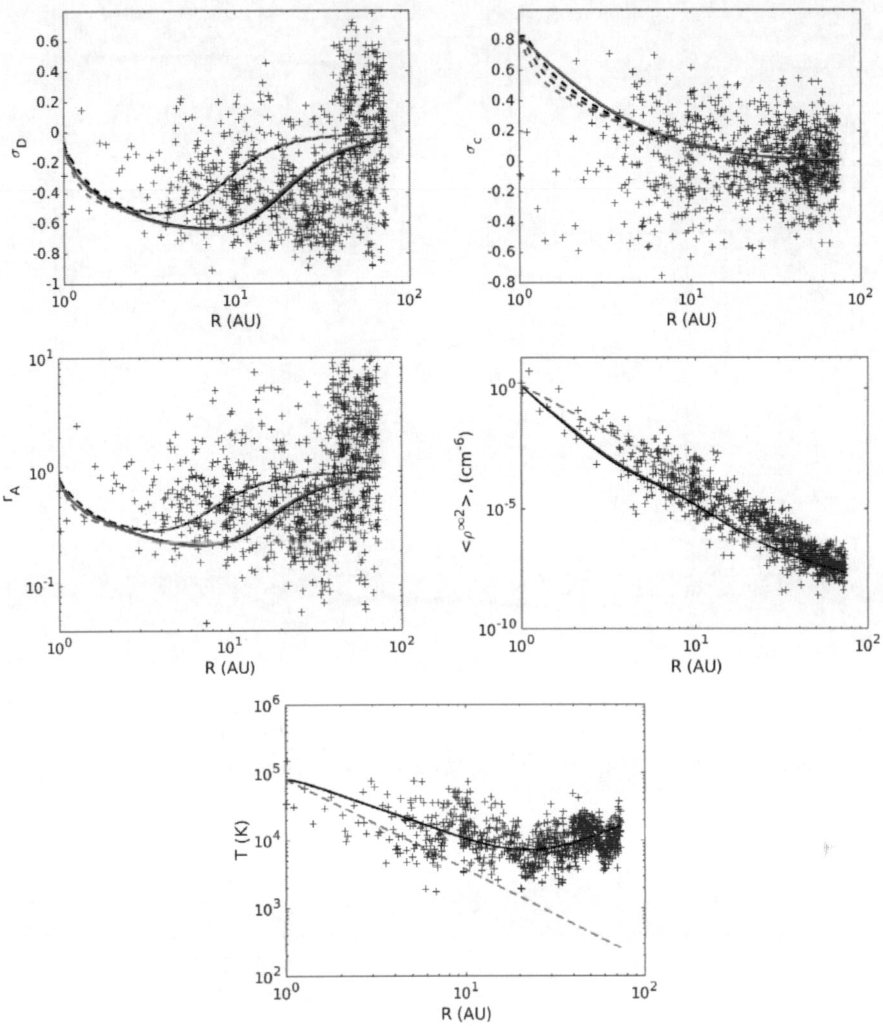

Figure 10. Comparison between the theoretical results derived from Table 4 and the observed values as a function of heliocentric distance. Description of the curves of the figure is similar to Figure 2.

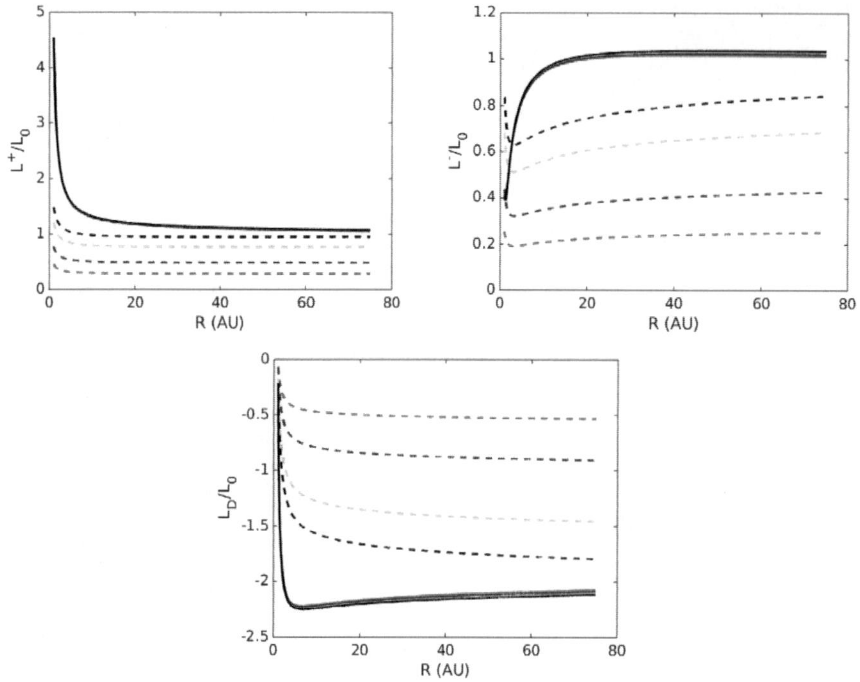

Figure 11. Normalized correlation functions derived from Table 4 as a function of heliocentric distance. Description of the figure is similar to Figure 3.

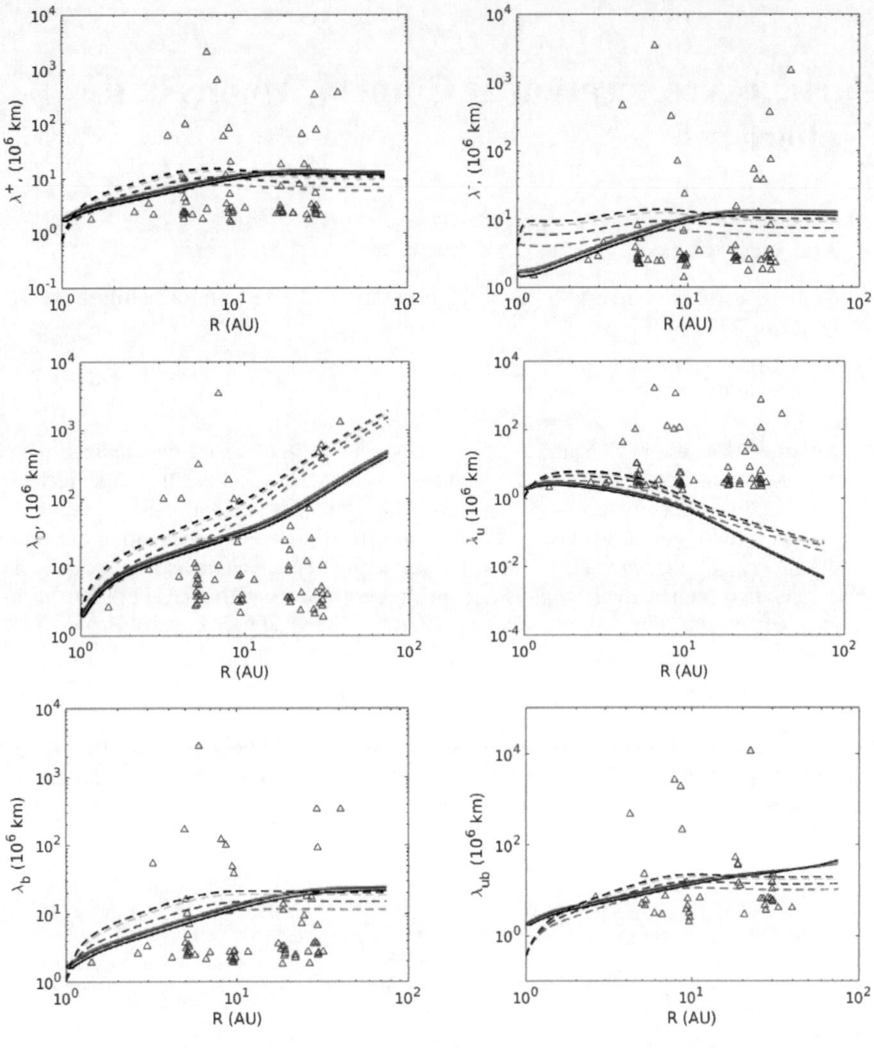

Figure 12. Comparison of the theoretical and observed correlation lengths derived from Table 4 with increasing heliocentric distance. Description of the figure is similar to Figure 4.

IOP Conf. Series: Journal of Physics: Conf. Series **900** (2017) 012002 doi:10.1088/1742-6596/900/1/012002

Pickup Particle Acceleration at Comets, Moons and Magnetospheres

Andrew J. Coates

Mullard Space Science Laboratory, University College London, Holmbury St Mary, Dorking RH5 6NT

a.coates@ucl.ac.uk

Abstract. Ionisation of neutrals plays a key role in the plasma interactions of comets and moons, and in magnetospheres. Following ionisation, ions and electrons are 'picked up' in the plasma flow, initially accelerating along the electric field and then gyrating around the magnetic field. For an ion with mass m (amu) this leads to an acceleration of up to 4m times the energy of the flowing plasma. Further scattering in pitch angle and energy may increase the acceleration further. Here, we illustrate the process at work with relevant plasma measurements from comets including Rosetta's comet 67P, from Titan and near Enceladus and Rhea.

1. Introduction

The pickup process begins when neutral particles from a comet or atmosphere are ionized by sunlight, electron impact or charge exchange. Neutral particles may come from a comet's coma, an atmosphere or interplanetary neutral particles. A summary of loss rates at solar system objects was given by Coates [1, see Figure 1].

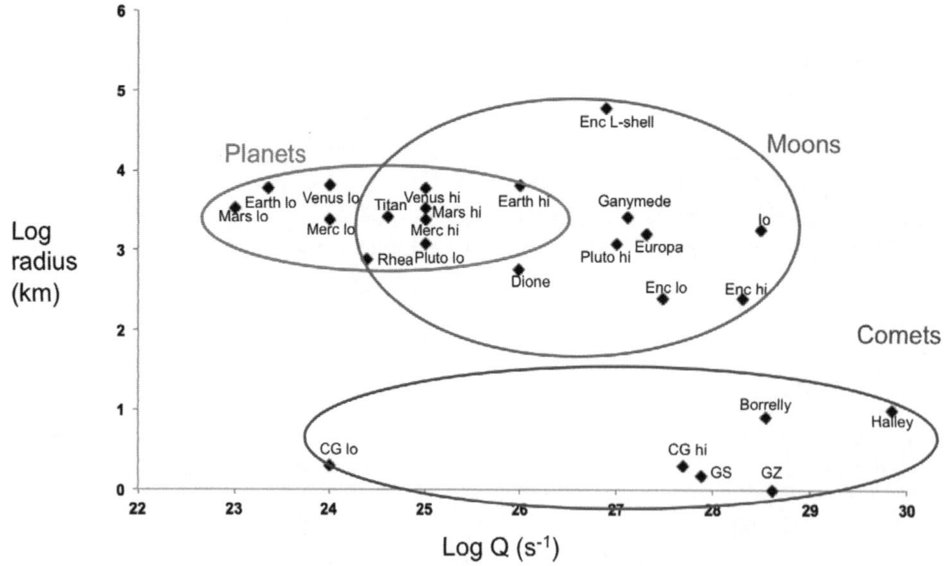

Figure 1 - Summary of neutral production rates (Q) for solar system objects compared to body radius (see [1] for details)

A summary of the pickup process is illustrated in Figure 2. In this paper, we will discuss the stages in the ion pickup process and illustrate each stage with examples from spacecraft data. We will also discuss the acceleration of particles as a result of pickup and from processes beyond pickup.

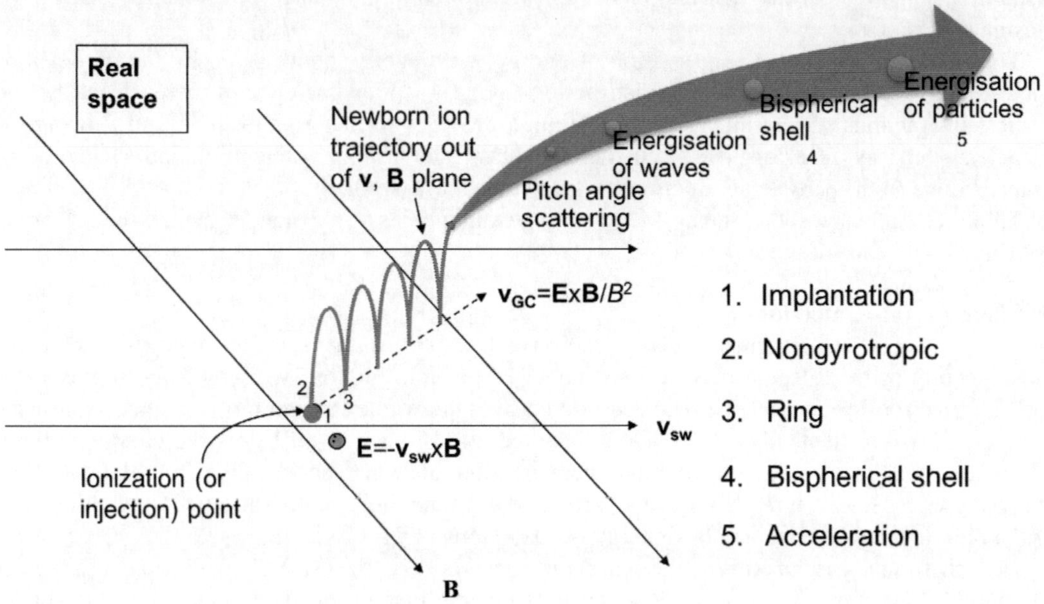

Figure 2 - Stages in the ion pickup process

2. Summary of stages in the ion pickup process

Following ionization, the particle is 'implanted' (Stage 1 in Table 1), and is then subject to the electric (**E**) and magnetic (**B**) fields in the collisionless flowing plasma. The particle is initially accelerated along the electric field, and in the ideal MHD approximation **E** = - **v** x **B**. The particle then gyrates around the magnetic field, with its initial motion as seen in **E** x **B** drift.

If there is a gradient in neutral particle density which is significant with respect to the particle's gyroradius, or at distances within ~2 gyroradii of the injection point, a 'non-gyrotropic' distribution of particles may occur (Stage 2), and has been observed at weak comets [2,3] and moons [4,5].

Table 1 – Stages in the ion pickup process, with indication of where each stage has been seen in spacecraft plasma observations. C=Comets, Ma=Mars, Mo=Moon, Msph=magnetospheres, V=Venus, Io=Io, E=Enceladus, T=Titan, R=Rhea, D=Dione, I=Interstellar

Stage in process	Timescale	Seen at
1. Implantation	<< gyroperiod (f_{ci})	C
2. Nongyrotropic ring	<gyroperiod	C, R, D, T
3. Ring	~gyroperiod	C, Ma, Mo, Msph, V, Io, E, T, I, R, D
4. (Bispherical) shell	~10 gyroperiods	C, Io?, E?, I?
5. Acceleration, shell filling	~100 gyroperiods	C
6. Maxwellian	?	?

Following this, and after at least a full gyration, the particle follows a cycloid in real space, corresponding to a ring in velocity space (Stage 3). The ring is unstable to plasma waves and instabilities, and the particle distribution becomes scattered in pitch angle, with energy being given to plasma waves.

The consequence of this conservation of energy, with energy being given by the particles to the waves, is a bispherical shell particle distribution (Stage 4) [6, and references therein]. Further wave-particle interactions and Fermi type II acceleration provides further energization, and slowing, of the particles – 'energy diffusion' (Stage 5). This process leads to both acceleration and shell filling, and finally (Stage 6) the particle distribution tends towards a Maxwellian.

Table 1 summarises the stages in the ion pickup process. We now consider the observations associated with each of the stages of pickup.

3. Stage 1 – Implanted ions

Newly implanted ions (or fresh pickup ions) have been seen in several cometary environments, but most recently by the Rosetta spacecraft at comet 67P [3,7]. In Figure 3 we show data from the Rosetta Ion and Electron Spectrometer instrument plotted over the whole Rosetta mission since orbit insertion in August 2014 to September 2016. The implanted ions are seen well below the energy of the solar wind in the second panel (ions). Although there are intervals where acceleration from the initial pickup energy is seen (Stage 2), many intervals particularly at large heliocentric distance (beginning and end of the plot) are characterized by low energy (<10eV) ions [3,7].

Note that in the case of Rosetta the spacecraft speed is very low (~cm s^{-1}) with respect to the comet. The initial, lowest energy at which Rosetta measures the pickup ions is dominated by the spacecraft potential [3]. During most of the interval shown the dominant plasma density is ionospheric and thus the spacecraft potential is a little negative, increasing their initial energy which is zero in the frame of the neutral particles. It is interesting that towards the beginning and end of the plot, the solar wind penetrates to Rosetta's position near to the cometary nucleus. There is therefore an electric field which can be seen accelerating the particles at times up to ~20-30 eV at heliocentric distances >~2.5 AU and frequently to 50-1000 eV for heliocentric distances <~2.5 AU (Stage 2).

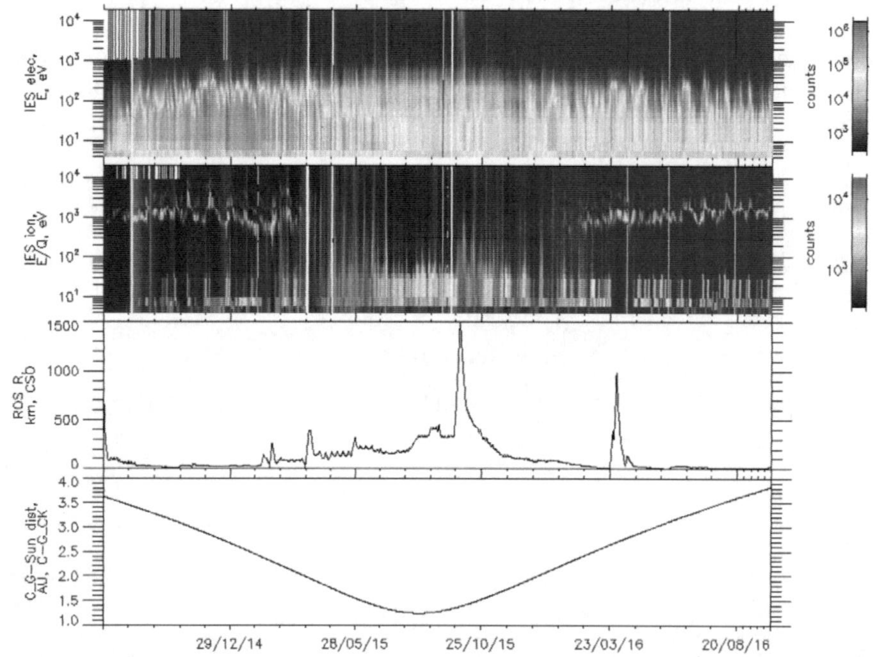

Figure 3 - RPC IES data from comet 67P. The top panel shows electrons, second panel shows ions, the third panel shows the Rosetta-comet distance and the 4th panel shows the comet-Sun distance.

IOP Conf. Series: Journal of Physics: Conf. Series **900** (2017) 012002 doi:10.1088/1742-6596/900/1/012002

4. Stage 2 – Nongyrotropic rings

The next stage of the pickup process beyond low energy (in the neutrals' frame) is the acceleration of the pickup ions along the electric field and the beginnings of gyration around the magnetic field. In the Rosetta data we already saw examples of such acceleration in Section 3. Further examples, and a comparison with pickup ions observed in the AMPTE neutral lithium and barium releases in 1984-5, were shown in [8]. The ion distributions observed in both the AMPTE and Rosetta data show a characteristic increasing energy due to following pickup ions around the early phase of the pickup ion trajectory. These constitute nongyrotropic distributions in velocity space [8].

Another way of producing a nongyrotropic distribution is at a more active, but still weak, comet such as Grigg-Skjellerup [2]. In this case, the ion production rate which decreases with distance from a cometary nucleus, varies between the cusps of the cycloid trajectory of pickup ions enough that the density changes with gyration angle. This was observed by Giotto at comet Grigg-Skjellerup [3] and led to the production of different instabilities [9,10].

An extreme version of this is the observation of pickup ions from moons with weak exospheres where pickup ions are observed originating from near the moon's surface. This was observed by the Cassini spacecraft at Rhea [4] and Dione [5], and has been seen as one of the pickup populations near Earth's moon [11].

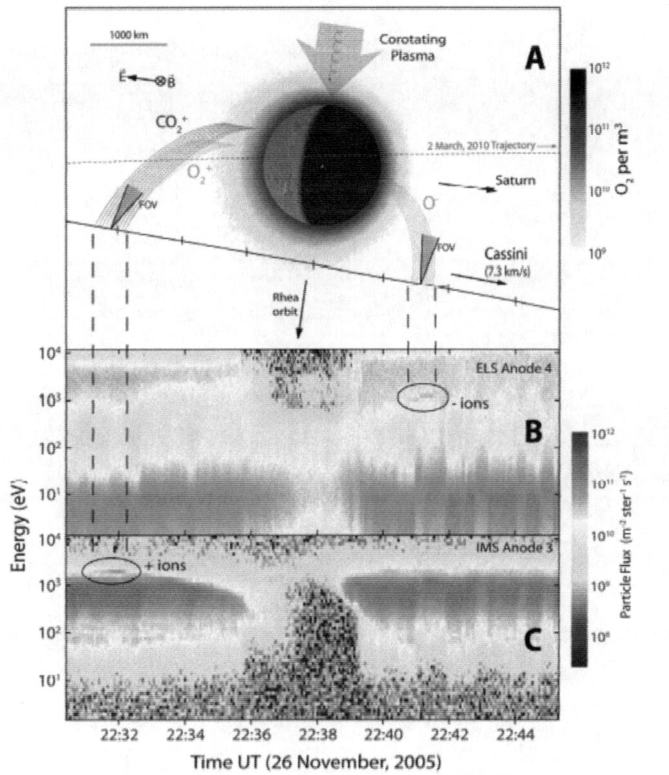

Figure 4 - Cassini CAPS observations of pickup ions near Rhea. Positive and negative pickup ions are seen, locating their source in a weak near-surface atmosphere (from [4])

Figure 4 shows observations by the CAPS instrument near to Saturn's moon Rhea. The bottom panel shows positive ions, with a narrow (in energy) pickup ion population seen before closest approach to the moon, likely O^+, superimposed on the corotating magnetospheric ion population before the spacecraft enters the plasma wake of Rhea. The middle panel shows data from the CAPS electron spectrometer, including cold magnetospheric electrons (<~20 eV), injection related hot

IOP Conf. Series: Journal of Physics: Conf. Series **900** (2017) 012002 doi:10.1088/1742-6596/900/1/012002

electrons (~1-5 keV) and the Rhea plasma wake. After closest approach, a narrow population of negative pickup ions is seen, likely O^-. Trajectory modelling of the early trajectory pickup ions locates the source near the moon and this, together with INMS measurements of neutral O_2 and CO_2, represents the discovery of Rhea's weak atmosphere. This is due to sputtering by magnetospheric ions, and is also present at Dione [5], and at Europa and Ganymede in the Jupiter system the target of the JUICE and Europa Clipper missions. For our purposes here, these are remarkable observations of Stage 2 positive and negative pickup ions.

Recently, pickup ion observations at Titan were re-examined [12,13] and the dominant direction of the early trajectory was confirmed to be along the E direction [13]. In addition, the loss rate due to pickup was found to be ~3.3×10^{23} s^{-1} [13], compared to about an order of magnitude higher rate from the ionosphere [14].

5. Stages 3,4 and 5 – Rings, bispherical shells, acceleration

The classical pickup process produces, after the early stages described above, rings in velocity space, bispherical shells and particle acceleration. After executing a complete cusp to cusp gyration in real space (early gyrations in Figure 2) the particle distribution function becomes a ring. These distribution functions are observed at comets (e.g. [1] and references therein) and are unstable to waves, which produce pitch angle scattering.

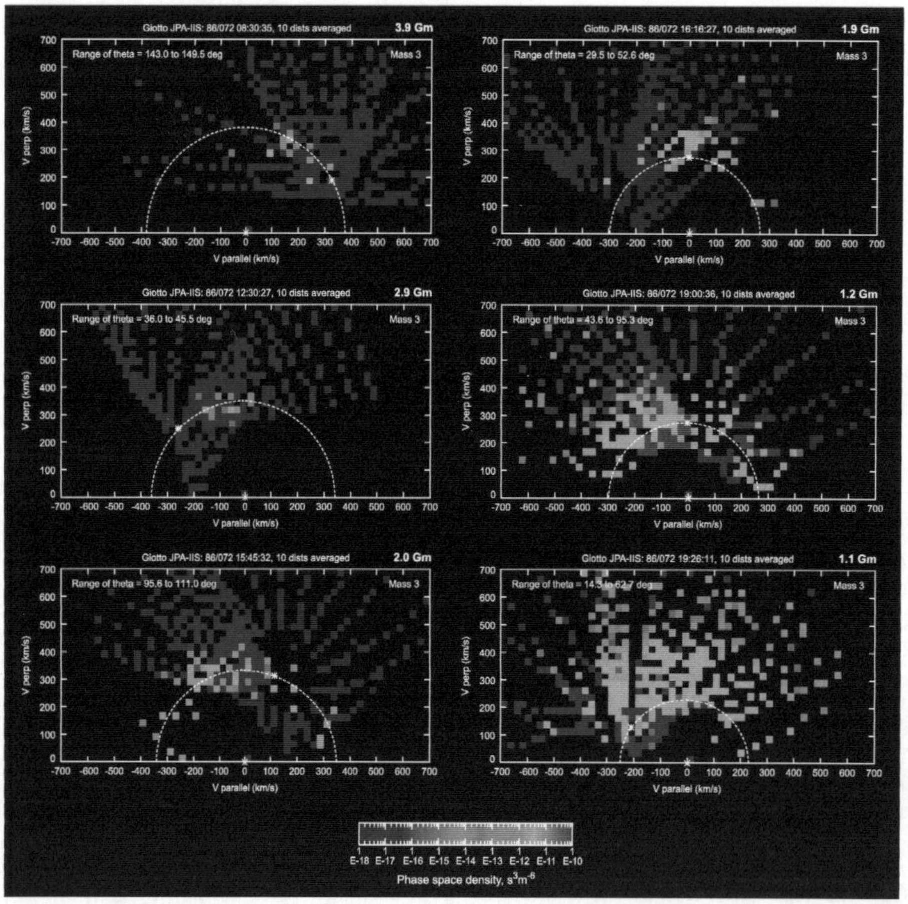

Figure 5 - Pickup water group ion distribution functions observed at comet Halley. The distribution follows the evolution from rings (upper left panels 1,2) to (bispherical) shell (panels 3,4) and shows particle acceleration upstream (panel 5) and downstream (panel 6) of the comet's bow shock (from [1], adapted from [15])

The maximum velocity of the ring with respect to the injection point (frame of the neutral particles) is $2|v|\sin\theta_{vB}$, where θ_{vB} is the angle between **v** and **B** and m is the ion mass in amu. Thus, the maximum energy of the ring is $2mv^2\sin^2\theta_{vB}$. This corresponds to $4mE_{sw}\sin^2\theta_{vB}$, where E_{sw} is the energy of the flowing plasma, for example the solar wind (see [1]).

Similarly, the maximum velocity of the bispherical shell is $2|v|$, thus the maximum energy is $4mE_{sw}$.

At the Moon, reflection of a fraction of neutral particles [11] gives additional acceleration and means that the maximum velocity of a bispherical shell is $3|v|$ in the lunar exosphere frame, and the maximum energy is $9mE_{sw}$ [11,1].

Stages 3,4 and 5 of the pickup process are well illustrated in Giotto data from the active comet Halley [15,16]; see Figure 5 (from [1]. In these plots the data have been transformed from the measured energy and angle dependent spectra in the spacecraft frame, to a magnetic field aligned solar wind frame [15], and arranged as $V_\perp$-$V_\parallel$ water group ion distributions. In this format, a ring distribution would appear at the point marked * on each plot, which denotes the injection point on the semicircle (representing the local solar wind velocity).

The first two panels are relatively narrow indicating limited scattering from the initial ring (*) around the semicircle, while the third and fourth panels show significant pitch angle scattering (almost around the entire semicircle). The bulk parameters of these populations (not shown) are consistent with bispherical shells centred on upstream and downstream propagating Alfven waves [16].

The fifth and sixth panels are samples immediately upstream and downstream of the cometary bow shock, and show significant acceleration beyond local pickup velocities. Although some of the pickup ions in these two panels are transported from upstream, the last panel in particular shows evidence of particle acceleration well beyond pickup. This is likely due second order Fermi processes with acceleration from the bispherical shell (stage 5) and first order Fermi acceleration associeated with approaching scattering centres near the bow shock [19,15]. The further accelerated ions are also seen in energetic particle detectors [20,21,22]. Pickup protons show somewhat similar behaviour [23].

Developed ring distributions are also seen in planetary magnetospheres. For example, in Saturn's magnetosphere near the orbit of Enceladus, the neutral particle density is about 2 orders of magnitude greater than the plasma density. Ring distributions of water group ions resulting from ionization of the neutrals were clearly seen there in the Cassini CAPS data [24]. The ions had ~90° pitch angle, consistent with the observation of early pickup ion rings in an almost dipolar magnetic field at ~4 Saturn radii.

6. Pickup ions dominate the comet-SW interaction

Here, we briefly summarise the solar wind-comet interaction and the relevant pickup ion processes. At weak comets, early phase implanted ions are important (Stage 1, section 3 above – see [25,8,3,7]). The mass loading associated with these slows the plasma enough that eventually a bow shock forms. Pickup ions play a key role at the bow shock [26,27,28].

The observed pickup water group ion distribution functions include Stage 1 implanted ions (Section 3), Stage 2 nongyrotropic rings [2], Stage 3 rings [15,23], and Stage 4 bispherical shells [16].

Acceleration beyond pickup (Stage 5) was also seen at comets Giacobini-Zinner [21,22], Halley [29,20,15,16] and at comet Grigg-Skjellerup [30,2,31,32].

7. Summary and conclusions

Plasma and magnetic field measurements are a sensitive indicator of cometary, moon and planetary activity in terms of neutral particle production rate, over many orders of magnitude of activity [1, see Figure 1]. Pickup ions result from the ionization of neutral particles in all these environments.

The ion pickup process includes different stages at weak, medium and strong comets, and we have summarized these in this paper (see Table 1 and sections 3-5). The different stages are clearly seen with Rosetta, Giotto and the other cometary missions.

The pickup process is also seen at moons (e.g., Rhea, Dione, Titan and Earth's moon). It is also seen in magnetospheres particularly where neutral particles dominate (e.g., Saturn's magnetosphere).

There is a clear need for particle and field measurements for activity detection at both weak and strong sources of neutral particles for future missions where activity and neutral particle production is important. One example of this is main belt comets [33,34].

Acknowledgments
We thank the organisers for the opportunity of giving this talk, and the STFC Consolidated Grant to UCL-MSSL for funding. I would also like to acknowledge the Giotto JPA, Cassini CAPS and Rosetta RPC teams for excellent collaboration for over 3 decades, and nearly 8 orders of magnitude in gas production rate. We thank Jim Burch and the RPC team for use of the data in Figure 3.

References

[1] Coates A J 2016 Plasma Measurements at Non-Magnetic Solar System Bodies *Magnetosphere-Ionosphere Coupling in the Solar System*, AGU Geophysical Monograph 222, ed C R Chappell, R W Schunk, P M Banks, J L Burch, and R M Thorne (Hoboken: Wiley) Chapter 21 pp259-276

[2] Coates A J, Johnstone A D, Wilken B and Neubauer F M 2003 Velocity space diffusion and non-gyrotropy of pickup water group ions at comet Grigg-Skjellerup *J. Geophys. Res.* **98** 20,985

[3] Goldstein R *et al* 2015 The Rosetta Ion and Electron Sensor (IES) measurement of the development of pickup ions from comet 67P/Churyumov-Gerasimenko *Geophys. Res. Lett.* 42, 3093-3099

[4] Teolis B D *et al* 2010 Cassini finds an oxygen–carbon dioxide atmosphere at Saturn's icy moon Rhea *Science* **330** 1813-1815

[5] Tokar R L, R E Johnson, M F Thomsen, E C Sittler, A J Coates, R J Wilson, F J Crary, D T Young, G H Jones 2012 Detection of Exospheric O_2^+ at Saturn's Moon Dione *Geophys. Res. Lett.* **39** L03105

[6] Coates A J, Wilken B, Johnstone A D, Jockers K, Glassmeier K-H and Huddleston D E 1990 Bulk properties and velocity distributions of water group ions at comet Halley: Giotto measurements *J.Geophys Res.* **95** 10249-10260

[7] Nilsson H *et al* 2015 Birth of a comet magnetosphere: A spring of water ions *Science* **347** aaa0571

[8] Coates A J, J L Burch, R Goldstein, H Nilsson, G Stenberg Wieser, E Behar and the RPC team 2015 Ion pickup observed at comet 67P with the Rosetta Plasma Consortium (RPC) particle sensors: similarities with previous observations and AMPTE releases, and effects of increasing activity *J. Phys.: Conf. Ser.* **642** 012005

[9] Motchmann U, Glassmeier, K-H 1993 Nongyrotropic distribution of pickup ions at comet P/Grigg-Skjellerup: A possible source of wave activity *J. Geophys. Res.* **98** 20,977-20,983

[10] Brinca, A L, Borda de Agua L, Winske, D 1993 On the stability of nongyrotropic ion populations - A first (analytic and simulation) assessment *J. Geophys. Res.* **98** 7549-7560

[11] Saito Y *et al* 2008 Solar wind proton reflection at the lunar surface: Low energy ion measurement by MAP-PACE onboard SELENE (KAGUYA) *Geophys. Res. Lett.* **35** L24205

[12] Hartle R E *et al.* 2006 Initial Interpretation of Titan Plasma Interaction as Observed by the Cassini Plasma Spectrometer: Comparisons with Voyager 1 *Planet Space Sci.* **54**, 1211-1224

[13] Regoli L H, A J Coates, M F Thomsen, G H Jones, E Roussos, J H Waite, N Krupp and G Cox 2016 Survey of pickup ion signatures in the vicinity of Titan using CAPS/IMS *J. Geophys. Res.* **121** 8317-8328

[14] Coates A J *et al.* 2012 Cassini in Titan's tail: CAPS observations of plasma escape *J. Geophys. Res.* **117**

[15] Coates, A J *et al* 1989 Velocity space diffusion of pick-up ions from the water group at Comet Halley *J. Geophys. Res.* **94** 9983

[16] Coates A J *et al* 1990 Bulk properties and velocity distributions of water group ions at comet Halley: Giotto measurements *J.Geophys Res* **95** 10249

[17] Thomsen M F *et al.* 1987, In-situ observations of a bi-modal ion distribution in the outer coma of Comet Halley, *Astron. Astrophys.* **187** 141

[18] Coates A.J. 1991 Cometary plasma energisation *Ann. Geophys.* **9** 158

[19] Terasawa T and M Scholer 1989 The heliosphere as an astrophysical laboratory for particle acceleration *Science* **244** 1050-1057

[20] McKenna-Lawlor *et al* 1986 Energetic ions in the environment of comet Halley Nature **321** 347

[21] Hynds R J *et al* 1985 Observations of energetic ions from comet Giacobini-Zinner *Science* **232** 361

[22] Gloeckler G *et al* 1986 Cometary pick-up ions observed near Giacobini-Zinner *Geophys. Res. Lett.* **13** 251

[23] Neugebauer M *et al* 1989 The velocity distributions of cometary protons picked up by the solar wind *J. Geophys. Res.* **94** 5227

[24] Tokar R L *et al.* 2008 Cassini detection of water-group pick-up ions in the Enceladus torus Geophys. Res. Lett. **35** L14202

[25] Glassmeier K H 2017 Interaction of the solar wind with comets: a Rosetta perspective *Phil. Trans. R. Soc. A* **375** 20160256

[26] Coates A J et al 1990 Plasma parameters near the comet Halley bow shock *J. Geophys. Res.* **95** 20,701

[27] Coates A J, Mazelle C and Neubauer F M 1997 Bow shock analysis at comets Halley and Grigg Skjellerup *J. Geophys. Res* **102** 7105

[28] Neubauer F M *et al* 1990 Giotto magnetic field observations at the outbound quasi-parallel bow shock of comet Halley *Ann. Geophys.* **8** 463

[29] Johnstone A D *et al* 1986 Ion flow at Halley's comet *Nature* **321** 344

[30] Johnstone A D, Coates A J, Huddleston D E, Jockers K, Wilken B, Borg H, Gurgiolo C, Winningham J D and Amata E 1993 Observations of the solar wind and cometary ions during the encounter between Giotto and comet Grigg-Skjellerup *Astron. Astrophys.* **273** L1-L4

[31] Coates, A J *et al.* 1993 Pickup water group ions at comet Grigg-Skjellerup *Geophys. Res. Lett.* **20** 483-486

[32] McKenna-Lawlor *et al* 1993 Energetic ions at Comet Grigg-Skjellerup measured from the Giotto spacecraft *Nature* **363** 326

[33] Jones, G H *et al.* 2017 Caroline - A search for the source of Earth's water *Adv Space Res.* submitted

[34] Snodgrass, C et al. 2017 The Castalia Mission to Main Belt Comet 133P/Elst-Pizarro *Adv Space Res* submitted

IOP Conf. Series: Journal of Physics: Conf. Series **900** (2017) 012003 doi:10.1088/1742-6596/900/1/012003

A Hilbert-Huang transform approach to space plasma turbulence at kinetic scales

G Consolini[1], T Alberti[2], E Yordanova[3], M F Marcucci[1], and M Echim[4,5]

[1] INAF-Istituto di Astrofisica e Planetologia Spaziali, Roma, Italy
[2] Dept. of Physics, University of Calabria, Rende, Italy
[3] Swedish Institute for Space Physics, Uppsala, Sweden
[4] Royal Belgian Institute for Space Aeronomy, Brussels, Belgium
[5] Institute of Space Science, Magurele, Romania

E-mail: giuseppe.consolini@iaps.inaf.it

Abstract. Heliospheric space plasmas are highly turbulent media and display multiscale fluctuations over a wide range of scales from the magnetohydrodynamic domain down to the kinetic one. The study of turbulence features is traditionally based on spectral and canonical structure function analysis. Here, we present an novel approach to the analysis of the multiscale nature of plasma turbulent fluctuations by means of Hilbert-Huang Transform (HHT). In particular we present a preliminary application of this technique to magnetic field fluctuations at kinetic scales in a fast solar wind stream as observed by Cluster mission. The HHT-energy spectrum reveals the intermittent and multiscale nature of fluctuation frequency at kinetic scales indicating that there are no-persistent and long standing frequencies.

1. Introduction

Several space plasma environments display complex dynamics characterized by multiscale processes, such as fluid and magnetoydrodynamic (MHD) turbulence. In the framework of heliospheric plasmas, turbulence is a phenomenon widely occurring in several different regions: solar wind and interplanetary medium, Earth's magnetosheath and tail central plasma sheet, etc.

Among these regions the solar wind and the interplanetary medium are surely those where the turbulence has been extensively investigated since the early era of space missions [1]. Indeed, although solar wind turbulence contains some features hardly classified within a general theoretical framework, this medium offers the best opportunity to study turbulence in collisionless plasmas, so that nowadays it is widely considered as a *natural laboratory* for the investigation and modeling of space plasma turbulence [1, 2, 3].

On the other side, the understanding and characterization of space plasma turbulence are fundamental to the modeling of plasma phenomena, such as plasma heating and acceleration, plasma transport across boundary regions, solar wind-magnetosphere coupling, etc. In the last years, it has been clearly understood that the understanding of such phenomena requires a detailed study of turbulent fluctuations of magnetic field and plasma parameters on different timescales from MHD down to kinetic domain.

Now, the study of the features of space plasma turbulence is essentially based on Fourier analysis and statistical approaches, being central issues the characterization of the spectral and scaling properties of both magnetic and velocity field increments/fluctuations. Both these approaches present some limitations and are based on some strong assumptions. For instance, Fourier analysis requires that the signals under investigation must be stationary, and that the increments/differences $\Delta x(\tau) = x(t+\tau) - x(t)$ at a timescale τ are representive of fluctuations at that scale.

However, there are several situations in which fluctuations appears to be non-stationary and, rigorously speaking, increments are only a proxy of fluctuations, since the fluctuations of a certain stochastic variable x are defined as the difference from the average value, i.e. $\delta x(t) = x(t) - \langle x(t) \rangle$, where the brackets mean some averaging procedure.

In 1998 Huang et al. [4] introduced a novel method of analysis, the *Hilbert-Huang Transform* (HHT), to characterize non-stationary signals resulting from the contribution of several oscillatory modes and embedded structures. The main feature of HHT technique stands in combining of a self-adaptive method, the *Empirical Mode Decomposition* (EMD), capable of identifying oscillatory modes and structures without using any *a priori* selected basis, with the standard Hilbert spectral analysis. The result of the application to a non stationary signal is that the signal is firstly decomposed into a set of zero-mean oscillatory modes, the *Intrinsic Mode Functions* (IMFs), which can be considered as the fluctuations at a given scale, to which successively is applied the Hilbert analysis, which allows to compute instantaneous frequency/spectrum. This allows to disentangle the traditional distinction between oscillatory modes and local fluctuations, solving the intrinsic problem of non-stationary fluctuations.

The aim of this work is to present a preliminary analysis of magnetic field fluctuations in the kinetic domain (i.e., below the ion-cyclotron frequency) using the HHT to investigate the potential of this method in characterizing the features of turbulent space plasmas down to non-MHD scales. To this end, we will present an analysis of an already studied solar wind period observed by ESA-Cluster mission.

2. The Hilbert-Huang Transform: a brief introduction

The Hilbert-Huang Transform was introduced by Huang et al. [4] in 1998 as a new tool for analysing nonlinear and non-stationary time series. HHT is indeed a very powerful method to investigate and analyze intermittent and amplitude-varying processes.

HHT consists of two successive steps: the *Empirical Model Decomposition* and the Hilbert spectral analysis [4, 5, 6]. The first step, the EMD, is indeed necessary to apply the Hilbert spectral analysis in such a way to remove possible fake results and is based on the assumption that a time series can be decomposed in a limited set of simple oscillatory modes, the *Intrinsic Mode Functions* (IMFs), which are characterized by significantly different average frequencies/periodicities. The set of IMFs is directly obtained by the time series without any *a priori* assumption via a self-adaptive method. Each IMF is characterized by two main features:

- the number of zero crossings and extrema must be the same or differ at most by one;
- the mean value of local maxima and local minima envelopes is zero.

As a result of the EMD, a time series $x(t)$ is decomposed as follows,

$$x(t) = \sum_{j}^{N} c_j(t) + res(t) \tag{1}$$

where $c_j(t)$ are the IMFs and $res(t)$ is a residue (the mean trend or a part of an oscillation characterized by a time scale longer than twice the time series under consideration). Nowadays, different methods have been developed to perform the EMD and to obtain the set of IMFs

IOP Conf. Series: Journal of Physics: Conf. Series **900** (2017) 012003 doi:10.1088/1742-6596/900/1/012003

[4, 5, 6, 7]. These methods are based on different stopping criteria of the sifting process [4] to get IMFs. Here, we make use of the Rilling et al. method [8].

Once the original time series is decomposed into the set of IMFs, it is, then, possible to apply the Hilbert spectral analysis to each IMF to compute the instantaneous frequency and amplitude. This consists in computing firstly the Hilbert transform of each IMF,

$$c_j^H(t) = \frac{1}{\pi}\mathcal{P}\int \frac{c_j(\tau)}{(t-\tau)}d\tau, \tag{2}$$

where $\mathcal{P}$ stands for the principal Cauchy value of the integral, and successively reconstructing the correspondent analytic signal $z_j(t)$, defined as:

$$z_j(t) = c_j(t) + ic_j^H(t) = A_j(t)e^{i\theta_j(t)}, \tag{3}$$

where $A_j(t)$ represents the instantaneous amplitude and $\theta_j(t)$ is the phase. The instantaneous angular frequency $\omega_j(t)$ (or frequency $f_j(t)$) is then computed by simply differentiating the phase $\theta_j(t)$, i.e.,

$$\omega_j(t) = \frac{d}{dt}\theta_j(t) \quad \rightarrow \quad f_j(t) = \frac{\omega_j(t)}{2\pi}. \tag{4}$$

Using the set of the instantaneous frequencies $f_j(t)$ and amplitudes $A_j(t)$ it is possible to compute the Hilbert spectrum, defined in terms of amplitudes, $H(t,f) = A(t,f)$, or squared amplitudes, $H(t,f) = A^2(t,f)$, by contouring the the instantaneous amplitude of each IMF as a function of both time and frequency. The Hilbert spectrum represents the amplitude (energy) in a time-frequency representation.

Other relevant quantities to characterize the fluctuation field in a time series can be defined from the Hilbert spectrum. For instance, it is possible to introduce a spectral density, named the *marginal spectrum* $H(f)$, which is equivalent to the Fourier power spectral density,

$$H(f) = \frac{1}{T}\int_0^T \frac{H(t,f)}{f}dt, \tag{5}$$

the *degree of stationarity* (DS) of a frequency,

$$DS(f) = \frac{1}{T}\int_0^T \left(1 - \frac{A(t,f)}{\bar{A}(f)}\right)^2 dt, \tag{6}$$

where $\bar{A}(f)$ is the mean amplitude of the frequency f, and the *persistency* (PS) of a frequency,

$$PS(f) = \frac{\Delta T(A(f,t) > 0)}{T}. \tag{7}$$

In the framework of space studies the HHT and, in particular, the EMD has been widely used to investigate the emergence of characteristic periodicities in studies related to both solar cycle and magnetosphere dynamics (see e.g., [9, 10, 11, 12, 13, 14]). Here, we apply the full HHT method to characterize some features of the magnetic field fluctuations at non-MHD/kinetic scales.

3. Data Description

To study the features of the magnetic fluctuations at the non-MHD scales we have consider a period of fast solar wind observed by ESA-Cluster mission on January, 20, 2007 from 12:00 UT to 13:15 UT. This period is a fast ($\bar{v} \sim 600$ km/s) high$-\beta$ ($\beta \sim 1.6$) solar wind stream characterized by nearly-stationary plasma conditions. Magnetic field data came from FGM and

IOP Conf. Series: Journal of Physics: Conf. Series **900** (2017) 012003 doi:10.1088/1742-6596/900/1/012003

STAFF experiments on-board of Cluster 3 satellite and are combined so to get a very high resolution (~ 450 samples per s).

Because the HHT procedure is very time consuming for very large datasets, the actual data have been down-sampled by a factor 4 so that the temporal resolution is of $\delta t \sim 8.9$ ms. Furthermore, data have been rotated from GSE reference system to the mean-field one.

Figure 1 shows the three magnetic field components for the selected time interval in the mean-field reference system. Looking to the traces plotted in Figure 1 it is evident how the magnetic field fluctuations are more pronounced in the perpendicular plane than in the parallel direction. This feature is confirmed by variance analysis ($\sigma_\perp^2/\sigma_{||}^2 \sim 26$) and suggests that the turbulent fluctuation field is essentially 2D.

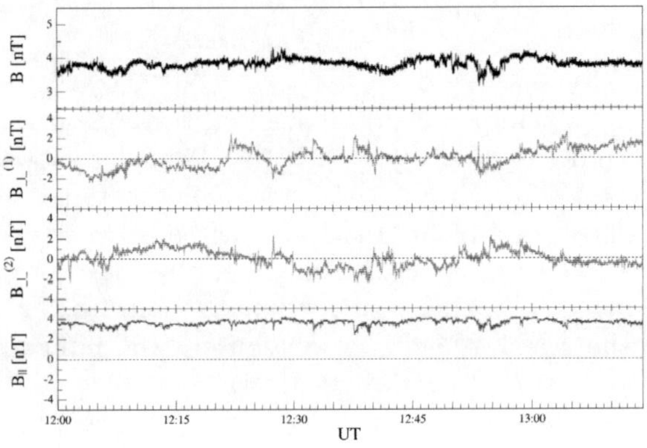

Figure 1. The magnetic field measurements relative to the period of fast solar wind observed by Cluster 3 satellite in the mean-field reference system.

4. Analysis and Results

As a first step of the HHT analysis, we apply the EMD to each time series of the magnetic field component and the magnetic field intensity. We get 18 and 19 IMFs for the magnetic field components and intensity, respectively. Each of this IMF is characterized by a mean characteristic time scale (periodicity) T_j that can be computed by measuring the average time interval between two successive zero-crossings of each IMFs. Figure 2 shows the characteristic periodicities T_j of the IMFs for the three components and the intensity. An interesting feature of EMD of the magnetic field components is that the components time series are decomposed into a set of IMFs characterized by the same periodicities.

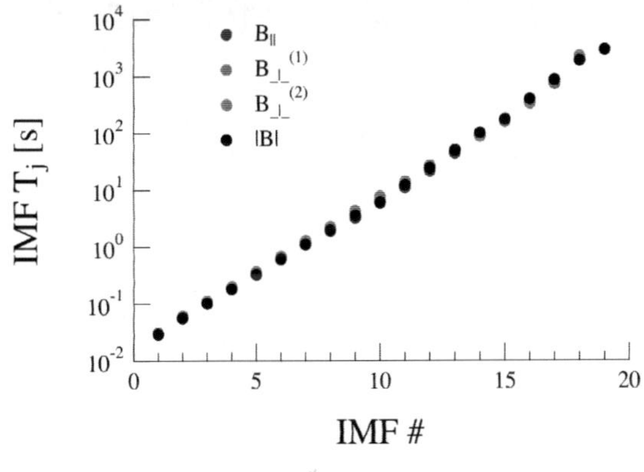

Figure 2. The characteristic periodicities T_j of the IMFs of the magnetic field components and intensity vs. the IMF mode number (IMF#).

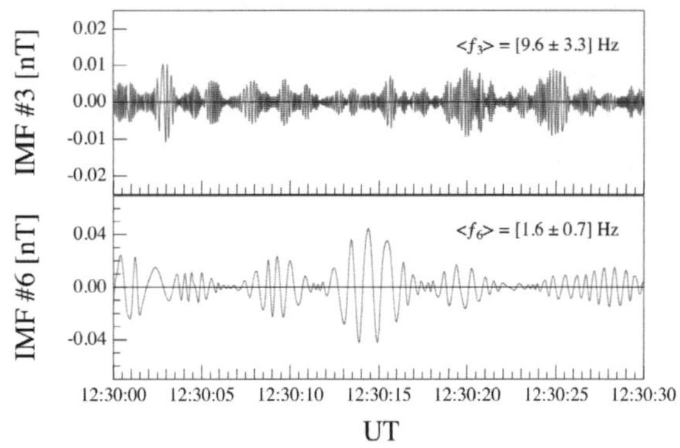

Figure 3. Samples of IMFs #3 and #6 of the magnetic field parallel component at timescales shorter than that corresponding to the ion-cyclotron frequency $f_\Omega \sim 0.06$ Hz. Here, $\langle f_j \rangle$ denotes the mean value of the local frequency $f_j(t)$ computed by Hilbert transform.

Figure 3 shows two samples of 30 s of the magnetic field parallel component at two timescales corresponding to frequencies above the ion-cyclotron frequency $f_\Omega \sim 0.06$ Hz. This plot shows the quasi-periodicity of the IMFs and the inherent intermittent character of fluctuations at these scales. Indeed, the IMFs show periods of relative small amplitude punctuated by intervals with a very large amplitude. However, although the traces seem nearly periodic, the local frequency $f_j(t)$, computed by Hilbert transform, is not constant from one point to another, but spread over a wide interval. For example, for the two IMFs shown in Figure 3 the interval $(\langle f \rangle - 2\sigma_f, \langle f \rangle + 2\sigma_f)$ is $f_3 \in (3.0, 16.2)$ Hz and $f_6 \in (0.2, 3.0)$ Hz, respectively.

Figure 4 shows the Hilbert energy spectrum of the parallel (left panel) and perpendicular (right panel) components for a short time interval of 30 min. The Hilbert spectrum of the perpendicular components is the total one computed as the sum of the two Hilbert spectra of the perpendicular components. The spectra show a singular character with non-persistent frequencies. This suggests that the inherent character of the fluctuation field seems to be more compatible with fluctuations instead of standing waves, as it should be for a regime of strong turbulence.

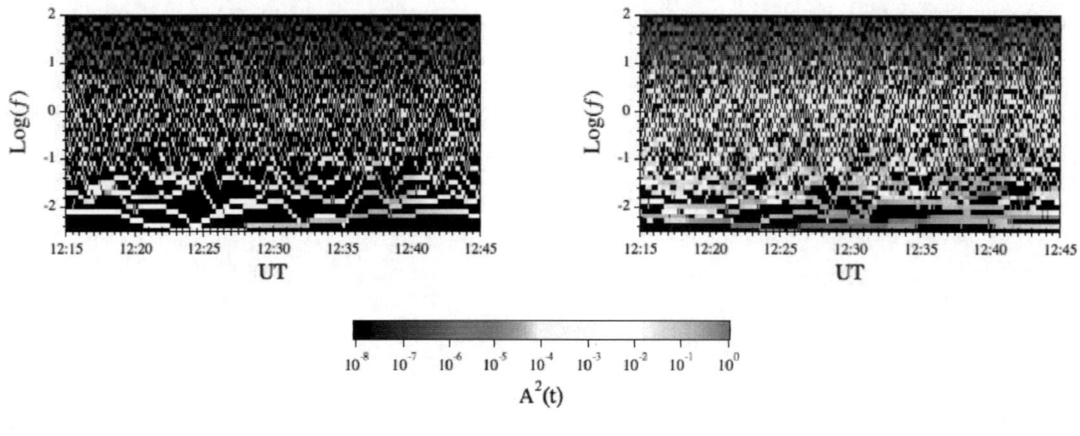

Figure 4. Hilbert energy spectra of the parallel (left panel) and perpendicular (right panel) components, respectively.

This point is confirmed by the degree of stationarity $DS(f)$ and the persistence degree $PS(f)$ as shown in Figure 5. Indeed, the persistence of the observed frequencies both in the MHD and kinetic domain do not exceed the 40%, clearly indicating that the fluctuation field is made by pulsations localized in time (or in the space assuming the Taylor's hypothesis).

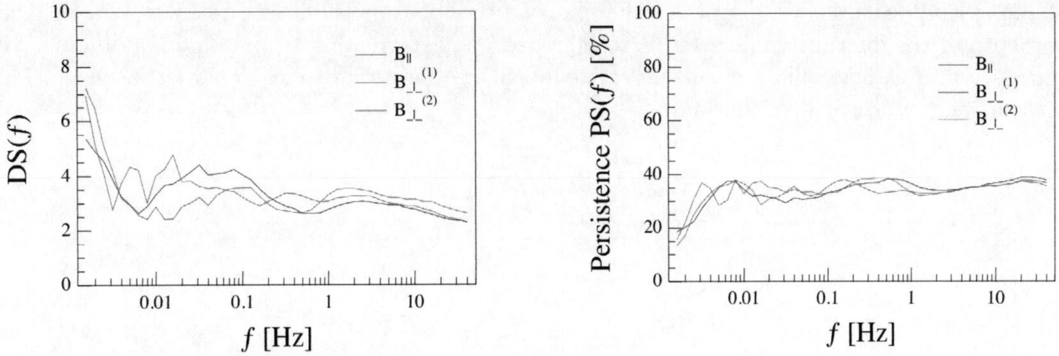

Figure 5. The stationary degree $DS(f)$ (left panel) and the persistence degree $PS(f)$ (right panel) for the parallel and perpendicular directions.

Moving from the Hilbert energy spectra, we evaluate the marginal energy spectra which is equivalent to the Fourier power spectrum. Figure 6 shows the marginal spectra in the parallel and perpendicular directions to the mean field.

Two different spectral regimes, characterized by different spectral exponents, can be identified above and below the spectral break $f_b \sim 0.3$ Hz. This frequency break f_b is compatible with the ion cyclotron frequency f_Ω considering the effect of Doppler shift. Furthermore, below the frequency break f_b all the spectra tend to the typical MHD Alfvénic turbulence spectrum ($S(f) \sim f^{-3/2}$), while above f_b, i.e., in the kinetic domain, the spectra are approximately $\sim f^{-8/3}$. This value seems to be compatible with several different kinetic turbulent regimes (e.g., compressible Hall-MHD turbulence, EMHD-turbulence, kinetic Alfvén wave turbulence), which predict a spectral slope near $\sim -7/3$.

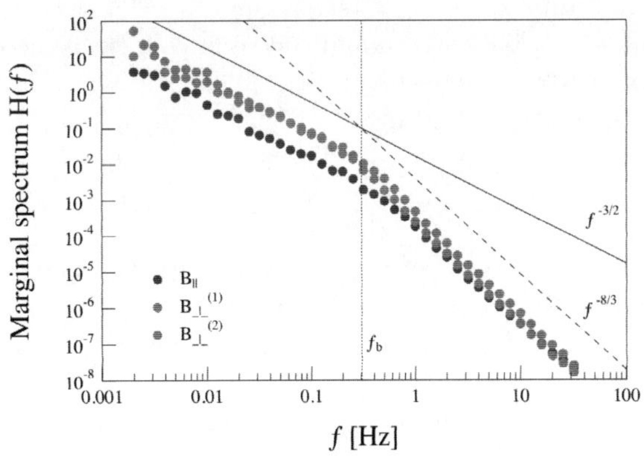

Figure 6. The marginal Hilbert energy spectra in the parallel and perpendicular directions to the mean field. The frequency break is $f_b \sim 0.3$ Hz.

An interesting feature emerging from the marginal spectra is that going to the kinetic scale the asymmetry between perpendicular and parallel spectra reduces. This can be evaluated introducing a sort of pseudo-dimension $D(f)$ defined as,

$$D(f) = \frac{1}{\max(H_j(f))} \sum_j H_j(f), \tag{8}$$

where j is the component index (i.e., parallel ($B_{||}$) and the two perpendicular components $B_{\perp}^{(1,2)}$) Figure 7 shows the behavior of this quantity computed using the marginal spectra. The most interesting feature emerging from this plot is how the dimensionality (or, if we prefer, the anisotropy) of the fluctuation field tends to increase (decrease) going to the kinetic domain. We underline how magnetic fluctuations tends to be 3D going toward electron inertial scale, i.e. towards a regime of isotropic fluctuation field.

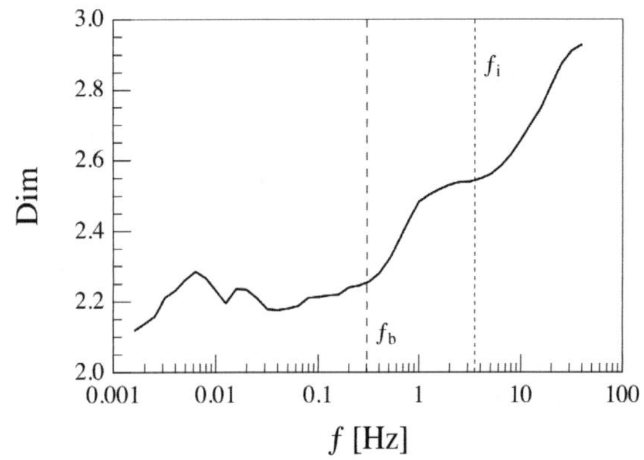

Figure 7. The pseudo-dimension $D(f)$ as a function of the frequency f. Here, f_b and f_i are the frequency break and the one corresponding to the ion-inertial scale η_i computed using the Taylor's hypothesis.

As last step of our analysis, we investigate the scaling features of the fluctuation field at the kinetic scale using the Hilbert spectrum. This can be done computing the qth-order pseudo-structure function, $S_q^*(f)$, using the frequency instead of the scale separation, i.e.,

$$S_q^*(f) = \langle A(t,f)^q \rangle_{\Omega_T} \sim f^{-\zeta(q)}, \tag{9}$$

where the averaging procedure is done over the time sub-interval, Ω_T, in which the frequency is present, i.e., $\Omega_T = \{\Delta t \in [0,T] \mid A(t,f) \neq 0\}$. The definition of the pseudo-structure function $S_q^*(f)$ is analogous to the standard structure function, $S_q(\tau) = \langle \mid x(t+\tau) - x(t) \mid^q \rangle$, once it is assumed that $A(t,f) \simeq \mid x(t+\tau) - x(t) \mid$ and $\tau \sim 1/f$. A similar procedure has been also adopted to analyze intermittency in homogeneous turbulence [15].

Figure 8 shows the behavior of the scaling exponent $\zeta(q)$ as a function of the moment order q. A linear trend is found for all the three directions, parallel and perpendicular to the mean field. Anyway, different scaling exponent are recovered indicating an anisotropy of the scaling feature parallel and perpendicular. The observed linear scaling of $\zeta(q)$ as a function of q suggests that at the kinetic scales the PDFs of Hilbert amplitudes, A_f, are characterized by mono-fractal scaling features. This is, indeed, the case as clearly shown in Figure 9 where we report the PDFs of the Hilbert amplitudes, A_f (where A_f stands for $= A(t,f)$), scaled to the standard deviation σ_{A_f} in the frequency range from 1 Hz to 20 Hz.

PDFs of scaled Hilbert amplitudes A_f collapse into a single master curve, which is compatible for the occurrence of mono-fractal scaling feature, i.e., there is no evidence of an intermittent cascading process which should be characterized by multi-fractal scaling features.

5. Summary and Conclusions

In this very preliminary work we have investigated the potentiality of the full Hilbert-Huang Transform approach in characterizing the properties of turbulent space plasmas. In particular, we have applied the full method and not limited our work only to the first step, the EMD analysis. This have allowed to characterize the intermittent nature of the fluctuation field clearly showing that there are no-standing frequencies at all the investigated MHD and kinetic scales.

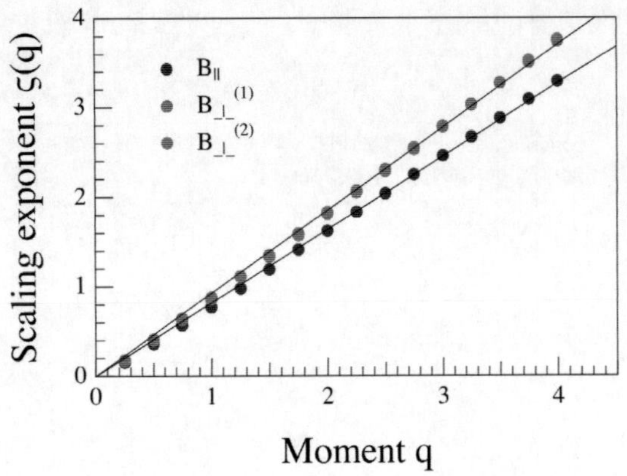

Figure 8. The scaling exponent $\zeta(q)$ as a function of the moment order q.

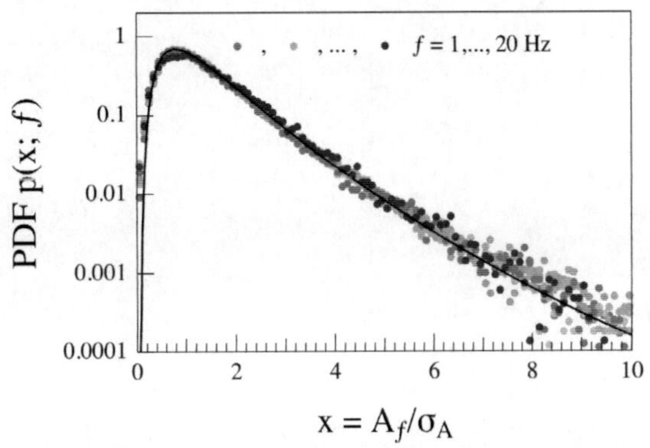

Figure 9. Data collapsing of the PDFs of the Hilbert amplitudes, A_f, scaled to the standard deviation σ_{A_f}. The black curve is a Log-Normal PDF.

The results suggest that the fluctuation field at kinetic scale is essentially characterized by a mono-fractal set of scale invariant fluctuations. Furthermore, we have found an increases of dimensionality from 2 to 3 going toward the electron inertial scale. In other words, we are in presence of an increase of isotropy of the fluctuation field toward the electron kinetic scales.

Clearly, this is only a first application of the HHT method, which has many other unexplored potentialities to investigate turbulence such as a multiscale variance analysis of the real fluctuation field.

Acknowledgments
We acknowledge the CLUSTER FGM and STAFF P.I.s and teams and the ESA-Cluster Active Archive team for making available the data used in this work. This research has received funding from the Italian Space Agency under Grant agreement No. ASI-INAF 2015-039-R.O "Missione M4 di ESA: Partecipazione Italiana alla fase di assessment della missione THOR".

References

[1] Bruno R and Carbone V 2013 *Liv. Rev. Solar Phys.* **10** 2
[2] Zhou Y, Matthaeus WH and Dmitruk P 2004 *Rev. Mod. Phys.* **76** 1015
[3] Carbone V and Pouquet A 2009 *Lect. Notes Phys.* **778** 71
[4] Huang NE, Shen Z, Long SR et al 1998 *Proc. Roy. Soc. Lond. Ser. A* **454** 903
[5] Huang NE, Wu M-LC, Long SR et al 2003 *Proc. Roy. Soc. Lond. Ser. A* **459** 2317
[6] Huang NE and Wu Z 2008 *Rev. Geophys.* **46** 1
[7] Flandrin P, Rilling G, and Goncalvès P 2004 *IEEE Sig. Proc. Lett.* **11** 112

IOP Conf. Series: Journal of Physics: Conf. Series **900** (2017) 012003 doi:10.1088/1742-6596/900/1/012003

[8] Rilling G, Flandrin P and Goncalves P 2003 *Proc. of 330 IEEE EURASIP workshop on nonlinear signal and image processing*
[9] Vecchio A et al. 2010 *Astrophys. J.* **709** L1
[10] Vecchio A et al 2012 *Astrophys. J.* **749** 27
[11] De Michelis P, Consolini G and Tozzi R 2012 *Nonlin. Processes Geophys* **19** 667
[12] Perri S et al 2012 *Phys. Rev. Lett.* **109** 245004 doi: 10.1103/PhysRevLett.109.245004
[13] Stangalini M et al 2014 *Astron & Astrophys* **569** A102
[14] Alberti T et al 2017 *J. Geophys Res* **122** doi:10.1002/2016JA023175
[15] Huang YX et al 2008 *EPL* **84** 40010

IOP Conf. Series: Journal of Physics: Conf. Series **900** (2017) 012004 doi:10.1088/1742-6596/900/1/012004

Heliosphere in a strong interstellar magnetic field

A Czechowski and J Grygorczuk

Space Research Centre PAS, ul.Bartycka 18A, Warsaw, POLAND

E-mail: ace@cbk.waw.pl

Abstract.

We present results of MHD simulations of the global structure of the heliosphere for a wide range of the interstellar magnetic field (ISMF) strengths (2-20 μG). For a strong interstellar field, the plasma flow in and around the heliosphere as well as the magnetic field draped over the heliopause have a simple structure, reminiscent of (though not identical to) the analytical model by Parker. We show how this structure evolves as the interstellar field strength is reduced down to the values in the observed range (few μG). We also present an example of using this evolution to understand the present observations of the heliosphere.

1. Introduction

In this paper we investigate the global structure of the heliosphere for the ISMF strengths B_∞ range from 2 μG up to 20 μG.

First in situ measurements of the ISMF were made by Voyager 1 in 2012 when the spacecraft crossed the heliopause and entered the outer heliosheath region [1]. Since that time the observed ISMF strengths vary from 4 to 6 μG as shown in Figure 1 [1, 2, 3, 4, 5]. As the spacecraft still traverses the region in which the ISMF is draped around the heliopause, we do not have yet direct information about the field undisturbed by the heliospheric obstacle. Several estimations of the pristine ISMF using global MHD simulations give magnitudes of 2-4 μG [6, 7, 8, 9, 10]. The latest determination from the IBEX ribbon gives B_∞=2.93 μG [11].

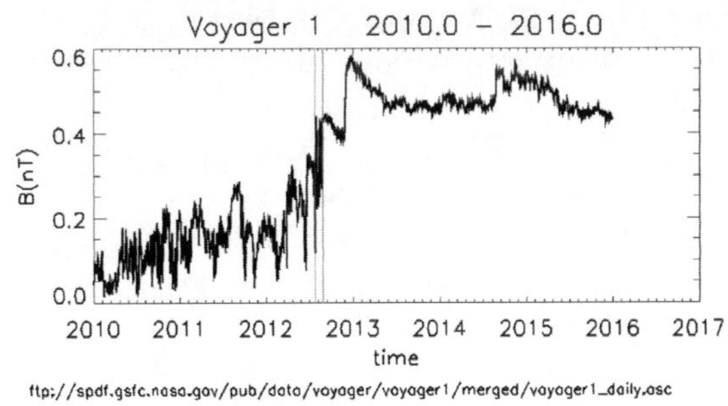

Figure 1. Daily averages of $|B|$ measured by Voyager 1 since 2010 (black line [2]). Two vertical red lines show DOY208 and DOY238 of 2012 (the last one corresponds to the date of the heliopause crossing by Voyager 1).

Why should we therefore consider the hypothetical heliosphere embedded in an unrealistically strongly magnetized (20 μG) interstellar plasma? Our reasons are as follows.

First is that the interaction of solar (stellar) wind with strong ISMF is of interest in itself as a theoretical problem.

IOP Conf. Series: Journal of Physics: Conf. Series **900** (2017) 012004 doi:10.1088/1742-6596/900/1/012004

Secondly, the range of the interstellar field strength which we consider (2 to 20 μG) is typical in the galactic disk and bulge [12]. Some of our results may therefore have application for astrospheres, including those of the solar like stars.

Third reason (our main motivation for this paper) is the observation (based on theoretical models [13, 14, 15]) that the large-scale structure of the heliosheath plasma flow and the magnetic field draped over the heliopause become simpler for high B_∞. By following the evolution of these structures with decreasing interstellar field strength one may then gain a better understanding of the heliosphere as actually observed, and explain some observations by Voyager 1 [1] and Interstellar Boundary Explorer (IBEX) [16].

To obtain global time-stationary models of the heliosphere, we use the 3D MHD code, adapted and corrected from [17] and already applied in [14, 15]. Except B_∞, other physical parameters of the local interstellar medium (ISM) as well as the parameters of the solar wind are kept close to the observed values, with the interstellar B_∞ direction towards the IBEX ribbon center [18], and the ISM velocity vector from [19]. The solar magnetic field is not included. The solar wind is either isotropic or two-component (slow/fast). The neutral hydrogen is approximated as a background with a constant flux parallel to the velocity of the undisturbed interstellar medium. To avoid overestimation of the effect of neutral hydrogen on the heliospheric plasma flow, we choose the constant value of the neutral hydrogen density to be n_H=0.1 cm^{-3}, close to the value observed inside the heliosphere [20], rather than the higher value expected in the interstellar medium. We use a fixed spherical grid with size $n_r \times n_\theta \times n_\phi = 348 \times 90 \times 180$ for most calculations.

The plan of our presentation is as follows. In Section 2 we discuss the analytical model by Parker and our expectations for the structure of the astrosphere of a moving star in a strongly magnetized medium. In Section 3 we concentrate of the two-stream structure of the plasma flow inside the astrosphere. Section 4 deals with the shape of the boundary of the astrosphere in the ($\vec{B}_\infty$,$\vec{V}_\infty$) plane. In Section 5 we present our results concerning the structure of the draped magnetic field and their application to Voyager 1 measurements near and beyond the heliopause. Section 6 presents briefly some of our results concerning the production of energetic neutral atoms in the heliosphere. In Section 7 we present our conclusions.

2. Star in a strong interstellar magnetic field

A simple analytical model of a stellar wind confined by the ISMF was proposed by Parker [13]. This model is applicable to the case of a star at rest relative to strongly magnetized ISM, with the magnetic field pressure much higher than the pressure of the interstellar plasma. The model astrosphere, from which the ISMF is excluded, consists of a central cavity connected with two outflow tunnels, parallel and anti-parallel to the interstellar field direction. After passing the (spherical) termination shock, the stellar wind plasma flows down these tunnels, forming two oppositely directed streams. The ISMF draped over the astropause has a simple form, described by an analytical formula [13].

Figure 2 shows the solution obtained from our simulations [15] for the case of the ISMF strength of 20 μG with the star at rest relative to the undisturbed medium, which includes also a neutral hydrogen component. The result is similar to the Parker model. Two streams (parallel and anti-parallel to the ISMF direction) as well as the tube-like shape of the heliopause are well visible. Note that the termination shock is not deformed by the magnetic field but approximately spherical, also in agreement with Parker model.

What happens when the star is moving through a strongly magnetized medium? There is no simple analytical model able to answer this question and we have to rely on qualitative arguments and numerical simulations.

The following qualitative argument suggests the form of the solution. The motion of the astrosphere through the magnetized medium induces the interstellar plasma flow, directed partly

IOP Conf. Series: Journal of Physics: Conf. Series **900** (2017) 012004 doi:10.1088/1742-6596/900/1/012004

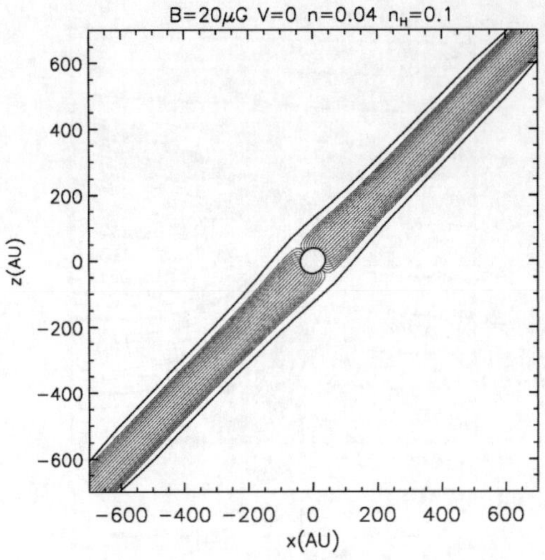

Figure 2. Flow lines of the solar plasma for the MHD solution corresponding to the hypothetical heliosphere with the Sun at rest (V_∞=0) relative to the ISM with plasma density n_{IS}=0.04 cm^{-3}, neutral hydrogen density n_H=0.1 cm^{-3} and very strong ISMF (B_∞=20 μG). The solar wind is spherically symmetric with the speed V_{SW}=750 km/s and the proton density at 1 AU $n_{SW,1AU}$=4.2 cm^{-3}. The solution is similar to the Parker model. The heliopause and the termination shock are shown by thick lines.

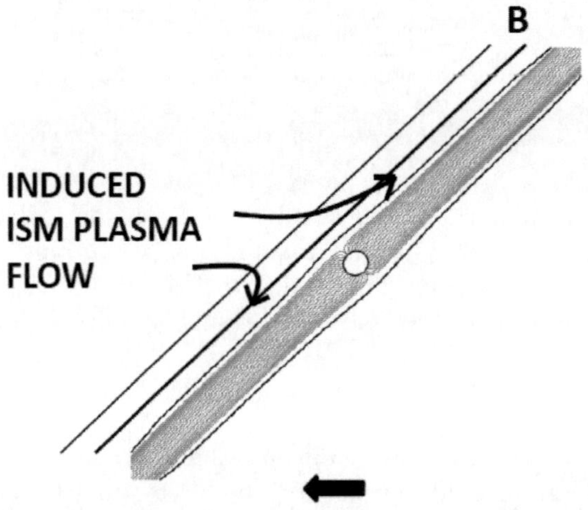

Figure 3. Schematic view of induced interstellar plasma flow in vicinity of an astrosphere when the star is moving through the strongly magnetized ISM. The induced ISM plasma flow moves partly along the magnetic field lines (denoted by B). The motion of the star (a thick arrow shows its direction) affects also the flow inside the astrosphere (see next Figure).

along the magnetic field lines, parallel and anti-parallel to the ISMF (see Figure 3). Combination of the motion along the magnetic field lines with the motion of the star would slightly deflect the flow, as seen in the rest frame of the star, from the magnetic field (and anti-field) direction. Apart from this deflection, the leading part of the boundary between the outer and the inner plasma would then be similar to the (slightly bent) tube-like structure of the Parker model. The flow inside the astrosphere would adjust to the boundary, suggesting that at least a part of the inner flow would also form two streams.

Our MHD simulation for strong ISMF (10-20 μG) is consistent with this simple picture (Figure 4, left panel). We conclude that the two-stream structure of the flow inside the astrosphere, obtained in the case of a star at rest, may survive to some extent also for a moving star.

Figure 4 also shows the effect of asymmetric pressure of the ISMF, which causes the astrosphere expansion in the ($\vec{B}_\infty$,$\vec{V}_\infty$) plane (upper panels) and contraction in the perpendicular direction (bottom panels). This effect is well known from many numerical simulations.

The boundary surface of the astrosphere (the astropause) in the Parker model has the topology of a tube, open on two sides. One could imagine a similar structure for the astropause

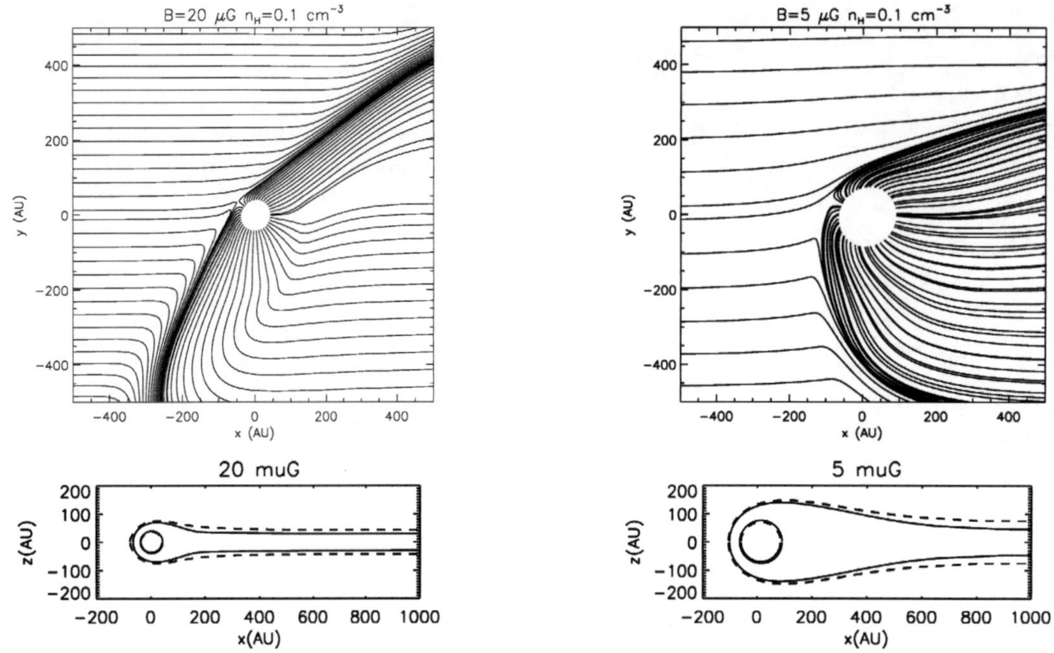

Figure 4. Plasma flow lines in the (x,y)≡($\vec{B}_\infty$,$\vec{V}_\infty$) plane (top panels) and the shape of the heliopause and the termination shock in the (x,z) plane (bottom panels) obtained from MHD simulations for B_∞=20 μG (left panels) and B_∞=5 μG (right panels), assuming V_∞=23.2 kms^{-1}, n_∞=0.06 cm^{-3}, n_H=0.1 cm^{-3}, and a spherically symmetric solar wind with V_{SW}=750 kms^{-1}, $n_{SW,1AU}$=4.2 cm^{-3}. The dashed lines in the bottom panels show the results obtained in a simulation using a coarse grid (95 × 50 × 100).

of a moving star, with the stellar plasma evacuated along the two tunnels rather than the single astrotail. Apart from the special case of the magnetic field parallel to the star motion, to our knowledge this possibility is not supported by numerical simulations. In particular, our results shown in Figure 4 indicate the presence of a single tail flattened by the magnetic field. The tube-like topology of the heliopause was, however, proposed in a different context [21, 22].

As shown in the bottom panel of Figure 4, our results for the thickness of the distant tail (although not for the shape of the heliosphere in the ($\vec{B}_\infty$,$\vec{V}_\infty$) plane) are sensitive to the choice of the grid. The full solution for the tail structure requires a separate study.

3. Two-stream structure of plasma flow

Figure 4 (top panels) shows plasma flow streamlines in the ($\vec{B}_\infty$,$\vec{V}_\infty$) plane for the case of the ISM moving relative to the Sun at the velocity V_∞=23.2 kms^{-1}. For the 20 μG case, the two-stream structure can be distinguished as two "bunches" of approximately straight streamlines near the forward boundary of the heliopause (which separates the interstellar and solar plasmas). Compared to Figure 2, the streams are deflected from the directions parallel and anti-parallel to ISMF.

After ~300-500 AU of approximately straight flow, the streams can be seen to commence turning tailward. This is caused by charge-exchange interaction between the plasma flow and the background neutral hydrogen. In the absence of this interaction, the two straight streams structure would be more extensive and prominent even for the 5 μG case.

Looking for a quantitative characterization of the two-stream structure, we examined different

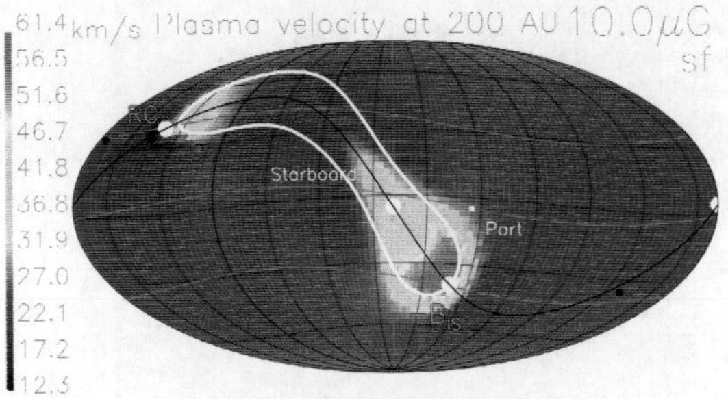

Figure 5. Directional distribution of the plasma speed at a distance of 200 AU from the Sun, obtained from MHD simulations for B_∞=10 μG. The all-sky Mollweide projection is centered on the anti-apex direction of the ISM flow (white dot). Also shown are the interstellar magnetic field and anti-field directions (white dots denoted by RC and $-B_\infty$, respectively), and the heliopause shape at the sphere shown here (white oval).

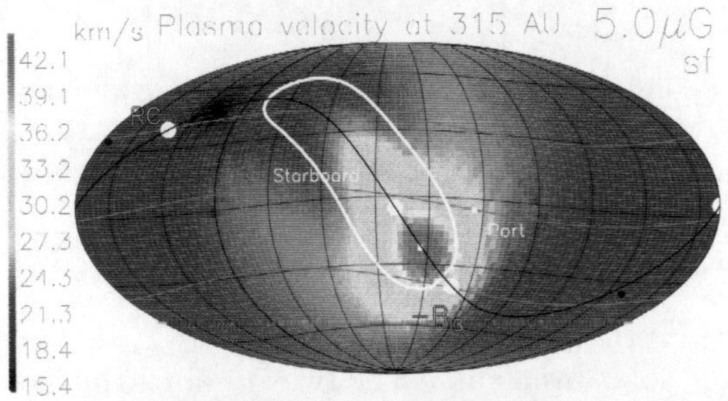

Figure 6. Directional distribution of the plasma speed at a distance of 315 AU from the Sun, obtained from MHD simulations for B_∞=5 μG. See Figure 5 caption for other descriptions.

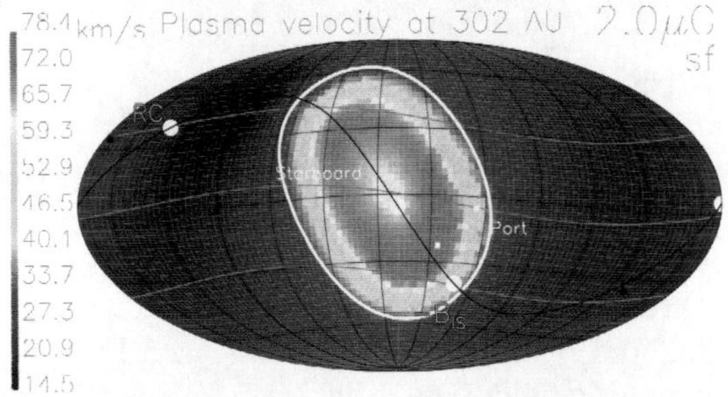

Figure 7. Directional distribution of the plasma speed at a distance of 302 AU from the Sun, obtained from MHD simulations for B_∞=2 μG. See Figure 5 caption for other descriptions.

possibilities. Figures 5, 6, and 7 show the calculated distributions of the plasma speed over the Sun-centered spheres of the radii 200 (B_∞=10 μG), 315 (B_∞=5 μG) and 302 AU (B_∞=2 μG). The heliopause is shown by the white ovals, with the enclosed region corresponding to the inside of the heliosphere. For strong ISMF (10 μG) the two streams appear as two separate high speed regions near the directions parallel and anti-parallel to the undisturbed ISMF. For 5 μG at 315 AU there is one high speed region, near the anti-field direction. However, two streams would be present at 200 AU (not shown). In the case of 2 μG, the speed distribution is approximately symmetric with respect to the center of the heliotail, which we interpret as the absence of the magnetic field-related stream structure.

Another way to illustrate the two-stream structure is to consider the energy flux distribution. Figure 8 shows the distribution of the radial component of the plasma energy flux in the ($\vec{B}_\infty$,$\vec{V}_\infty$) plane at the heliocentric distance 200 AU (20 μG) and 300 AU (5 μG), plotted as a function of the angle counted from the ISM inflow direction. Again we can see two maxima near the field and anti-field direction appearing for the strong magnetic field 20 μG and only one surviving for the 5 μG case at 300 AU.

Another characterization of the stream structure can be provided by the distribution of the plasma mass flux across the sphere surrounding the Sun at the chosen distance. The plasma

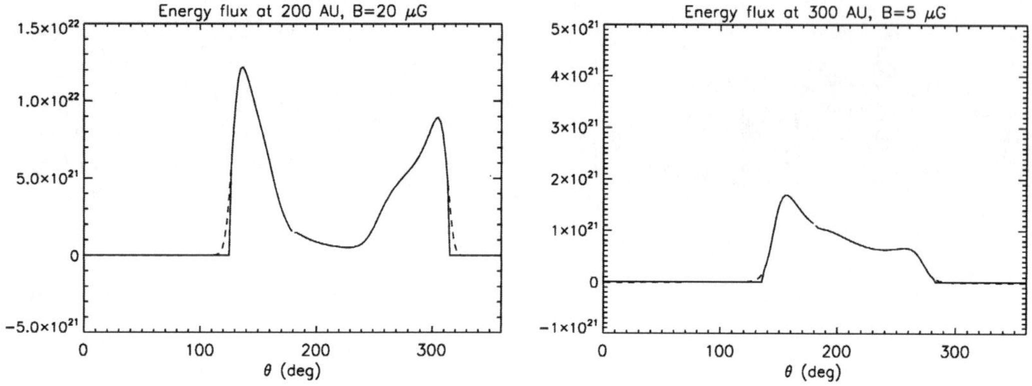

Figure 8. Energy flux distribution at fixed radial distance from the Sun in ($\vec{B}_\infty$,$\vec{V}_\infty$) plane, versus angle from ISM inflow direction (θ). Shown are results of MHD simulations for B_∞=20 μG (200 AU, left panel) and B_∞=5 μG (300 AU, right panel). For 20 μG two streams are present, for 5 μG only one survives.

mass flux is, however, dominated by the region outside the heliopause.

The above results were obtained for the symmetric solar wind case. The two-stream structure of the heliospheric flow is present also when the two-component solar wind is considered. For our simulations of this case see [15].

4. Shape of the heliopause: Alfvén wings

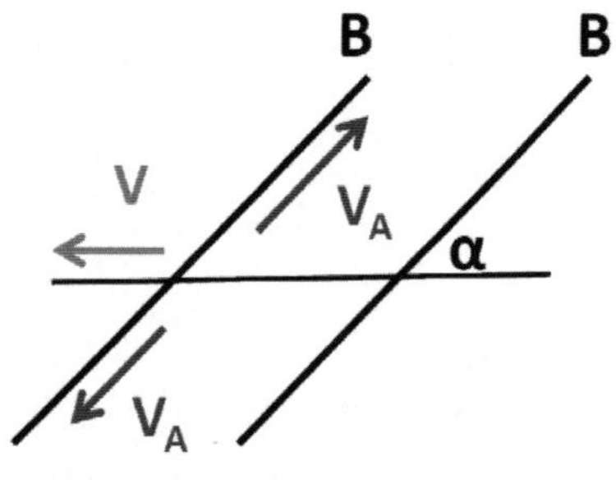

Figure 9. Schematic formation of Alfvén wings arising from the interaction of a conducting or mass loading source moving through a magnetized plasma. The source speed is denoted by V, its direction is shown by red arrow. The magnetic field lines direction is inclined to the source velocity vector by angle α. The signal propagates with the Alfvén speed (V_A) along B lines (as shown by blue arrows). As a source moves, Alfvén waves launched away from the source form Alfvén wings (not shown).

For strong B_∞ (20 μG and, if n_H is small, even for 5 μG) we find that the forward part of the heliopause in the ($\vec{B}_\infty$,$\vec{V}_\infty$) plane has the form of straight "wings". This is reminiscent of the Alfvén wings [23, 24], which appear when a conducting body (like a satellite or a planetary magnetosphere) is moving through a magnetized plasma (Figure 9). The angle α between the Alfvén "wings" and the x axis (the V_∞ direction) is given by:

$$\tan\alpha = \frac{\tan\gamma}{1 \pm V_\infty/V_A \cos\gamma}, \qquad (1)$$

where γ is the angle between the interstellar magnetic field and the x axis, and V_A is the Alfvén speed in the interstellar medium.

The Alfvén wings shape of the heliopause in the ($\vec{B}_\infty$,$\vec{V}_\infty$) plane was derived in a simplified model [25] using the Newtonian approximation [26]. The heliosphere of this form was also recently proposed by [27].

IOP Conf. Series: Journal of Physics: Conf. Series **900** (2017) 012004 doi:10.1088/1742-6596/900/1/012004

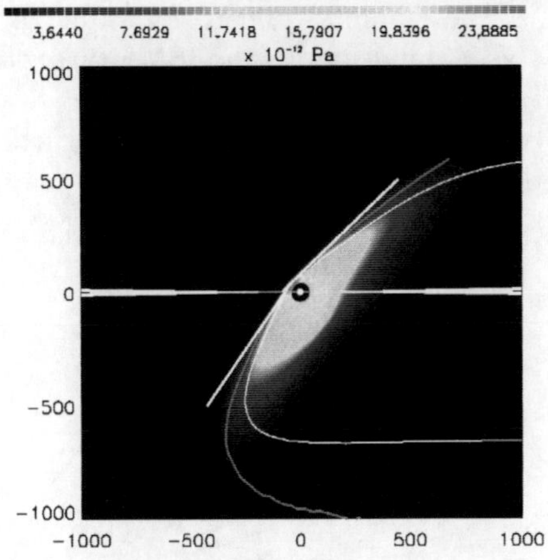

Figure 10. Heliopause (white curve) in ($\vec{B}_\infty$,$\vec{V}_\infty$) plane versus Alfvén wings (white straight lines) over thermal pressure map (color scaled) obtained from MHD simulations for B_∞=20 μG and V_∞=23.2 kms^{-1}. Also shown is estimated outer limit of heliopause transition region (red curve).

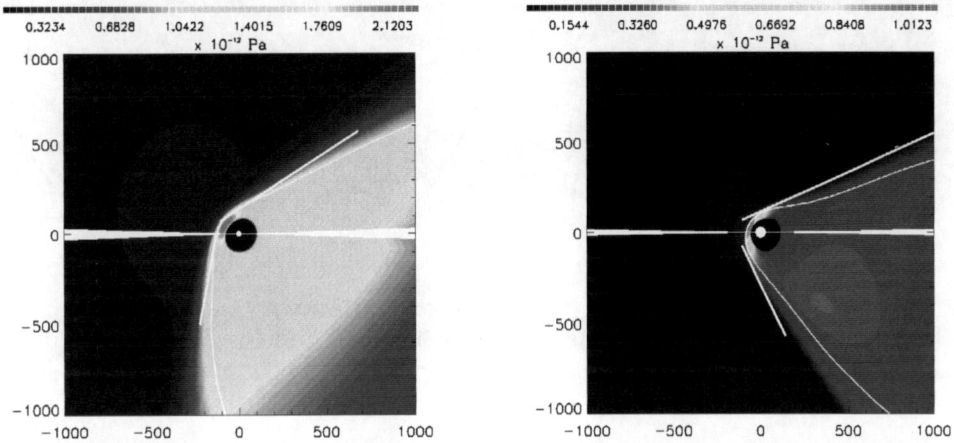

Figure 11. Similar to Figure 10. Results obtained for B_∞=5 μG and V_∞=23.2 kms^{-1} (left panel) or V_∞=46.4 kms^{-1} (right panel).

Although our results confirm the presence of straight "wings", the angle between each wing and the ISM velocity direction is not in agreement with Eq. 1. In Figures 10 and 11, the directions of the Alfvén wings are shown by straight white lines.

Note that the definition of the heliopause used by us (the boundary between the flow lines of the inner and the outer plasma flow), although correct for the exact solution, might be not the optimal definition for a numerical calculation. In a numerical simulation, the heliopause does not appear as a boundary surface, but rather as a finite width transition region between the inner and the outer plasmas. We estimated the outer boundary of the transition region in our simulation (the thin red curve in Figure 10), and found that the the Alfvén wing lies outside this region.

5. Draped interstellar magnetic field structure

We found it convenient to present our discussion in terms of the magnetic field lines projected on a celestial sphere (sky projections for short).

Figure 12 (left panel) shows a set of magnetic field lines passing through selected points just

IOP Conf. Series: Journal of Physics: Conf. Series **900** (2017) 012004 doi:10.1088/1742-6596/900/1/012004

above the heliopause (the change from black to red occurs at these points). Their sky projections are shown in the right panel. The lines result from our MHD simulation for the ISMF direction towards the IBEX ribbon center (RC, [18]), and the field magnitude $B_\infty = 3\ \mu$G.

Away from the heliopause, the field lines must finally become parallel to the undisturbed field direction. Their sky projections must therefore approach the point corresponding to the $\vec{B}_\infty$ direction (here: RC) (Figure 12, right panel).

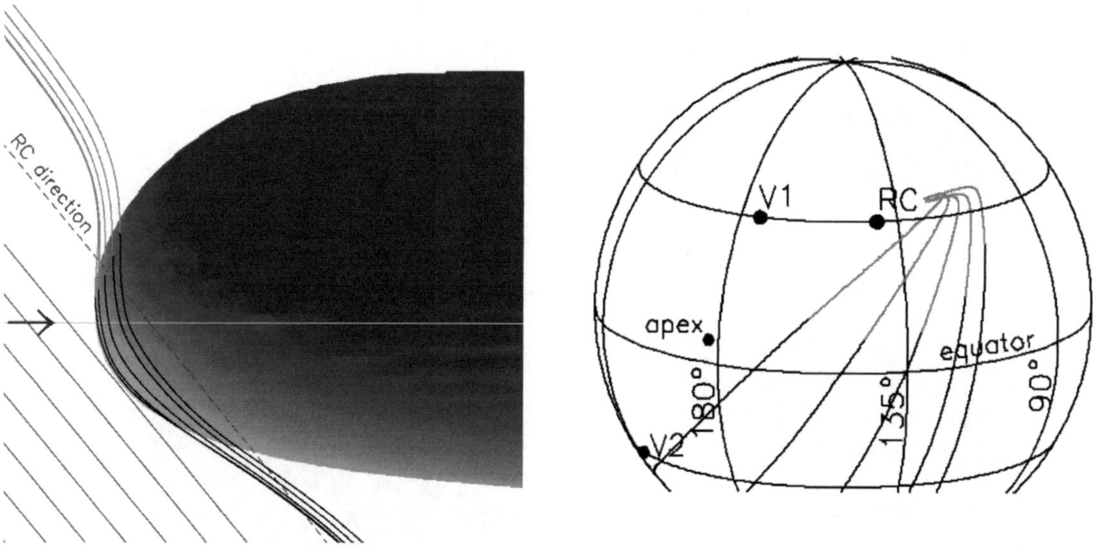

Figure 12. Magnetic field lines passing through selected starting points just outside the heliopause (black/red lines, color changes at starting point) shown as the lines draped around the heliopause (gray shaded surface in the left panel) and in projection on the celestial sphere (right panel).

In the analytical Parker model [13], the magnetic field lines in this projection have the form of great circles crossing at the points corresponding to the directions parallel and antiparallel to the undisturbed magnetic field. From our calculations we found [14] that, for a strong interstellar field (20 μG), the magnetic field draped around the moving heliosphere still has a simple structure similar to the Parker model. The weaker the field, the more the sky projections of its lines differ from great circles, especially in vicinity of the undisturbed field direction (Figure 12, right panel).

5.1. Direction of the ISMF observed by Voyager 1 just beyond the heliopause

Our analysis of evolution of the magnetic field draped over the heliopause for the decreasing interstellar field magnitude helped us in our attempt [14] to understand the observations by Voyager 1 just beyond the heliopause.

Voyager 1 crossed the heliopause and entered the interstellar medium in August, 2012 [1]. Observed before the crossing, the interplanetary magnetic field direction was close to the expected Parker spiral (i.e. lying in the spacecraft's heliographic parallel plane). Unexpectedly, there was no significant (in fact: $\sim 20°$) change in the field direction during crossing of the boundary [1].

In [14] we proposed a simple explanation of this fact. We noticed a special coincidence between the directions of Voyager 1 trajectory and of the undisturbed ISMF (assumed to be given by the IBEX ribbon center). They share approximately the same heliolatitude ($\sim 35°.5$) and are not far separated in longitude (see Figure 12, right panel).

In a sky projection, the Parker spiral of solar magnetic field just ahead of the heliopause approximately follows the heliographic parallel. The projection of the interstellar magnetic field line, observed by Voyager 1 just beyond the heliopause, should connect Voyager 1 trajectory direction with the direction of the undisturbed interstellar field [14].

Figure 13 illustrates schematically (left panel) and basing on our simulation for B_∞=3 μG (right panel) such projections for two cases: strong ISMF (solid red line) and weak ISMF (dashed red line). V1 and RC denote the projections of Voyager 1 trajectory and the direction of B_∞ (IBEX ribbon center) which share the same heliographic latitude. The solid red line is the arc of great circle which corresponds to sky projection of the magnetic field line in the analytical Parker model. The angle (denoted by ϕ) between the projections and the heliographic parallel (which is approximately the projection of the solar magnetic field line) for the case of strong field is small, but may be bigger for the weaker field. However, our simulations (Figure 13, right panel) give the value of this angle which is small enough [14] to be consistent with Voyager 1 observations.

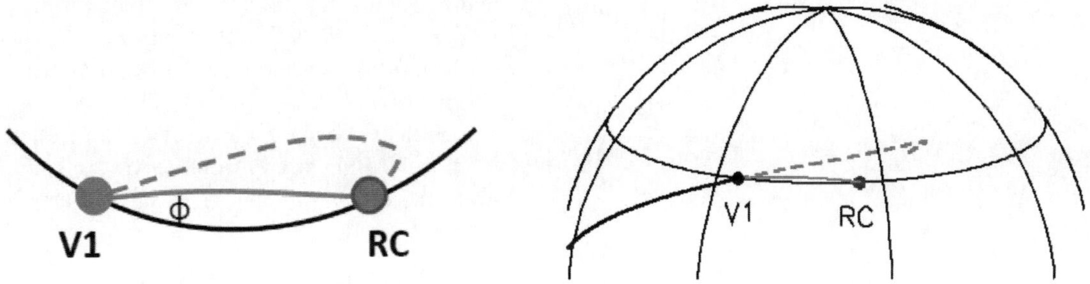

Figure 13. Sky projection of magnetic field lines which cross the Voyager 1 trajectory just beyond the heliopause, and approach RC far from the heliopause (left panel: schematic view, right panel: our simulation results). Solid red line (arc of a great circle) corresponds to the analytical Parker model for a strong ISMF, dashed red line to a weak field case. The (dashed red / solid black) line shown in the right panel is obtained from the same simulation as in Figure 12 (B_∞ of 3 μG).

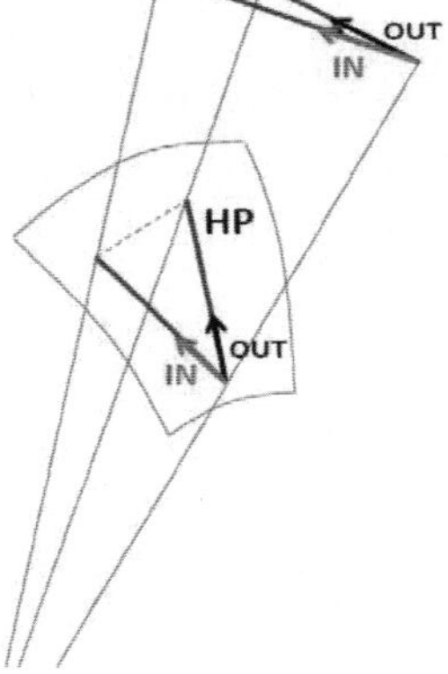

Figure 14. Geometrical construction showing the segments of the magnetic field lines draped over the inner (IN) the outer (OUT) side of the surface element of the heliopause (HP), and their sky projections (the top of the figure). The angle between the field lines is not equal to the angle between their sky projections, but the difference is small provided that the surface element of the heliopause over which the field lines are draped is approximately perpendicular to the heliocentric radial direction.

Note that the angle between the sky projected vectors is not the same as the angle between the vectors draped on both sides of the heliopause (Figure 14). However, the difference will be small it the surface element of the heliopause is not far from perpendicular to the heliocentric radial direction. This condition was met by our MHD simulation.

5.2. Evolution of magnetic field direction beyond the heliopause along Voyager 1 trajectory

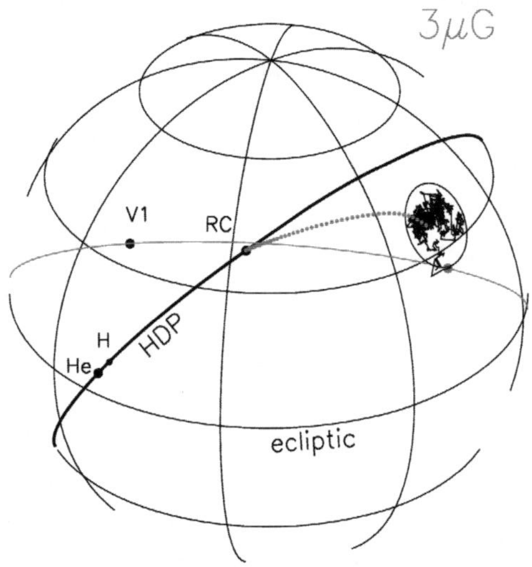

Figure 15. Celestial sphere in orthogonal projection and solar ecliptic coordinates. Shown are the magnetic field directions observed by Voyager 1 just beyond the heliopause (blue dot [1]) and then daily averages up to 2016.0 [2] (black curve inside a circle of ∼9° radius). The evolution of magnetic field direction from the heliopause to ∼5000 AU distance from the Sun at Voyager 1 direction, obtained from our MHD simulations for B_∞=3 μG, and projected on the celestial sphere is shown by red dots.

The interstellar magnetic field directions observed by Voyager 1 up to 2016 [2, 5] are shown in Figure 15 (black curve inside the circle). The observed field was disturbed by several shocks of solar origin [5]. The draped ISMF directions obtained from our simulations for B_∞ from 2.7 to 3 μG ([14], Table 1), but not for stronger field, fall inside this region.

Our simulations permit us to follow the variation of the magnetic field direction at Voyager 1 with the increasing distance from the Sun (Figure 15, red dotted curve). The curve links the region of observed field directions with the undisturbed field (RC). The shape and position of the curve depends on the parameters assumed in the simulation, in particular on B_∞ (in Figure 15, 3 μG).

6. Simulated ENA flux distribution originated in the inner heliosheath

A detailed discussion of production of the energetic neutral atoms (ENA) in the heliosphere for a wide range of ISMF strengths was presented in [15]. In the present work we limit ourselves to a brief presentation of our recent results based on a new model (Zank et al. [28]) of the termination shock energetic ion spectrum.

We calculate the heliospheric ENA flux assuming that the ENA derive from the energetic ions accelerated at the termination shock and transported from there to the ENA production site by plasma convection (spatial diffusion is unimportant for the ions in the IBEX energy range). ENA flux from a given line-of-sight (LOS) is then determined by the energetic ion spectrum at those points of the termination shock which are linked to the line of sight by plasma streamlines (Figure 17).

The main factors which shape the directional dependence of the ENA flux are: (1) the energetic ion flux intensity in the relevant region at the termination shock, (2) the length of the line-of-sight interval within the heliosphere and (3) the transport effects on the energetic ion

IOP Conf. Series: Journal of Physics: Conf. Series **900** (2017) 012004 doi:10.1088/1742-6596/900/1/012004

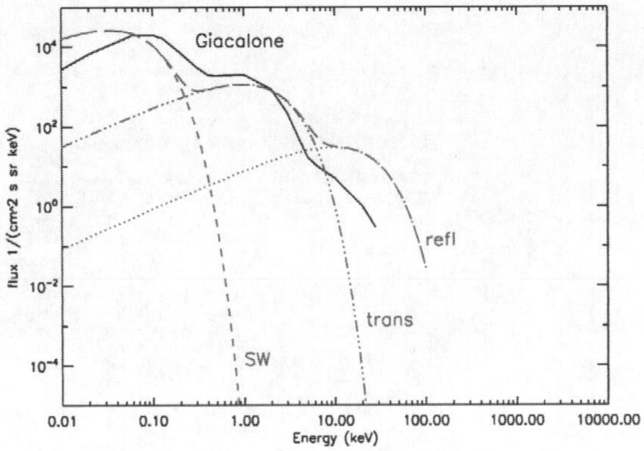

Figure 16. Model of energetic ion spectrum at Voyager 2 location on the termination shock. Shown are result of hybrid simulation by Giacalone et al. [31] (blue solid line), and obtained from theoretical model Zank et al. [28] (red broken line). The theoretical ion flux consists of three components corresponding to solar wind (SW, dashed line), transited ions (dash-dot line), and ions reflected from the shock (dotted line).

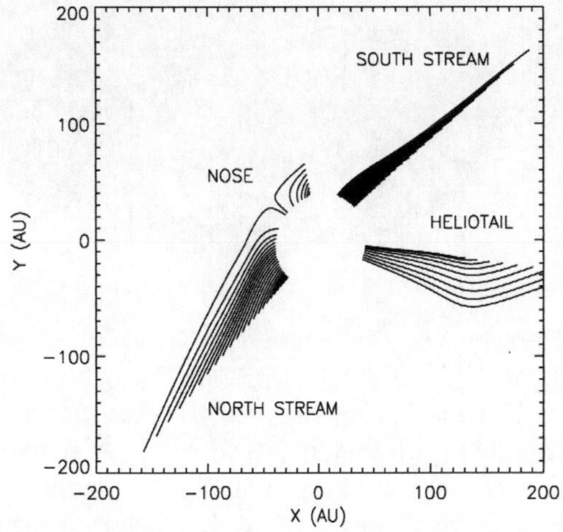

Figure 17. Streamlines ending at four selected lines-of-sight Two directions correspond to peaks of ENA flux associated with plasma streams, one direction to nose (forward part of the heliosphere), and one to heliotail direction. View in $(\vec{B}_\infty,\vec{V}_\infty)$ plane.

distribution, mainly the neutralization loss due to charge-exchange with the background neutral hydrogen.

For the ion spectrum at the termination shock we use the model of Zank et al. [28], based on the ideas of Zank et al. [29] and Lee [30] (in our previous work [15] we have used a simpler trial model). Figure 16 shows a comparison of the energetic proton spectrum which we derived using the model of Zank et al. [28] for the case of the termination shock crossing by Voyager 2, with the result of hybrid simulation by Giacalone and Decker [31]. The latter was shown to approximately match the termination shock particle spectra measured by Voyagers [32].

For the "comet-like" heliosphere expected for the case of weak interstellar magnetic field, the large length of LOS occurs for the heliotail direction (Figure 17). However, this LOS is fed by the energetic ions coming from the rear parts of the termination shock, where the energetic ion flux intensity is relatively low, due to low pick-up ion density. In result, the ENA flux from the heliotail direction can be lower than from the ISM apex region. For higher energy (above $\sim$50 keV), the flux from the tail direction may become higher than from the apex because of lower neutralization losses.

For the "two-stream", strong interstellar magnetic field heliosphere, large LOS length occur also for the parts of the heliosphere corresponding to the streams. These are fed by the energetic ions from the forward and flank parts of the termination shock. The pick-up ion density in these regions can be high, and therefore one may expect a high value of the ENA flux from the two stream directions. Our results (Figures 18, 19) confirm this conclusion.

We wish to stress the difference between the stream directions (corresponding to large LOS length and high ENA flux) and the heliotail direction (large LOS length but low ENA flux). By the above argument, the models of the heliosphere similar to the Parker model would correspond to the ENA intensity map with two peaks (and not two low flux regions) near the stream (tunnel)

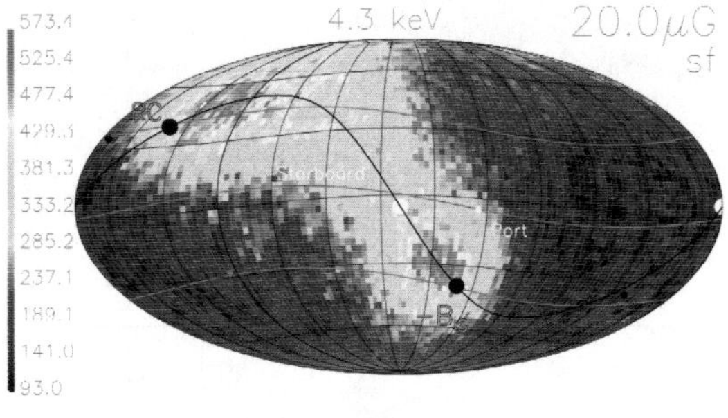

Figure 18. Directional distribution of the 4.3 keV ENA flux for the model heliosphere with spherically symmetric solar wind in the 20 μG interstellar field directed towards the IBEX ribbon center. The Mollweide projection is centered at the heliotail (ISM anti-apex) direction.

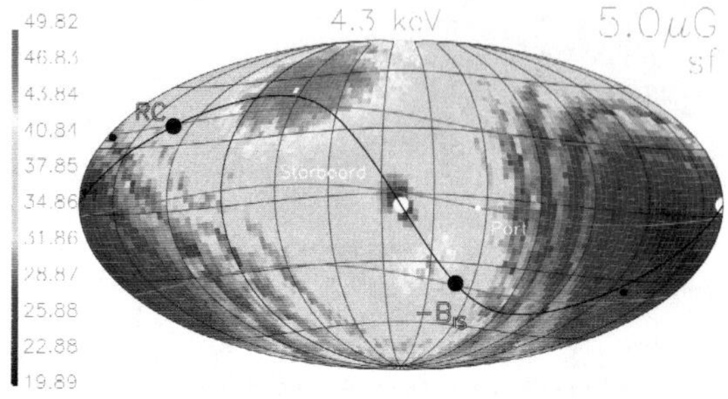

Figure 19. As Figure 18, but for the case of 5 μG interstellar field.

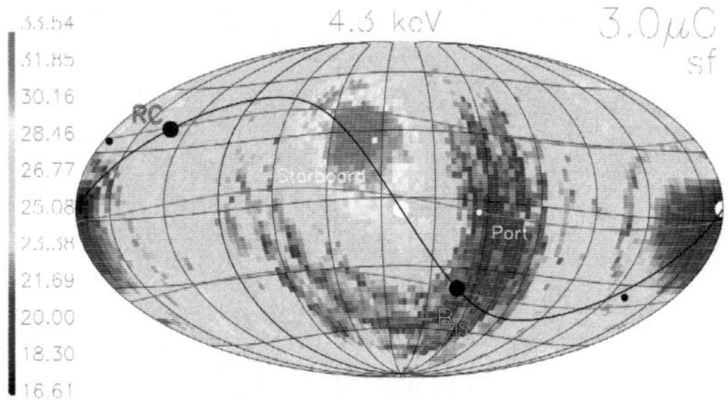

Figure 20. As Figure 18, but for the case of 3 μG interstellar field.

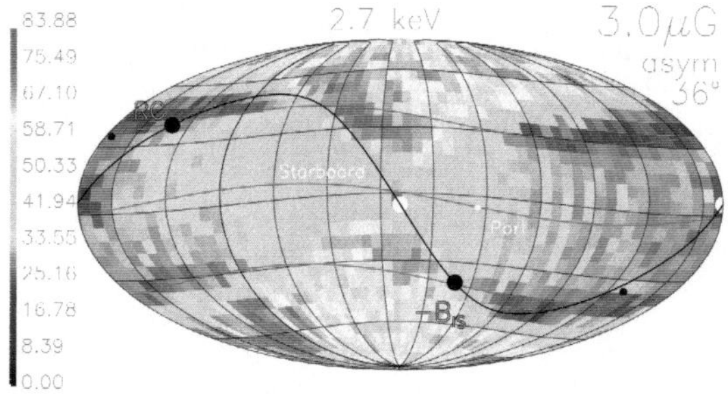

Figure 21. As Figure 20, but for the ENA with energy 2.7 keV and the asymmetric (fast/slow) solar wind.

directions. This observation may be relevant in view of some authors referring to Parker-like models of the heliosphere in connection with ENA observations [33, 34].

Figures 20, 21 show the ENA flux distributions calculated for the case of ISMF strength of 3 μG, which can be compared with IBEX observations. The solar wind is assumed to be either spherically symmetric (Figure 20) or two-component, with the fast wind restricted to within $\pm 36^\circ$ from the solar equator (Figure 21), respectively. For asymmetric solar wind, the ENA distribution near the heliotail corresponds to the "split-tail" structure, with two low ENA flux lobes flanking the heliotail direction. This "split-tail" effect was already obtained in [15], using a different termination shock spectrum. A similar structure was recently derived by [35].

Figure 21 shows some similarity to the ENA distribution observed by IBEX [36]. Some of the differences can be attributed to our assumption of constant neutral hydrogen background.

7. Conclusions

We constructed a range of time-stationary MHD models of the heliosphere for different ISMF strengths, from 2 to 20 μG. The ISMF direction (the IBEX ribbon center) and the parameters of the solar wind and the ISM were assumed to be close to the observed values.

Comparing the models for different ISMF, we were able to follow the evolution of the global structure of the heliosphere, with the ISMF decreasing from unrealistically high to observed values. In particular, we found that the two-stream structure of the heliospheric plasma flow may survive down to $\sim$5 μG ISMF.

As an example of application of this evolution to understand the actual heliosphere, we review our suggested explanation [14] of the observations by Voyager 1, indicating an unexpectedly small change in direction of the magnetic field during crossing of the heliopause.

We also used our models to obtain the directional distribution of the heliospheric ENA in the IBEX energy range. Basing on our results for high B_∞, we point out that a heliosphere open at two ends would correspond to two areas of high ENA flux coming from the directions of the tunnels, rather than two low flux areas. Differently from our previous work [15], our ENA calculations presented here use the new model of the termination shock spectrum proposed by Zank [28]. The "split tail" effect, which we found in [15] for another termination shock spectrum, is still present when we use the model of Zank.

Acknowledgments

We thank Space Science Data Facility (SPDF) Data and Orbit services (http://spdf.gsfc.nasa.gov/) for data usage [2]. The early version of the MHD code used in this work was provided to us by R. Ratkiewicz.

References

[1] Burlaga L F, Ness N F and Stone E C 2013 *Science* **341** 147
[2] Space Physics Data Facility Data and Orbit services file **ftp**: *//spdf.gsfc.nasa.gov/pub/data/voyager/voyager1/merged/voyager1_daily.asc*
[3] Burlaga L F and Ness N F 2014 *Astrophys. J.* **784** 146
[4] Burlaga L F and Ness N F 2014 *Astrophys. J. Lett.* **795** L19
[5] Burlaga L F and Ness N F 2016 *Astrophys. J.* **829** 134
[6] Florinski V, Pogorelov N V, Zank G P, Wood B E and Cox D P 2004 *Astrophys. J.* **604** 700
[7] Grygorczuk J, Ratkiewicz R, Strumik M and Grzedzielski S 2011 *ApJL* **727** L48
[8] Heerikhuisen J and Pogorelov N 2011 *Astrophys. J.* **738** 29
[9] Zank G P, Heerikhuisen J, Wood B E, Pogorelov N V, Zirnstein E and McComas D J 2013 *Astrophys. J.* **763** 20
[10] Heerikhuisen J, Zirnstein E J, Funsten H O, Pogorelov N V and Zank G P 2014 *Astrophys. J.* **784** 73
[11] Zirnstein E J, Heerikhuisen J, Funsten H O, Livadiotis G, McComas D J and Pogorelov N V 2016 *ApJL* **818** L18
[12] van Marle A J, Meliani Z and Marcowith A 2015 *Astron. & Astrophys.* **854** A49
[13] Parker E N 1961 *Astrophys. J.* **134** 20
[14] Grygorczuk J, Czechowski A and Grzedzielski S. 2014 *Astrophys. J.* **789** L43
[15] Czechowski A, Grygorczuk J and McComas D J 2015 *ArXiv e-prints* **arXiv:1507.00540**
[16] McComas D J et al. 2013 *Astrophys. J.* **771** 77
[17] Ratkiewicz R, Ben-Jaffel L and Grygorczuk J 2008 *Astronomical Society of the Pacific Conference Series* **385** 189-194
[18] Funsten H O et al. 2009 *Science* **326** 964
[19] McComas D J et al. 2012 *Science* **336** 1291
[20] Bzowski M, Möbius E, Tarnopolski S, Izmodenov V and Gloeckler G 2008 *Astron. and Astrophysics* **491** 7
[21] Drake J F, Swisdak M and Opher M 2015 *ApJL* **808** L44
[22] Golikov E A, Izmodenov V V, Alexashov D B and Belov N A 2017 *MNRAS* **464** 1065
[23] Drell S D, Foley H M and Ruderman M A 1965 *J. Geophys. Res.* **70** 3131
[24] Neubauer F M 1980 *J. Geophys. Res.* **85** 1171
[25] Czechowski A and Grzedzielski S 1998 *Geophys. Res. Lett.* **25** 1855
[26] Fahr H J, Grzedzielski S, Ratkiewicz-Landowska R. 1988 *Ann. Geophys.* **6** 337
[27] Kivelson M G and Jia X 2013 *J. Geophys. Res.: Space Phys.* **118** 1-8
[28] Zank G P, Heerikhuisen J, Pogorelov N V, Burrows R and McComas D 2010 *Astrophys. J.* **708** 1092-1106
[29] Zank G P, Pauls H L, Cairns I H and Webb G M 1996 *J. of Geophys. Res.* **101** 457-477
[30] Lee M A, Shapiro V D and Sagdeev R Z 1996 *J. Geophys. Res.* **101** 4777-4789
[31] Giacalone J and Decker R 2010 *Astrophys. J.* **710** 91-96
[32] Hsieh K C, Giacalone J, Czechowski A, Hilchenbach M, Grzedzielski S and Kota J 2010 *ApJL* **718** L185-L188
[33] McComas D J et al. 2009 *Science* **326** 326

[34] Krimigis S M, Mitchell D G, Roelof E C, Hsieh K C and McComas D J 2009 *Science* **326** 971
[35] Zirnstein E J, Heerikhuisen J, Zank G P, Pogorelov N V, Funsten H O, McComas D J, Reisenfeld D B and Schwadron N A 2017 *Astrophys. J.* **836** 238
[36] Schwadron N A et al. 2014 *Astrophys. J. Suppl. Series* **215** 13

Response times of Cassini/INCA > 5.2 keV ENAs and Voyager ions in the heliosheath over the solar cycle

K. Dialynas[1], S. M. Krimigis[1,2], D. G. Mitchell[2], R. B. Decker[2] and E. C. Roelof[2]

[1]Office of Space Research and Technology, Academy of Athens, Athens 10679, Greece.

[2]Applied Physics Laboratory, The Johns Hopkins University, Laurel, Maryland 20723, USA.

Abstract. Both a magnetosphere-like tail and a bubble model of the heliosphere were posited by E. N. Parker in 1961. Recently, we showed that heliosheath ions are the source of > 5.2 keV Energetic Neutral Atoms (ENA), whose images of the heliosphere exhibit a rough nose to anti-nose (tail) global symmetry that resembles a diamagnetic bubble. The comparison between energetic neutral atom (ENA) global images of the helioshphere obtained with the Ion and Neutral Camera (INCA) on board Cassini and ions measured in-situ by the Low Energy Charged Particle experiment (LECP) on board Voyager 1 and 2 (V1/V2) in overlapping energy bands over an 11-year period shows that the heliosphere responds promptly, within ~2-3 years, to outward propagating solar wind changes in both the nose and tail directions. Here we focus on the recovery of solar cycle 24 and the response times of > 5.2 keV ENAs to show that this ~2-3-year time delay is consistent with a "tail" of ~80-120 AU. This preliminary rough calculation is generally consistent with lower energy ENA data (E < 6 keV, from the IBEX-Lo and IBEX-Hi) and is supported by recent modelling of the heliosphere.

1. Introduction

The Sun's atmosphere is not static, but expands in the form of a magnetized fluid called solar wind, reaching to distances of potentially hundreds of Astronomical Units (1 AU = 1.5×10^8 km), shaping our local bubble, called "heliosphere", through its interaction with the Local Interstellar Medium. This process occurs with a 22-year periodicity (two full Solar Cycle), during which the general properties of the solar wind change drastically [1]. The question of how the heliosheath responds to the variability of solar wind conditions and in what manner this response is connected to the solar activity through the solar cycle is of paramount importance for understanding the physics of the Heliosphere and has been the subject of many theoretical models (e.g. [2],[3],[4],[5],[6]). Voyagers 1 and 2 (V1, V2) discovered the reservoir of ions and electrons that constitute the heliosheath after crossing the termination shock [7],[8] north and south of the ecliptic plane at 94 [9],[10],[11] and 84 AU [12], respectively. Owing to the coexistence of ions and neutrals in the heliosphere that allows the charge-exchange reaction to take place, the local measurements by each Voyager were placed in a global context by remote sensing images using ENAs obtained with the Ion and Neutral Camera (INCA) on-board Cassini [8],[13] orbiting Saturn (~10 AU away from the Sun).

ENA imaging of the heliosphere was long anticipated [14] and the recent ENA observations from eV (e.g. Interstellar Boundary Explorer-IBEX) to keV (e.g. Ion and Neutral Camera-INCA) energy range have already revolutionized our understanding about the formation and interactions of the global heliosphere, providing insights on the plasma processes at ~100 AU that were substantially at variance with previous theories and models. For example, the narrow bright ENA stripe known as IBEX-ribbon (E < 5.6 keV) [15] that is most likely formed outside the heliopause [16] is surrounded by the Globally Distributed Flux (GDF), probably of heliosheath origin [17],[18],[19].

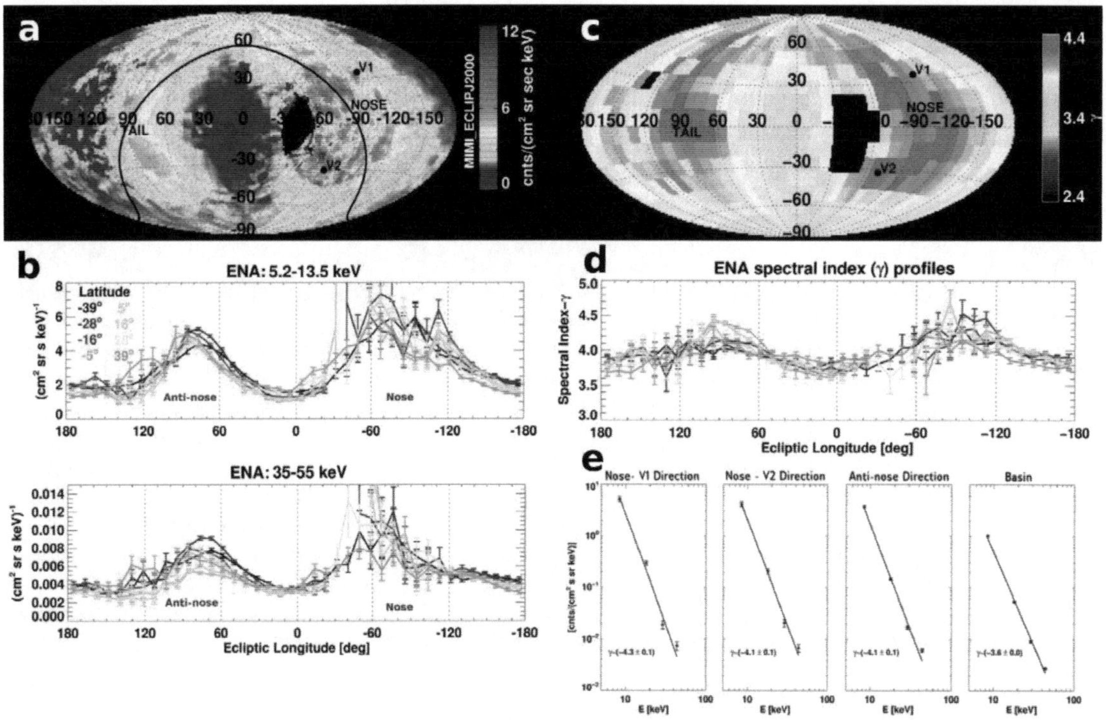

Figure 1. a) A combination of 320x160 pixel INCA/ENA images (5.2–13.5 keV) organized in ecliptic coordinates, averaged over the 2003-2009 time period, showing the high-intensity belt and the regions of ENA minima (basins) that reside in the center of the image and reappear beyond the belt at ≥ 120°. The solid black line represents the galactic equator. **b)** ENA longitudinal profiles for various latitudes in the energy ranges of 5.2-13.5 keV (top) and 35-55 keV (bottom), averaged over 11x9 (latitude, longitude) in ecliptic coordinates. **c)** The projection of the spectral index γ from fits of the 5.2-55 keV ENA energy spectra (e.g. panel e) to a power law form. **d)** Latitude and longitude dependence of the deduced γ-indices that can be directly compared with the intensity profiles in panel-c. **e)** ENA energy spectra sampled from four selected heliosphere regions. The black lines represent the power law fits.

In fact, a rough comparison of IBEX ENAs with INCA [20] led us to propose that the globally distributed flux and the ribbon are distinct features that originate from different source plasma populations and that [...] *different energies in IBEX correspond to different layers of the heliospheric structure starting from outside the heliopause and moving inward to the heliosheath at the high energy channel (E < 5 keV) [...]*. A very interesting recent analysis of the proton velocity distributions in the inner heliosheath [21] provides a coherent picture: [...] *Consequently, the energy spectrum of at least the background ENAs [22] observed from a given direction is fed differently from different regions across the inner heliosheath. This offers the opportunity to indirectly probe the spatial structure of the latter. Depending on the source region of the ribbon ENAs [for a review see [23]] this applies also to these particles. [...]*.

At higher energies (E > 5.2 keV), first images of the heliosphere with Cassini/INCA [24], revealed the existence of a prodigious (yet unexpected) broad band of emission in the sky, called

IOP Conf. Series: Journal of Physics: Conf. Series **900** (2017) 012005 doi:10.1088/1742-6596/900/1/012005

"Belt" (Figure 1a) that wraps around the celestial sphere in ecliptic coordinates, passing though the heliospheric nose, tail (henceforth called "anti-nose") and the ecliptic poles [8],[20]. Recently, we have shown that the heliosheath ions are the source of INCA/ENA, whereas the combined use of > 5.2 keV ENAs and in-situ LECP ions, have demonstrated that the heliosphere responds promptly, within ~2-3 years, to outward propagating solar wind changes in both the nose and anti-nose directions [25]. Taking into account the V1 measurement of a ~0.5 nT [26] interstellar magnetic field (ISMF) and the high plasma beta in the heliosheath [13], we have concluded that the heliosphere behaves as a diamagnetic bubble with little substantial tail-like features, consistent with recent modelling [27],[28],[29],[30]. Although different interpretations and models of the V1 measurements exist, showing that the spacecraft may not have exited to interstellar space [31], [32], a fact that can possibly point to a heliosheath with different magnetic properties [33] other than a "diamagnetic bubble", such a view does not preclude the rough nose-"anti-nose" symmetry shown in our recent analysis [25].

Given the line-of-sight integral that determines the ENA intensity, assessing any proposed interpretation of the shape and properties of the heliosheath requires the evaluation of the physical properties of the underlying source proton population in the heliosphere in the broad context of the solar wind. A significant advantage of the INCA/ENAs is that they correspond to a time period that partly overlaps with solar cycle 23 and the ongoing solar cycle 24 and at the same time they can be directly contrasted with in-situ ion measurements from the heliosheath sampled by V1 and V2 in overlapping energy bands. The aforementioned recent developments in this direction [25] verified our previous interpretations [8],[13],[20],[34],[35],[36], and partly addressed key questions; the Belt was previously identified as a relatively stable feature as a function of energy, corresponding to a reservoir of particles that exist within the heliosheath, constantly replenished by new particles from the solar wind [20]. Consequently, the Belt ENAs are expected to be associated with a region of enhanced particle pressure that is formed between the termination shock and the heliopause and contributes to balancing the pressure of the ISMF [8],[13],[34], forming a roughly symmetric time-dependent obstacle to the inward interstellar flow; an 11-year "breathing mode" of the global heliosphere, manifested by the outward propagating changes of the solar wind through the solar cycle.

The present paper is organized as follows: Section 2 provides the necessary "background" information concerning the INCA experiment and the ENA measurements presented in this study. Section 3 presents the > 5.2 keV ENA measurements over the 2003-2014 time period and compares the "nose" and "anti-nose" ENA intensities as a function of time together with in-situ ions in the heliosheath and solar wind energy input. Section 4 extends the interpretations shown in our recent analysis by focusing on the recovery of ENA intensities after the onset of solar cycle 24. In Section 5 we present a summary of our results and discuss the features that need to be explored in our future analyses.

2. Data and experiment details

The INCA imager is part of the Magnetospheric Imaging Instrument (MIMI) [24] on board the Cassini spacecraft in orbit around Saturn's (~10 AU away from the Sun) since 1st July of 2004. INCA utilizes a large geometry factor of ~2.4 (cm^2 sr) and a broad field of view (FOV) of 90° in the azimuthal direction and 120° in the perpendicular direction; nominal angular resolution is ~5° for Hydrogen. The detection of ENAs is then based on the time-of-flight (TOF) technique, and several discrete energy passbands are defined within the energy range of ~5.2 to ~200 keV.

The ENA data used in this study correspond to four TOF channels within the energy range of 5.2-55 keV (defined by each energy channel passbands as $E = \sqrt{E_i \cdot E_{i+1}}$; ~8.4, 18.0, 29.0, 43.9 keV) and to both "spin" and "stare" imaging periods, i.e. a rotation around the spacecraft Z-axis (spin axis) that results in a scan over 360° across the sky, and also periods of staring in a fixed direction in the sky sphere, respectively. Previous analyses (e.g. [8],[20]) have shown that possible "background" sources are insignificant when compared with the measured foreground INCA ENAs, whereas all time periods where INCA was surveying Saturn's plasma sheet or pointing in the general direction of the Sun were excluded from our analysis (e.g.[20]).

The resulting INCA images (Figure 1a) represent a set of 320x160 pixel maps and, to the extent that our pixel exclusion criteria (as described above and in [8],[20],[25]) are successfully applied, we asses that the ENA data are consistent with Poisson counting statistics: all measured counts (*C*) that are accumulated in the INCA detector per each pixel are bounded to an uncertainty that is equal to $\sqrt{C}$. The ENA emissions in the keV energy range are relatively low differential intensity events ($j = C/(cm^2 \cdot sr \cdot sec \cdot keV)$), resulting in relatively high uncertainties ($\delta j = \delta C/(cm^2 \cdot sr \cdot sec \cdot keV)$). Thus, to ensure credible statistics we average the measured intensities over several pixels in both the longitude and latitude direction, as explained in the captions of each of the Figures 1, 2, 3.

Figure 1b illustrates the longitudinal dependence of the combined 2003-2009 ENA intensities for various latitudes, verifying that the peak intensities -on average- between the nose and anti-nose directions are nearly similar throughout the INCA energy range (although only the lower, 5.2-13.5 keV and higher, 35-55 keV energy channels are included in this Figure). As shown before [20] and also in Figure 1d, the 5.2-55 keV ENA spectra are fairly consistent with a single power law form in energy ($j \sim E^{-\gamma}$). The INCA spectra (Figure 1c,d) are softer ($4.0 < \gamma < 4.4$) roughly within the belt region and harder ($3.5 < \gamma < 4.0$) in the surrounding regions (basins, where the ENA minima occur), following the ENA intensity changes shown in Figure 1a,b. The ENA spectral index shows no noticeable asymmetry between the nose and anti-nose directions, while its spatial dependence with latitude is nearly similar (within error bars). The consistency of the ENA intensities, taken together with the convergence of the ENA energy spectra between the nose and anti-nose directions, provide indisputable evidence that the global heliosphere must be roughly symmetric. However, we must keep in mind that, as explained in [25], a perfect symmetry not only would require a negligible ram pressure of the interstellar flow [8], but also, models including Hubble Space Telescope observations of H-Lyα absorption have been interpreted as showing a nose–tail asymmetry [37]. Thus, although this nose-tail asymmetry can be in fact attributed to the non-negligible anisotropic ram pressure of the interstellar medium, we stress that such distortion would differ substantially from any heliosheath structure that includes a very prolonged tail [16].

3. Global ENA emissions over Solar Cycles 23 & 24

A quantitative examination of ENA histories in both the anti-nose as well as the V1 and V2 directions (towards the nose) during the 11-year period (Figure 2) reveals a substantial decay of the belt intensities (by a factor of ~3 among all TOF channels) that is not "symptomatic", but closely correlated with the declining phase of solar cycle 23. The gray-scaled lines in Figure 2 illustrate the time profiles of ENAs at ~80-100 deg in ecl. longitude (where the approximate center of the anti-nose belt lies) and for various latitudes above and below the ecliptic equator (20°x20° averaging).

The first Cassini/INCA image during 2003 was taken three years after the onset of the declining phase of SC23 (where solar activity was high enough, SSN~70), that exhibited a late solar minimum ("deep" solar minimum) during late 2009 to early-2010, where the solar activity was effectively minimized (SSN < 20) and onset of SC24 occurred. As shown in Figure 2, the year 2003 is associated with high ENA intensities emanating from the anti-nose region, that decrease in following years to levels found in the basins by 2011-2012. However, the rapid recovery of ENA intensities beyond 2012 is associated with the onset of solar cycle 24 where solar activity is gradually restored towards the new solar maximum. Note that solar cycle 24 is expected to be among the smallest cycles in this century [38].

These observations have important implications for our understanding of the global heliosphere, especially if we take into account the following factors:

a) The decay of ENAs in the nose and anti-nose directions present similar time dependence: The blue and red lines in Figure 2 illustrate the time profiles of ENAs in the V1 and V2 directions respectively, taking into account the displacement of both spacecraft over the observation period: V1 and V2 travel through the HS with velocities of ~3.6 AU/yr and ~3.2 AU/yr respectively. From 2003 up to 2015, V1 slowly ascends in ecliptic latitude with a speed of ~0.03°/yr (and ~0.25°/yr in ecliptic longitude), i.e. over an 11-year time period the displacement of V1 in the ecliptic plane is ~0.33° in latitude and ~2.75° in longitude, which is well below the 5°x5° pixel averaging that we perform in our INCA/ENA measurements (and roughly comparable, in the longitude direction, with the INCA pixel width). By contrast, V2 descends in ecliptic latitude with a speed of ~0.6°/yr and ~0.17°/yr in ecliptic longitude. This results in an apparent (and more important, compared with V1) movement of V2 in the ecliptic sphere of ~6.6° in latitude and ~1.9° in longitude over an 11-year time period. The 5°x5° pixel averaging that we perform in our ENA intensities ensures enough counts to produce statistically significant results (reflected in the error bars of Figure 2) and at the same time defines a relatively narrow region around both V1 and V2 for a meaningful comparison between the LECP ions and INCA ENAs (which will be explained in point b). To avoid possible "spatial effects" (especially in the V2 direction) over the 11-year period that we analyze here, our averaged 5°x5° ENA intensities take into account the displacement of both spacecraft in ecliptic coordinates, by always keeping the V1 and V2 positions at the center of the averaging bins (henceforth "V1" and/or "V2 pixels"). With the above in mind, Figure 2 shows that although the decrease of ENA intensities in V1 and V2 pixels persists up to the year 2012/2013, the overall profiles between the nose and anti-nose ENAs are very similar within error bars (a 1-year uncertainty should be taken into account here due to the time averaging that we perform in the ENA intensities). Therefore, the ENA intensities between the nose and anti-nose regions are not only consistent in an average sense (Figure 1), but exhibit a consistent time variability.

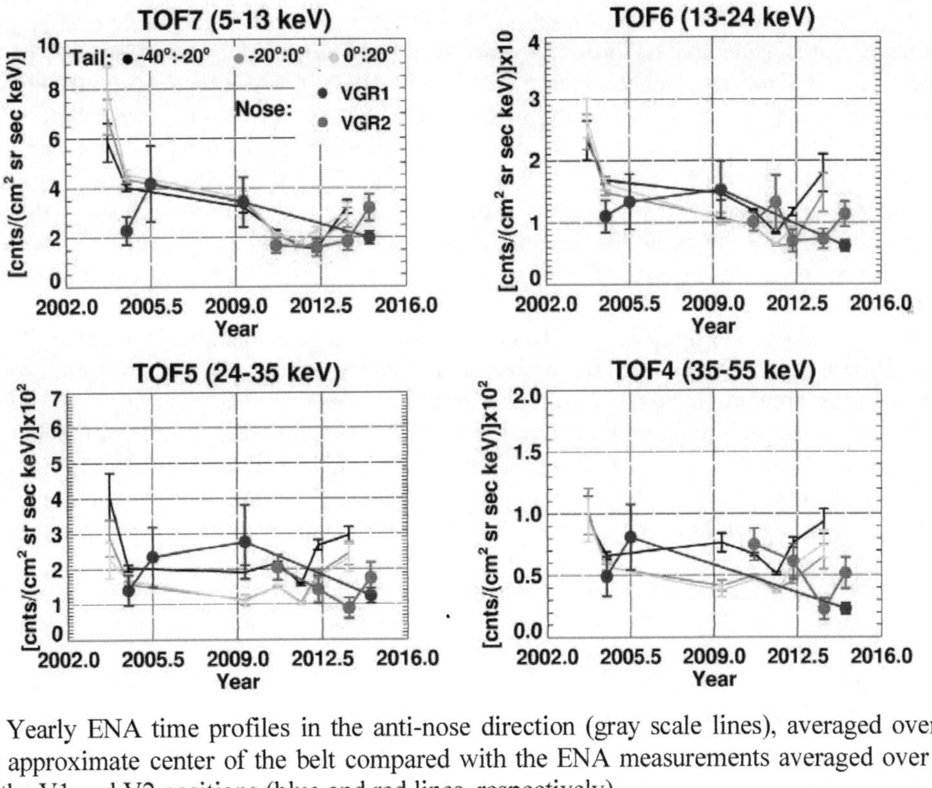

Figure 2. Yearly ENA time profiles in the anti-nose direction (gray scale lines), averaged over 20x 20° in the approximate center of the belt compared with the ENA measurements averaged over 5x5° enclosing the V1 and V2 positions (blue and red lines, respectively).

b) In-situ ions and remote-sensed keV ENAs in the heliosphere are directly comparable and the source of > 5.2 keV ENAs is the heliosheath: V1 crossed the TS at ~94 AU [9],[10],[11] while it recently crossed the HP at a distance of ~122 AU (where the complete evacuation of keV ions and simultaneous increase of galactic cosmic rays occurred) [35]. Below the ecliptic equator, V2 crossed the TS at a distance of ~84 AU [12] and still surveys the heliosheath. As shown in [25], the > 5.2 keV

ENA intensities in the directions of both V1 and V2 have undergone a significant decrease during the time period 2009 to 2013 (by a factor of ~2.3), very similar to the > 30 keV in-situ ion decrease (by a factor of ~2.7) at V2 (V1 crossed the HP on 2012.65). Furthermore, the period after 2013 is characterized by a turn-up in both ENA (tail and in V2 direction towards the nose, by a factor of ~2) and ion intensities (~1.5). Also, as explained in our recent publication [25], despite the ~138 AU separation between the two Voyagers (~35° above and ~34° below the ecliptic equator, respectively), the ion intensity histories and spectra at both spacecraft are very similar in both shape and numbers. Furthermore, they are closely correlated with the ENA intensities at the two different latitudes of both V1 and V2, a fact which indicates that the modulation of superthermal ions during the solar cycle is global throughout the heliosheath. Most importantly, however, the coherence in the overlapping energy ranges of ions and ENA at E > 28 keV in the variability measured in-situ by V1,2 and remotely by Cassini, shows that the source of ENA is definitely the heliosheath. Given the similarity in the overall appearance of the images throughout the INCA energy range (~5-55 keV), we infer that the source of ENAs at <30 keV (i.e. below the V1,2 ion threshold) to also be the heliosheath. In contrast, the source of the IBEX-ribbon at lower energies (<6 keV) is not clear [16] and could lie beyond the HS due to processes yet to be revealed. It is likely, however, that the IBEX-defined Globally Distributed Flux-(GDF)[17] is also the HS [19]. For example, despite a 10-70% ENA reduction in the 1.1-4.29 keV pole measurements [39] over the 2008-2011 time period, the 0.71 keV IBEX channel showed no decrease in ENAs during that same period. This is yet another indication that the GDF and the Ribbon are possibly two distinct features that originate from different source plasma populations (heliosheath and outside the heliopause, respectively) as proposed earlier [20]. We should note, however, that a solar cycle connection to the N–S pole IBEX ENA variations (north and south pole estimates of L_{HS}~210 AU and L_{HS}~160 AU, respectively) has recently been reported [40].

c) The intensity profile of ENA and ion measurements at ~100 AU and beyond are directly comparable to the solar wind energy input at 1 AU: As explained above, the SSN can be used as proxy for the solar cycle phases and we have used these measurements to connect solar activity with the ENAs and ions in the heliosheath. Recent studies [41] have undertaken a significant effort to calculate the solar wind parameters as a function of time for the in-ecliptic solar wind data at 1 AU from the OMNI database. The solar wind energy flux and density at 1 AU show a coherent decay with time (peak to basin ratio of ~2) up to early 2010 and a recovery in both quantities thereafter (e.g. Figure 9 in [41]). This, as expected (e.g. see [1]), is consistent with the measured SSN and overall solar activity explained in the previous paragraphs over the declining phase of solar cycle 23 and onset of solar cycle 24. Consequently, the plasma output from the sun at 1 AU is directly contrasted and connected with the intensity changes in both ENA and ions at ~100 AU and beyond, i.e. inside the heliosheath. The ~2-3 years time difference between all aforementioned quantities will be discussed in detail in paragraph 4 where we elaborate on the travel times of ENAs and ions in the heliosphere and provide a quantitative calculation of the width of the heliosheath based on those travel times.

As explained elsewhere [34], the transition of the high ENA intensities in the tail to the low ENA intensities in the Basin, that occurs at ~60° to 30° in ecl. longitude (previously defined as "transition region"), is identified in all energy ranges and serves as a relatively smooth boundary, with a spatial width of ~30° deg in ecl. longitude, between the very low and high ENA emissions in the heliosheath. The ENA intensities in this region diminish with an average spatial rate of ~2.4% /degree, that is nearly constant as a function of energy. Here we note that by contrast to the belt, the ENA intensities in the basins remain essentially stable with time throughout the INCA energy range.

Figure 3 illustrates the evolution of the global ENA heliosphere over the time frame of an 11-year observation period (2003-2014), that involves a general reduction of the Belt in both intensities and width, up to at least the year 2011/2012 and a recovery thereafter, that is clearly shown in both the 5.2-13.5 and 35-55 keV channels, above and below the ecliptic equator. Thus, the rate through which the transition region is diminished is also variable with time (as expected), presenting its minimum value over the years 2011-2012 where the Belt ENA intensities are effectively minimized. This observation may act as an indirect assessment (which will be examined in our future work) of the fact that the heliosphere, independently of its shape (even if it is "bubble-shaped" as we conclude) cannot be a closed and unaltered system through time, since the solar wind energy input has to be evacuated somehow.

First, the solar wind is the source population of heliosheath plasma but, due to the measured high plasma beta in the heliosheath contrasted with the relatively low stagnation pressure exerted from the interstellar flow, the heliosphere is expected to exhibit a diamagnetic behavior [25]. Therefore, the dynamic properties of ENA from the heliosheath are apparently strongly related to the dynamic properties of the solar wind over the solar cycle and can quantitatively relate to the properties of the heliosphere interactions with the local interstellar medium. Secondly, the ENA intensity gradient in the transition region translates to a partial pressure gradient in the same region, which in turn produces a variable "shielding current" as a function of time. Therefore, in combining the above two points, we infer that the heliosphere bubble may remain diamagnetic throughout the solar cycle phases, but it is not expected to remain a perfectly symmetric and unaltered sphere, as the pressures, pressure gradients and -in turn- the "shielding currents" in the heliosheath are in fact time-depended.

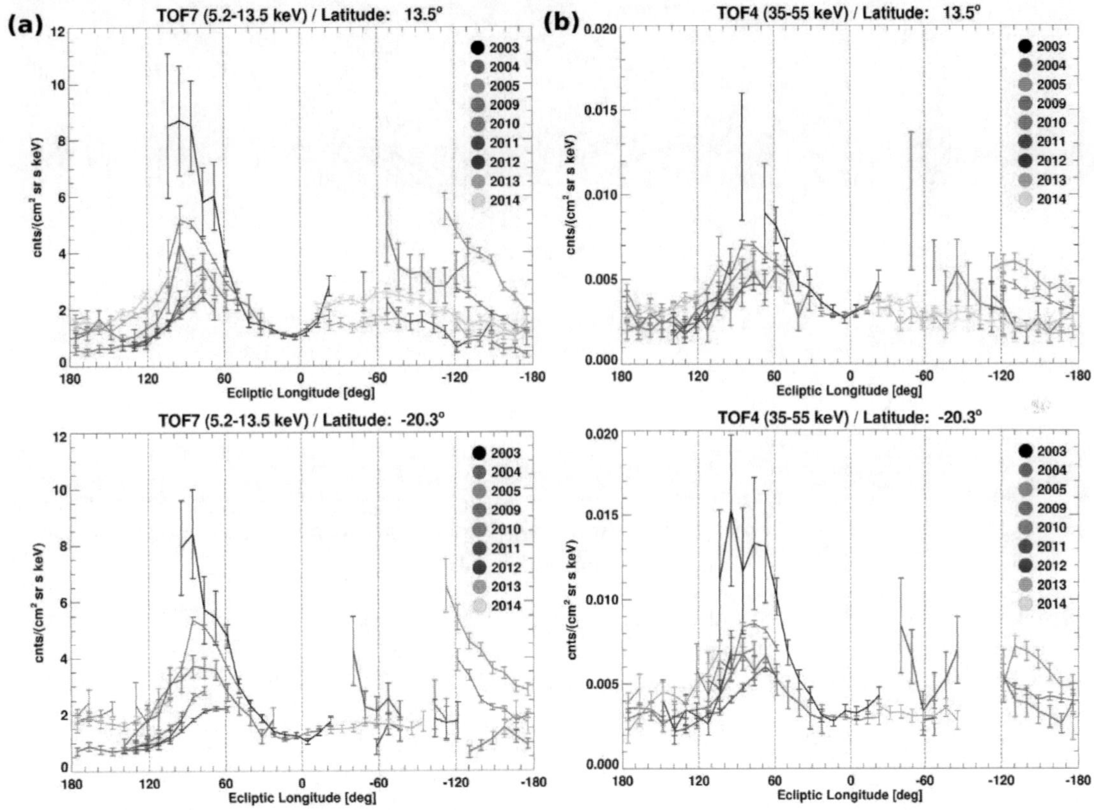

Figure 3. a) Longitudinal distributions of the 5.2-13.5 keV, averaged over 11x9 (Latitude, longitude) in ecliptic coordinates for both above (top) and bellow (bottom) the ecliptic equator for the time period of 2003-2014 following the declining phase of solar cycle 23 and the onset of Solar cycle 24. **b)** the same as in (a) for the 35-55 keV ENAs.

Consequently, the heliosphere bubble can inflate with time in either the anti-nose direction (as "tail" models suggest) or along the direction of the interstellar magnetic field (as in the Parker-Bubble model). The recent model that shows a view of the heliosphere with two prominent polar jets [27],[28], provides one of the possible ways through which the solar wind input can be evacuated from the heliosphere.

4. Recovery times of ENAs over the onset of solar cycle 24 and the width of the heliosheath.

We now draw our attention to the recovery times of ENAs in the heliosphere. Figures 2 and 3 show a clear, but explainable, mismatch of ~2-3 years between the observed ENA minimum and the minimum of cycle 23 as measured by the SSN and the solar wind energy input at 1 AU (described in paragraph 3 of the present paper). In fact, given our results, this time delay enables us to re-think the following question, which goes to the heart of our analysis: *If the time profile of V1 and V2 LECP >28 keV ions are indicative of the solar wind changes through the SC inside the HS, when should V2 be expected to "measure" the solar minimum of solar cycle 23 (occurred in early 2010) and detect the recovery at the onset of solar cycle 24?*

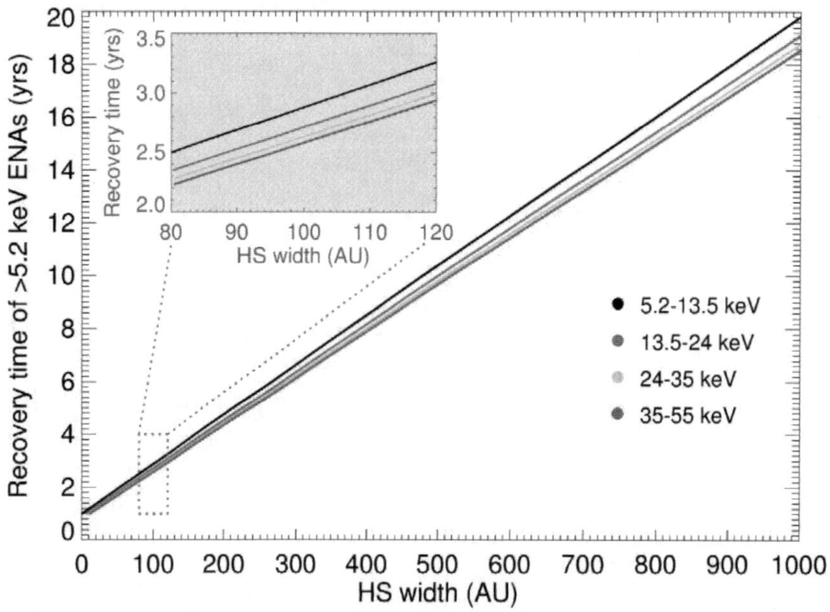

Figure 4. Recovery times of ENAs of different energies (5.2-55 keV), as a function of the heliosheath width using Eq.1 and the parameters explained in the text. The inset focuses on the ~2-3 year recovery times of >5.2 keV ENAs, for which a heliosheath of ~80-120 AU width is obtained.

V2 crossed the termination shock at L_{TS} ~84 AU, whereas its position in early 2013 was at L~102 AU (-34° ecliptic latitude). Both V1 and V2 measured that the field aligned streaming velocities of ions cannot be more than ~100 km/s, typically around V_{plasma} ~60-80 km/s or even lower. The solar wind at the nearly ecliptic equator travels with a speed of V_{SW}~400-500 km/s. Therefore, it takes the solar wind $t_{SW} = L_{TS}/V_{SW}$ = 0.7 to 1 year to reach the termination shock in the V2 direction. From this point on it would require an additional time of $t_{plasma} = (L-L_{TS})/V_{plasma}$ ~1 to 1.4 yrs for the plasma to reach the V2 position. Consequently, as the minimum intensity in keV ions towards the nose (V2 direction) occurred during the first days of 2013, a time delay of > 2.4 years between the solar cycle parameters and the measured ions at V2 (and V1) can be in fact regarded as indicative of the time difference between the actual minimum on the Sun and its manifestation in the heliosheath; this is simply due to the solar wind propagation up to the termination shock and the subsequent plasma propagation with relatively low speeds in the heliosheath. This simple calculation together with the observed consistency between ENAs and ions in the heliosheath implies that no source regions, other than the heliosheath itself, and no sources other than the solar wind itself are needed to explain the time variation of the measured > 5.2 keV ENAs.

The above picture is also consistent with the rapid recoveries in both ENA intensities from the V2 pixel and V2/LECP (in-situ) ion measurements after 2013. So, the question here is: *can we estimate the thickness of the heliosheath given the recovery times of ENAs?* Before getting into the details of this question, we need to be reminded that ENAs and ions are consistent in both the V1 and V2 pixels, whereas the ENA spectra are found to be in excellent agreement between the nose and anti-nose regions. Therefore, we can -at this point- estimate the width of the global heliosheath by providing simple calculations, although the details of this will be examined in future analyses.

Here we can make the reasonable assumption that the termination shock is roughly symmetric around the sun, located at L_{TS}~90 AU, which is also consistent with recent modelling that depicts a view of the heliosphere which includes a symmetric termination shock [30]. Again, we can adopt an average solar wind speed V_{SW}~450 km/s. Although the plasma speed inside the heliosheath (in the V1 and V2 pixels) was found to be typically V_{plamsa}<100 km/sec (e.g. [42], [43]), recent models show that the average magnetosonic speed towards the tail can be as low as ~50 km/sec (e.g. [27]) and as high as ~140 km/sec (e.g. [22]).

Figure 4 shows the recovery times of ENAs as a function of heliosheath width after incorporating the numbers explained in the aforementioned paragraph in the following simple equation:

$$T_{Recovery} = L_{TS}/V_{SW} + L/V_{plasma} + (L_{TS} + L)/V_{ENA}(E) \quad (1)$$

where $V_{ENA}(E)$ is the energy-dependent speed of ENAs (e.g. a 10 keV H ENA travels with a speed of ~1400 km/sec whereas a 45 keV is twice as fast, i.e. ~2800 km/sec), and L is the spatial range in which the charge exchange reaction takes place. Since we have concluded that our ENA measurements are produced by the interaction between the ions and neutrals from inside the heliosheath, we can in fact match this parameter (L) to the actual width of the heliosheath. Here we also need to be reminded that the ENA emission is optically thin (that is, ENA intensity cannot be significantly reduced between the termination shock and INCA, located at ~10 AU from the Sun). Further, although, ENAs will be generated throughout the whole distance from L_{TS} to $L_{TS} + L$, which means that Eq. 1 would yield a wide range of recovery times for ENAs of a given energy, we have selected to adopt a method shown elsewhere [40], making the simplifying assumption that charge exchange will take place near $L/2$ (so, replace L with $L/2$ in Eq. 1).

With the above assumptions, Figure 4 shows that the >5.2 keV ENAs would be consistent with a heliosheath width of 80 - 120 AU towards the tail (due to the ~2-3 yrs recovery after solar minimum). It is noted that the differences in the travel times of ~10 keV ENAs compared to 45 keV ENAs from a heliosheath that is as thick as ~100 AU is only a few months (~3 months), whereas the differences between higher energy ENAs (>20 keV) are practically insignificant (see inset of Figure 4). ENAs of lower energies (e.g. at ~1 keV, not shown here) would require ~3.5 to 4.3 years to travel back from a HS with a width of ~80 to 120 AU (and the differences in travel times of low energy ENAs are much more important). This view and the above calculations are consistent with the fits of the IBEX-Hi/ENA measurements from the North (N) and South (S) poles ([40]), where the inner heliosheath thickness is found to be ~210 AU in the north and ~160 AU in the south. This is also consistent with the heliosheath tail thickness of ~ (220±110) AU, inferred by IBEX-Lo measurements ([44]).

5. Summary and Conclusions

After our most recent work [25], we have used an extended set of >5.2 keV INCA/ENA measurements over the time period of 2003-2014 to re-evaluate our previous assessments and further provide preliminary calculations to show that the ~2-3 yrs time delay between the SSN and solar wind

IOP Conf. Series: Journal of Physics: Conf. Series **900** (2017) 012005 doi:10.1088/1742-6596/900/1/012005

energy input at ~1 AU with the measured ENAs at ~10 AU from Cassini is consistent with a heliosheath with a thickness of ~ 80-120 AU in all directions.

As it becomes transparent in recent publications, using a variety of different ENA measurements from eV (IBEX-Lo, IBEX-Hi) to 10s of keV (INCA) , but most importantly due to the comparison of INCA/ENAs with in-situ LECP ions, together with the recent V1 crossing of the heliospause, we seem to obtain the same (or similar) result, depicted in Figure 5: A roughly symmetric heliosheath that can be ~29 AU in the V1 direction [35], up to 70 AU in the V2 direction [36], ~160 and 210 AU towards the N and S poles respectively [40] and ~100 to 220 AU towards the tail[25],[44].

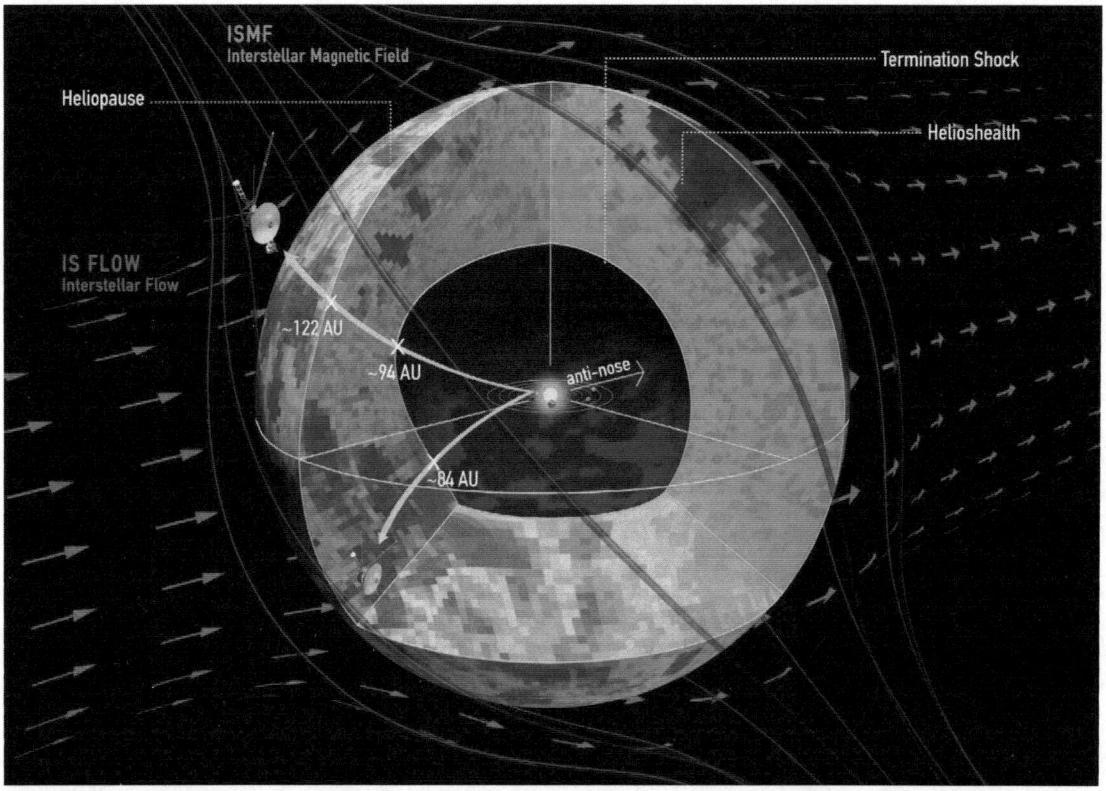

Figure 5. A composite, conceptual representation of the global heliosphere summarizing its basic properties in three dimensions, based on both remote ENA and *in-situ* ion measurements from Cassini/INCA and LECP/V1 and V2, respectively (published in [25]). The heliosheath includes a belt of varying ENA intensities (colour-coded with relative scale ranging from 1 (blue) to 12 (red)) surrounding the termination shock and extending to the heliopause. The asymmetric termination shock crossings from V1 and V2, above and below the ecliptic equator respectively, together with the heliopause crossing of V1 are shown. The heliosheath is possibly ~30–50% thicker towards the V2 direction (compared to the V1 direction), inconsistent with a compressed heliosheath in the Southern hemisphere. The interstellar flow (red arrows) impinges the nose and is deflected around a roughly symmetric nose to anti-nose heliosheath that interacts directly with the strong interstellar magnetic field (gray lines).

At this point we should also be reminded of two very important observations: The first is that the plasma beta inside the heliosheath, as measured along the trajectories of both Voyager 1 and Voyager 2, showed that the particle pressure was persistently much higher than the magnetic pressure at all times [13]. Second, after Voyager 1 exited the heliopause it measured a strong interstellar magnetic field of ~0.5 nT [45] (much stronger than was anticipated in the past), as was predicted by the joint use of INCA and Voyager measurements [13]. In combining all the aforementioned points, we regard our measurements to be consistent with a roughly symmetric, bubble-like heliosphere that exhibits diamagnetic behavior. This notion, as far as the broad context of the heliosphere interaction with the local interstellar medium is concerned, is consistent with the Parker-bubble model [46] (compared with the "magnetosphere-type" Parker model). A modified version of the Parker-bubble was suggested as a plausible interpretation of the "early" INCA measurements [8], and was proved in our recent analysis [25].

Together with the above recent developments, modern heliosphere modelling seems to provide a similar view: Magnetohydrodynamic modelling using the mini magnetosphere of Ganymede embedded in the sub-Alfvenic flow of Jupiter's magnetospheric plasma [29], was used as a proxy for the heliosphere and resulted in a "tailless" structure with no bow shock. The sensitivity of MHD models to seemingly "small" changes in the interstellar magnetic field was clearly demonstrated [47], from a heliosphere that includes a strong bow shock when B~0.2 nT, to no bow shock when B~0.4 nT. In other recent studies [30] the "Parker-bubble" was tested and a heliosphere with a symmetric termination shock was concluded when using an interstellar magnetic field close enough (0.44 nT), comparable to the measured one (~0.5 nT). Finally, a very recent and challenging model [27],[28] depicts a view of the heliosphere with turbulent jets deflected into the tail region by the interstellar wind (a "croissant-type" heliosphere), but not strong enough to drive a "comet-type" tail.

References

[1] Richardson, J. D., Wang, C., Paularena, K. I. 2001, *Adv. Space Res.* **27**, No. 3, pp. 471-479.
[2] Izmodenov, V. V., Malama, Y. G., Ruderman, M. S. 2008, *Adv. Space Res.*, **41**, 2, p318-324.
[3] Opher, M., Liewer, P. C., Velli, M., Bettarini, L., Gombosi, T. I., Manchester, W., DeZeeuw, D. L., Toth, G., Sokolov, I. 2004, *Astrophys. J.*, **611**, 1, pp. 575-586.
[4] Zank, G., and Muller, H.-R. 2003, *J. Geophys. Res.*, **108**, A6, pp. SSH 7-1.
[5] Washimi, H., Zank, G. P., Hu, Q., Tanaka, T., Munakata, K. 2007, *Astrophys. J.*, **670**, 2, pp. L139-L142.
[6] Pogorelov, N. V., Suess, S. T., Borovikov, S. N., Ebert, R. W., McComas, D. J., Zank, G. P. 2013, *Astrophys. J.*, **772**, 1, 17 pp.
[7] Decker, R. B., Krimigis, S. M., Roelof, E. C., Hill, M. E. 2010, *Proceedings of the 9th Annual International Astrophysics Conference, AIP Conference Proceedings*, **1302**, pp. 51-57.
[8] Krimigis, S. M., Mitchell, D. G., Roelof, E. C., Hsieh, K. C., McComas, D. J. 2009 *Science*, **326**, 971-973.
[9] Decker, R. B., Krimigis, S. M., Roelof, E. C., Hill, M. E., Armstrong, T. P., Gloeckler, G., Hamilton, D. C., Lanzerotti, L. J. 2005, *Science*, **309** (5743):2020-4.
[10] Burlaga, L. F., Ness, N. F., Acuña, M. H., Lepping, R. P., Connerney, J. E. P., Stone, E. C., McDonald, F. B. 2005, *Science*, **309**, 5743, pp 2020-2024.
[11] Stone, E. C., Cummings, A. C., McDonald, F. B., Heikkila, B. C., Lal, N., Webber, W. R. 2005, *Science*, **309**, 5743, pp 2012-2020.
[12] Decker, R. B., Krimigis, S. M., Roelof, E. C., Hill, M. E., Armstrong, T. P., Gloeckler, G., Hamilton, D. C., Lanzerotti, L. J. 2008, *Nature,* **454**, pp 67-70.
[13] Krimigis, S. M., Mitchell, D. G., Roelof, E. C., Decker, R. B. 2010, *Proceedings of the 9th Annual International Astrophysics Conference. AIP Conference Proceedings*, **1302**, 79.
[14] Fahr, H-J, Fichtner, H., Scherer, K. 2007, *Rev. Geophys.*, **45**, 1-38.
[15] McComas D. J. et. al 2009, *Space Sci. Rev.* **146**, 11.
[16] McComas, D. J., Dayeh, M. A., Funsten, H. O., Livadiotis, G., Schwadron, N. A. 2013, Astrophys. J., 771, 77, 9pp.
[17] Livadiotis, G., McComas, D. J., Dayeh, M. A., Funsten, H. O., Schwadron, N. A. 2011, *Astrophys. J.,* **734**:1, pp19.

[18] Heerikhuisen, J., Pogorelov, N. V., Zank, G. P., Crew, G. B., Frisch, P. C., Funsten, H. O., Janzen, P. H., McComas, D. J., Reisenfeld, D. B., Schwadron, N. A. 2010, *Astrophys. J. Lett.*, **708**, L126.
[19] Dayeh, M. A., McComas, D. J., Livadiotis, G., Ebert, R. W., Funsten, H. O., Janzen, P., Reisenfeld, D. B., Schwadron, N. A. 2011, *Astrophys. J.*, **734**, 29.
[20] Dialynas, K., Krimigis, S. M., Mitchell, D. G., Roelof, E. C., Decker, R. B. 2013, *Astrophys. J.*, **778**, 1, 13 pp.
[21] Fahr, H.-J., Sylla, A., Fichtner, H., Scherer, K. 2016, *J. Geophys. Res. Space Physics*, **121**, 8203-8214, doi:10.1002/2016JA022561.
[22] Schwadron N. A. et al. 2011, *Astrophys. J.* **731**, 56, 22pp.
[23] McComas, D. J., et al. 2014, *Astrophys. J. (Supp.)*, **213**, 20, doi:10.1088/0067-0049/213/2/20.
[24] Krimigis, S. M., et al. 2004, *Space Sci. Rev.*, **114**, pp. 233-329.
[25] Dialynas, K., Krimigis, S. M., Mitchell, D. G., Decker, R. B., Roelof, E. C. 2017, *Nat. Astron.*, **1**, No. 0115, doi:10.1038/s41550-017-0115.
[26] Burlaga, L. F., Ness, N. F., Stone, E. C. 2013, *Science*, **341**, 147-150.
[27] Opher, M., Drake, J. F., Zieger, B., Gombosi, T. I. 2015, *Astrophys. J. Lett.*, **800**, 7 pp.
[28] Drake, J. F., Swisdak, M., Opher, M. 2015, *Astrophys. J. Lett.*, **808**, 6.
[29] Kivelson, M. G. and Jia, X. 2013, *J. Geophys. Res.*, **118**, 6839-6846.
[30] Izmodenov, V. V. and Alexashov, D. B. 2015, *Astrophys. J. Suppl. S.*, **220**, 14.
[31] Gloeckler, G., and L. A. Fisk 2014, *Geophys. Res. Lett.*, **41**, 5325–5330.
[32] Gloeckler, G., and L. A. Fisk 2016, *Astrophys. J.*, **833**:290 (5pp).
[33] Gloeckler, G., & Fisk, L. A. 2014, in *ASP Conf. Ser.* **484**, Outstanding Problems in Heliophysics: From Coronal Heating to the Edge of the Heliosphere, ed. Q. Hu & G. P. Zank.
[34] Dialynas, K., Krimigis, S. M., Mitchell, D. G., Roelof, E. C. 2015, *J. Phys.: Conf. Ser.* **577,** 012007.
[35] Krimigis, S. M. Decker, R. B., Roelof, E. C., Hill, M. E., Armstrong, T. P., Gloeckler, G., Hamilton, D. C., Lanzerotti, L. J. 2013, *Science*, **341**, 144-147.
[36] Roelof, E. C., Krimigis, S. M., Mitchell, D. G., Decker, R. B., Dialynas, K. 2012, *AIP Conf. Proc.,* **1436**, 239.
[37] Wood, B. E., Izmodenov, V. V., Alexashov, D. B., Redfield, S., Edelman, E. 2014, *Astrophys. J.,* **780**, 12.
[38] McComas, D. J.Angold, N., Elliott, H. A., Livadiotis, G., Schwadron, N. A., Skoug, R. M., Smith, C. W. 2013, *Astrophys. J.,* **779**, 10.
[39] Reisenfeld, D. B. et al., 2012, *Astrophys. J.*, **747**:110, 14pp.
[40] Reisenfeld, D. B. et al., 2016, *Astrophys. J.,* **833**:277, 15pp.
[41] Sokół, J. M., Swaczyna, P., Bzowski, M., Tokumaru, M. 2015, *Solar Phys.,* **290**, 2589-2615.
[42] Decker, R. B., Krimigis, S. M., Roelof, E. C., Hill, M. A. 2012, *Nature*, **489**, 124-127.
[43] Richardson, J. D. and R. B. Decker, 2014, *Astrophys. J.* **792**:126 (5pp).
[44] Galli, A. et al. 2016, *Astrophys. J.,* **821**:107, 10pp.
[45] Burlaga, L. F., Ness, N. F., Stone, E. C. 2013, *Science* **341**, 147-150.
[46] Parker, E. N. 1961, *Astrophys. J.,* 134, 20.
[47] Heerikhuisen, J., Zirnstein, E. J., Funsten, H. O., Pogorelov, N. V. & Zank, G. P. 2014, *Astrophys. J.* **784**, 73.

Acknowledgments
This work was supported at JHU/APL by NASA under contract NAS5 97271 and NNX07AJ69G and by subcontract at the Office for Space Research and Technology. The authors are grateful to M. Kusterer for software development and assistance with the INCA data processing.

IOP Conf. Series: Journal of Physics: Conf. Series **900** (2017) 012006 doi:10.1088/1742-6596/900/1/012006

The pump acceleration mechanism

L A Fisk and G Gloeckler

Department of Climate & Space, University of Michigan
2455 Hayward Street, Ann Arbor, MI, 48109-2143, USA

Email: lafisk@umich.edu

Abstract. Starting in the early 2000s, Fisk and Gloeckler developed a theory for a pump acceleration mechanism to account in detail for the ubiquitously observed, so-called, -5 spectra, a distribution function that is a power law in particle speed with spectral index of -5. To date, this theory has seen no general use in explaining data and relatively little constructive discussion. In this paper, we will make another attempt to interest experimentalists and theorists in using and expanding on the pump acceleration mechanism. The inevitability of the pump acceleration occurring in conditions that are common in space plasmas will be discussed, along with supporting numerical studies and observations.

1. Introduction

In the last decade, detailed observations have been made of the spectra of lower energy particles accelerated both in the inner heliosphere, from the Ulysses and ACE spacecraft, and in the heliosheath beyond the termination shock of the solar wind, from the Voyager spacecraft. In both cases, in specific events in the inner heliosphere, and throughout the heliosheath, the spectra are the same: a power law in particle speed with a spectral index of -5, when the spectrum is expressed as a distribution function (also referred to as phase space density); or equivalently, a differential intensity spectrum that is a power law in energy with a spectral index of -1.5. This common spectrum generally has an exponential rollover at higher energies, indicating the maximum energy particles acquire in the acceleration process.

To explain the common spectrum, Fisk and Gloeckler [1-6] developed a new acceleration mechanism, a pump acceleration mechanism. However, despite the fact that the pump acceleration mechanism can account for the common spectrum in all the different circumstances where it is observed, the theory has seen no general use, other than the applications of Fisk and Gloeckler, and relatively little constructive discussion. In this paper we will make one more attempt to interest experimentalists and theorists in using and expanding on the pump acceleration mechanism.

We provide a step-by-step, concise presentation of the logic behind the pump acceleration mechanism; we then discuss examples of where it is successful; and ways in which we can test the pump acceleration mechanism. All the arguments and conclusions summarized in this paper, as well as supporting observational evidence, are discussed in detail in Fisk and Gloeckler [5,6].

2. The theory of the pump acceleration mechanism

The following sections describe the essential features of the pump acceleration mechanism.

2.1. The plasma conditions in which the pump acceleration mechanism operates
The following are the plasma conditions in which the pump acceleration mechanism is expected to operate and accelerate particles:

- The plasma contains an embedded mean magnetic field. The energy in the particles is greater than that of the magnetic field, which has as its only function to couple the motion of the particles.
- The plasma can be considered to be in a fixed volume—i.e. on applicable scales we are not concerned with the expansion of the solar wind—and is thermally isolated, which means simply that there is no net influx or outflow of particles and energy across the boundaries of the volume; i.e., no large-scale spatial gradients.
- The plasma contains mesoscale compression and expansion regions. These compressions and expansions occur normal to the mean magnetic field, and the regions are of size large compared to the gyroradii of the particles that are to be accelerated in the plasma, but small compared to the total volume of the plasma.

The pump acceleration mechanism should apply in very common conditions in the solar wind. In particular, the required conditions are likely to occur downstream from shocks, which are subsonic regions with ample compressions and expansions.

2.2. The proper description of particle behavior in the pump acceleration mechanism
The particles to be accelerated have sufficiently high speeds so that they can spatially diffuse, which we define as: in the presence of a sufficient spatial gradient, they can escape from a compression region within the time that a compression region will become an expansion region. Recall each location is the site of alternating compression and expansion regions.

The compression and expansion regions have a cross section much larger than the particle gyroradii, and so the best description of the behavior of the particles within each compression and expansion region is the Parker [7] transport equation, written in terms of the distribution function, f:

$$\frac{\partial f}{\partial t} + \delta\mathbf{u}\cdot\nabla f = \frac{\nabla\cdot\delta\mathbf{u}}{3} v \frac{\partial f}{\partial v} + \nabla\cdot\left(\tilde{\kappa}\cdot\nabla f\right). \tag{1}$$

Here, v is particle speed; t is time; $\delta\mathbf{u}$ is the mean velocity of the plasma, and is responsible for the compressions and expansions thru the term $\nabla\cdot\delta\mathbf{u}$; the spatial diffusion tensor is $\tilde{\kappa}$.

The particle behavior in the pump acceleration mechanism can be described by the Parker transport equation, which is the standard equation that has been used successfully to describe galactic cosmic ray modulation for over 50 years, i.e., we treat each compression and expansion region as a small modulation region.

2.3. The consequences for using the Parker equation for the total particle energy
Particles gain or lose energy in the Parker transport equation only by an adiabatic compression or expansion; e.g. the adiabatic cooling of galactic cosmic rays in the expanding solar wind. In an adiabatic compression or expansion, no energy is added to or removed from the particles; no work is done on or by the particles. All that happens is there is a change in the volume experienced by the particles. Since the total volume is constant, the compressions and expansions do not result in a net change in the total energy of all the particles in our plasma.

In the pump acceleration mechanism, the sum of the energy of the particles that are being compressed or expanded, plus the energy of the particles that are responsible for the compression or expansion, is constant.

2.4. *Why a high-energy tail develops and the spectrum of the tail*

We have a thermally isolated system in which particles gain or lose energy by adiabatic compressions and expansions. Total particle energy and number of particles are constants. We have an irreversible process, the spatial diffusion. Under these circumstances, the system will evolve to a state of maximum entropy, and in doing so will create a high-energy tail.

In a state of maximum entropy, the high-energy tail must be isentropic in each compression and expansion region, or equivalently, the pressure in the high-energy tail must behave according to the pressure equation for an adiabatic compression or expansion. Thus, the pressure, P_t, in the tail between particle speeds v_1 and v_2 must satisfy

$$\frac{\partial P_t}{\partial t} + \delta\mathbf{u}\bullet\nabla P_t + \frac{5}{3}(\nabla\bullet\delta\mathbf{u})P_t = 0 . \qquad (2)$$

We can find the spectral form of the high-energy tail that satisfies equation (2) by first noting that in a fully developed high-energy tail, spatial gradients between adjacent compression and expansion regions are small, and thus there is no loss or gain of particles by spatial diffusion, in which case, from equation (1), the particle behavior is described by

$$\frac{\partial f}{\partial t} + \delta\mathbf{u}\bullet\nabla f - \frac{\nabla\bullet\delta\mathbf{u}}{3} v \frac{\partial f}{\partial v} = 0 . \qquad (3)$$

We can then integrate equation (3) over particle speed to determine an equation for the particle pressure, and note that if $f = f'v^{-5}$, where f' is a normalization constant, the resulting pressure equation is equation (2).

$$\frac{4\pi m}{3}\int_{v_1}^{v_2} v^4 dv \left[\frac{\partial(f'v^{-5})}{\partial t} + \delta\mathbf{u}\bullet\nabla(f'v^{-5}) - \frac{(\nabla\bullet\delta\mathbf{u})}{3} v \frac{\partial(f'v^{-5})}{\partial v}\right] = 0$$
$$\frac{\partial P_t}{\partial t} + \delta\mathbf{u}\bullet\nabla P_t + \frac{5}{3}(\nabla\bullet\delta\mathbf{u})P_t = 0 \qquad (4)$$

where m is particle mass.

In the pump acceleration mechanism, the high-energy tail that results when a system with adiabatic compressions and expansions, and irreversible spatial diffusion, evolves to a state of maximum entropy is a distribution function with a spectral index of -5.

2.5. *The source of the energy that creates the high-energy tail*

In our system, particle energy is conserved. There is no change in the sum of the energy of the particles that are being compressed or expanded, plus the energy of the particles that are responsible for the compression or expansion. It follows therefore that the energy that creates the high-energy tail comes from lower energy particles.

The consequence of the conservation of energy was illustrated by Bykov [8], who modeled the time evolution of the spectra of particles accelerated in superbubbles. The resulting nonrelativistic spectra are shown in the left panel of figure 1, taken from Bykov [8]. The acceleration is due to mesoscale turbulence generated by shocks. There is a nonlinear coupling between the accelerated particles and the turbulence, such that the energy in the total system is conserved. Noting that $v^2 f$ is plotted, the time asymptotic spectrum (#5) is a power law with spectral index of -5.

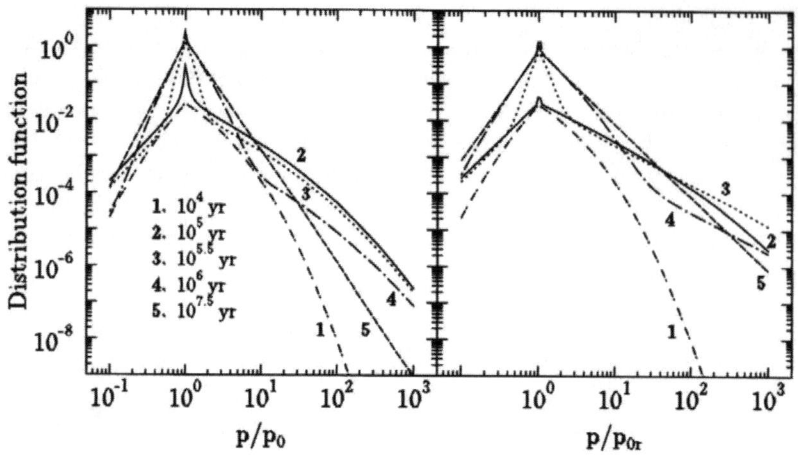

Figure 1. The temporal evolution of the distribution function ($v^2 f$) in the Superbubble of radius 220 pc with the standard interstellar magnetic field. Monoenergetic injection was assumed with the injection energies 10 keV (left panel) and 10 GeV (right panel) (from [8]).

In the pump acceleration mechanism, total particle energy is a constant, in which case the energy that creates the high-energy tail must come from the lower energy particles that are responsible for the compressions and expansions. Moreover, the Bykov [8] model confirms the conclusion that a system with mesoscale compressions and expansions, in which energy is conserved and there is an irreversible coupling between particles of different energies, results in a distribution function with a tail that is a power law with spectral index of -5.

2.6. The time evolution of the high-energy tail

Fisk and Gloeckler [5,6] carefully derived an equation that describes the time evolution of the spectrum of particles accelerated by the pump acceleration mechanism. The derivation is in terms of a nonstandard form for the mean distribution function, f_o, which is defined as the distribution function that determines the mean pressure in the accelerated particles, or

$$\langle P \rangle = \frac{4\pi m}{3} \int_{v_1}^{\infty} v^4 f_o dv \equiv \int_{v_1}^{\infty} P_v dv \,. \tag{5}$$

This choice for the mean distribution function allows Fisk and Gloeckler to correctly capture the collective behavior of the particles within a compression or expansion region. The resulting acceleration that describes the time evolution of f_o is then

$$\frac{\partial f_o}{\partial t} = \frac{1}{v^4}\frac{\partial}{\partial v}\left(\frac{\delta u^2}{9\kappa} v \frac{\partial}{\partial v}\left(v^5 f_o\right)\right) . \tag{6}$$

Here, δu^2 is the mean squared variation in the mean plasma speed and κ is the spatial diffusion coefficient for diffusion across the mean magnetic field.

A typical solution to equation (6) is

$$f_o \propto v^{-5} \exp\left[-9\kappa / \left(\left(1+\alpha\right)^2 \delta u^2 t\right)\right] . \tag{7}$$

In this case, the diffusion coefficient is particle speed times a power law in particle rigidity, with spectral index of α. The solution evolves into a power law with spectral index of -5 and an exponential rollover that increases with time.

Technically, f_o is not the observed mean distribution function. What is observed is the time average of the distribution function at a given particle speed, not an integral over particle speed, as in equation (5). However, Fisk and Gloeckler [5,6] show that in the -5 portion of the spectrum, f_0 is identical to the time-averaged distribution function, and there are only small deviations between f_o and the time-averaged distribution function in the roll-over region of the spectrum.

The time evolution of the spectrum in the pump acceleration mechanism is well described by equation (6) and the resulting solutions, e.g. equation (7), and these solutions provide excellent fits to spectra observed in the many instances where the pump acceleration mechanism is expected to apply.

2.7. The rate of acceleration of the high-energy tail

We can determine the rate of acceleration in the pump mechanism by rewriting equation (6) in terms of the differential pressure, P_v, in equation (5):

$$\frac{\partial P_v}{\partial t} = \frac{\partial}{\partial v}\left(\frac{\delta u^2 v^2}{9\kappa}\frac{\partial P_v}{\partial v}\right) + \frac{\partial}{\partial v}\left(\frac{\delta u^2 v}{9\kappa}P_v\right). \tag{8}$$

The first term on the right in equation (8) describes the diffusion of particle energy to higher energies. The second term describes a first-order acceleration.

The pump acceleration mechanism is a first-order acceleration, which will always accelerate particles more rapidly than any stochastic acceleration mechanism, and there are many cases where this first-order acceleration is comparable to or exceeds the first-order acceleration in diffusive shock acceleration.

2.8. Conclusions about the theory for the pump acceleration mechanism

If you have a plasma with the following conditions, which you can find in many locations in the solar wind:

1. *The plasma is in a fixed volume that can reasonably be considered to be thermally isolated, with no net influx or outflow of particles and particle energy across the boundaries of the volume.*
2. *The plasma contains mesoscale compression and expansion regions, where the compressions and expansions are normal to the mean magnetic field that causes all particles within a region to experience either a compression or expansion.*
3. *There are particles with sufficiently high speeds, and thus sufficiently large gyroradii that they can spatially diffuse out of a compression region, into an adjacent expansion region, thereby irreversibly coupling the compression and expansion regions.*

Then, a high-energy tail, a power law with a spectral index of -5, will develop.

4. *The tail will develop rapidly because the pump acceleration mechanism responsible for the high-energy tail is a first-order acceleration.*

3. Applications of the pump acceleration mechanism

The following are some particularly compelling applications of the pump acceleration mechanism.

3.1. The heliosheath

The heliosheath is an ideal location for the pump acceleration mechanism. It is a subsonic region, with dominant pressure in the interstellar pickup ions, and ample mesoscale compression and expansion regions.

Indeed, when Voyager 1 and Voyager 2 crossed the termination shock of the solar wind, each, after some initial fluctuations in the spectra, locked onto the differential intensity spectra shown in figure 2, a power law with spectral index of -1.5, which is equivalent to a distribution function in particle speed, with spectral index of -5, the spectrum produced by the pump acceleration mechanism.

Of even more importance, when Voyager 1 reached ~120 AU it observed the full ACR spectrum shown in figure 3, a distribution function that is a power law with spectral index of -5, extending up to 10s of MeV/nucleon in energy.

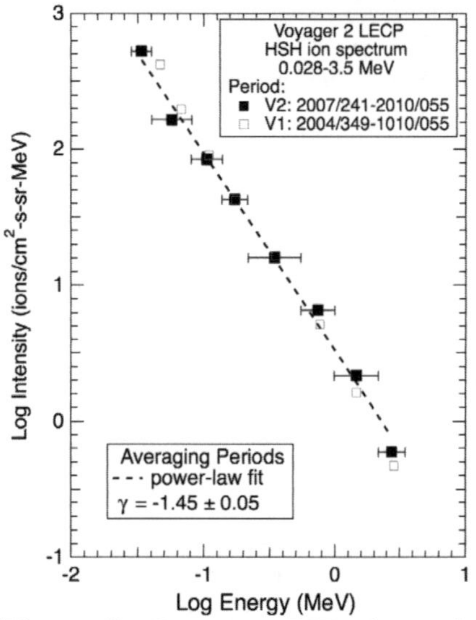

Figure 2. Voyager LECP observations taken from Decker et al. [12]; Gloeckler et al. [13].

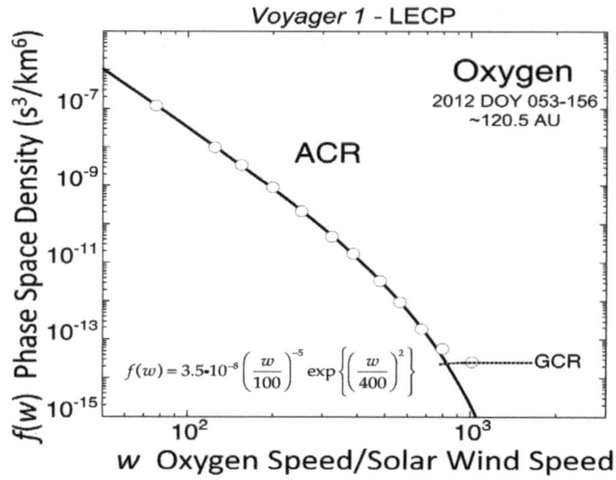

Figure 3. Phase space density of ACR O in the prime acceleration region of the heliosheath at ~117 AU. The equation shown in the figure is the formula for the spectrum produced by the pump acceleration mechanism and provides an excellent fit to the observations (from [5]).

As is discussed in detail in Fisk and Gloeckler [9], the acceleration of the ACRs in the heliosheath must be due to the pump acceleration mechanism, and cannot be due to diffusive shock acceleration at the termination shock. The pressure in the ACRs in the heliosheath is observed to be about half the total particle pressure in the heliosheath [10]. If ACRs originate at the termination shock on the flanks, as suggested by McComas and Schwadron (e.g., [11]), the ACR pressure would be added to the pressure in the pickup ions that are convected radially downstream from the termination shock with the solar wind. Pressure would then not be constant, as is required in the subsonic heliosheath, unless the pickup ion pressure could be reduced, which seems highly unlikely. Rather, it is necessary to accelerate the ACRs out of the pickup ions using the pump acceleration mechanism, which yields the correct spectral shape for the accelerated ACRs.

3.2. The interstellar medium

Superbubbles in the interstellar medium are also a possible location for application of the pump acceleration mechanism; the problem that Bykov [8] considered. There is ample energy in low-energy particles and supernovae to excite mesoscale compression and expansion regions.

Fisk and Gloeckler [14] applied the pump acceleration mechanism to the acceleration of all GCRs in the interstellar medium. The scale sizes of the turbulence were chosen realistically; particles were assumed to diffuse across the magnetic field by field line random walk; and the problem was done relativistically. The resulting spectrum is shown in figure 4. The slope below the knee, which occurs when the particle gyroradii equals the scale size of the turbulence, is fixed by the choice of parameters. The slope above the knee is a fixed multiple of the slope below the knee and agrees well with observations.

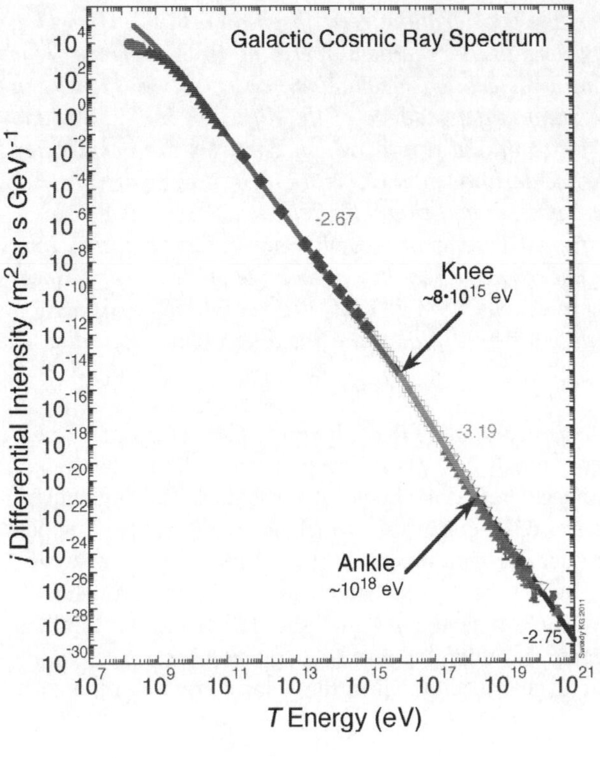

Figure 4. Differential intensity, j, versus kinetic energy, T, of galactic cosmic ray protons (symbols) and fits to the spectrum computed by Fisk and Gloeckler [14] (red curve). The relation $j = p^2 f$ was used to convert the distribution function, f, to j. The spectrum above the ankle ($T \geq 10^{18}$ eV), assumed to be of extragalactic origin, was fit using a power law with spectral index of -2.75. Filled circles are Voyager 1 measurements [15], while other data are a compilation of various measurements by Swordy [16].

The spectrum above the knee is a problem for diffusive shock acceleration, which becomes ineffective when the particle gyroradii exceed the scale size of the shocks. It is not a problem for the pump acceleration mechanism, which uses the entire interstellar medium to accelerated particles and only becomes ineffective when the particle gyroradii exceed the scale size of the galaxy.

3.3. The solar wind at 1 AU

The pump acceleration mechanism is also important at 1 AU, and yields suprathermal tails with spectral index of -5. The acceleration frequently occurs in the subsonic regions immediately downstream of shocks, although it can also occur in the absence of shocks. Shown in figure 5 are typical observations at 1 AU. The observed spectrum has three distinct parts: The thermal solar wind, labeled the core in figure 5; a steep extension on the core labeled the halo; and finally the classic -5 tail, which given the small scale sizes at 1 AU only extends up to ~4 times the solar wind speed, where the spectrum rolls over.

Fisk and Gloeckler [6] developed a detailed theory for the halo portion of the spectrum, referred to as 'the inefficient pump mechanism'. Particles in this speed range are unable to readily escape from a compression region and will experience a smaller than average compression and gain less energy. As a result the halo has a steeper spectrum, which Fisk and Gloeckler [6] demonstrate varies with the rigidity dependence of the diffusion coefficient.

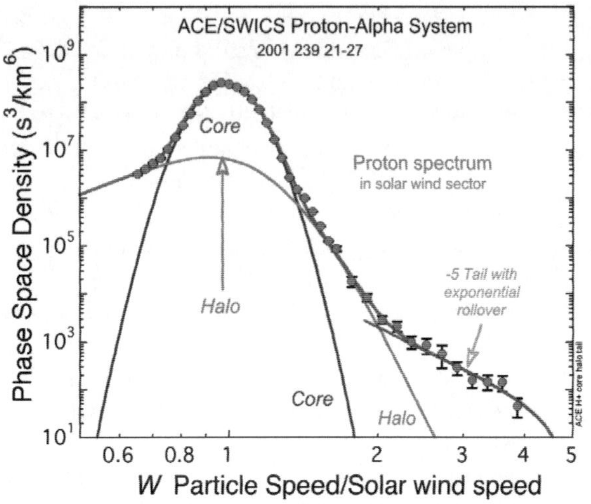

Figure 5. A typical spectrum of particles that are accelerated in the solar wind at 1 AU. The core, thermal solar wind is shown in blue, the steep halo in red, and the common spectrum, with spectral index of -5 and an exponential rollover, in green.

There is a wealth of information available on particle acceleration from observations at 1 AU, which allows us to determine the dominant particle acceleration mechanism operating in the various solar wind plasma conditions. In the case of the pump acceleration mechanism, observations such as in figure 5 can, from the halo, reveal the required rigidity dependence of the diffusion coefficient, and from the particle speed at the roll-over of the spectrum, the rate at which particles are accelerated and/or the escape of the particles from the prime acceleration site. Comparison of the requirements of the pump acceleration mechanism with concurrent plasma and magnetic field measurements provide a full test of the pump acceleration mechanism. It must be emphasized, however, that very high time resolution measurements at all relevant particle energies are required to completely test every aspect of the pump mechanism. The scale sizes involved in acceleration at 1 AU are very small, compression regions with dimensions only a few times larger than the gyroradii of the particles being accelerated.

4. Concluding remarks

The acceleration mechanism of choice for particle acceleration in the solar wind has, for decades, been diffusive shock acceleration, which has been employed to explain acceleration to high energies at shocks associated with CMEs in the solar corona; acceleration at shocks at 1 AU; at CIRs; at planetary bow shocks, etc. In principle, the pump acceleration mechanism, which as with diffusive shock acceleration is first order, could be the dominant acceleration mechanism in many of these observed acceleration events. As we learn more about the pump acceleration mechanism, and in particular we verify all aspects of the pump mechanism with observations at 1 AU, we should expand our thinking and explore whether the pump acceleration mechanism is dominant over diffusive shock acceleration in some or even many of the events where particles are accelerated in the solar corona and the solar wind.

Acknowledgments

This work was supported in part by NSF Grant AGS-1043012.

References

[1] Fisk L A and Gloeckler G 2007 *Proc. Nat. Acad. Sci.* **104** 5749
[2] Fisk L A and Gloeckler G 2008 *Astrophys. J.* **686** 1466
[3] Fisk L A, Gloeckler G and Schwadron N A 2010 *Astrophys. J.* **720** 533
[4] Fisk L A and Gloeckler G 2012a *Space Sci. Rev.* **173** 433
[5] Fisk L A and Gloeckler G 2014 *J. Geophys. Res* **119** 8733
[6] Fisk L A and Gloeckler G 2017 Common spectrum of particles accelerated in the heliosphere: Observations and a mechanism *Kappa Distributions: Theory and Applications in Plasmas*, ed G Livadiotis (Netherlands: Elsevier)
[7] Parker E N 1965 *Planet. Space Sci.* **13** 9
[8] Bykov A M 2001 *Space Sci. Rev.* **99** 317
[9] Fisk L A and Gloeckler G 2013 *Astrophys. J.* **776** 79
[10] Gloeckler G and Fisk L A 2010 *Proc. 9th Annual Intl. Astrophysics Conf.* vol 1302 (Melville, NY: AIP) p 110
[11] McComas D J and Schwadron N A 2006 *Geophys. Res. Lttrs.* **33**, doi: 10.1029/2005GL025437
[12] Decker R B, Roelof E C, Krimigis S M and Hill M E 2006 *Physics of the Inner Heliosheath* vol 858 (Melville, NY: AIP) p 73
[13] Gloeckler G, Fisk L A, Mason G M and Hill M E 2008 *Particle Acceleration and Transport in the Heliosheath and Beyond* vol 1039 (Melville, NY: AIP) p 367
[14] Fisk L A and Gloeckler G 2012b *Astrophys. J.* **744** 127
[15] Cummings, A C, Stone E C, McDonald F B, Heikkila B C, Lal N and Webber W R 2008 *Particle Acceleration and Transport in the Heliosphere and Beyond* vol 1039 (Melville, NY: AIP) p 343
[16] Swordy S P 2001 *Space Sci Rev* **99** 85

IOP Conf. Series: Journal of Physics: Conf. Series **900** (2017) 012007 doi:10.1088/1742-6596/900/1/012007

Reconnection at the Heliopause: Predictions for Voyager 2

S. A. Fuselier[1,2] and I. H. Cairns[3]

[1]Space Science and Engineering Division, Southwest Research Institute, San Antonio, TX 78228, USA

[2]Department of Physics and Astronomy, University of Texas at San Antonio, San Antonio, TX 78249, USA

[3]School of Physics, University of Sydney, NSW 2006, Australia

sfuselier@swri.edu

Abstract. Predicted and observed properties of the inner and outer heliosheath were recently used to assess whether magnetic reconnection was occurring at the Voyager 1 crossing of the heliopause. It was concluded that reconnection may not have been occurring locally, but may have been occurring at a location remote from Voyager 1. Here observations of 37 to 70 keV electrons measured by the LECP instrument on Voyager 1 are interpreted as possible evidence for remote reconnection and the study is extended to the heliopause near the projected crossing location of Voyager 2, where the plasma depletion layer (PDL) should be significantly stronger. The predicted plasma properties are used to determine if local reconnection is possible at this projected heliopause crossing.

1. Introduction

Voyager 1 crossed the heliopause in August 2012 and was the first spacecraft to directly sample the outer heliosheath/local interstellar medium (LISM) near the boundary [1,2,3]. Over the course of almost a year, the plasma density increased and the magnetic field magnitude decreased (except during the passage of shocks in the LISM). These changes in the plasma parameters are characteristic of a plasma depletion layer [4]. The plasma depletion layer near the heliopause was predicted [5] and is analogous to the plasma depletion layer at the Earth's magnetopause.

A plasma depletion layer forms because draping of the interstellar magnetic field against the heliopause increases the magnetic field magnitude and the perpendicular temperature of the plasma. These quantities increase until the perpendicular to parallel temperature anisotropy of the ions exceeds the threshold of the electromagnetic ion cyclotron (EMIC) instability. The waves grow rapidly to saturation and pitch-angle scatter ions into the parallel direction, reducing the temperature anisotropy. The scattered ions stream along the magnetic field and leave the region, resulting in a density decrease. The temperature anisotropy and the plasma density decrease are bounded throughout the plasma depletion layer by the marginal stability threshold of the EMIC instability [6,7]. The PDL is an example of a micro-kinetic process that has global-scale consequences in a plasma. Since this process is kinetic in nature, it is not captured in global, isotropic magnetohydrodynamic (MHD) computer simulations.

Magnetic reconnection is another inherently kinetic process that has global consequences at the Earth's magnetopause. In particular, reconnection can mitigate the effect of magnetic field draping at the magnetopause. Reconnected field lines may convect north and south over the magnetic poles instead of slipping east or west around the flanks of the magnetopause. A similar effect may occur at the heliopause if there is reconnection at the boundary.

Reconnection at the Earth's magnetopause is asymmetric and often involves a substantial guide field (i.e., component reconnection). The asymmetry occurs because there is a large change in density and a smaller change in magnetic field magnitude across the magnetopause. The magnetosheath density at the magnetopause is of the order of 20 cm^{-3} while the outer magnetosphere density is of the order of 0.1 cm^{-3}, resulting in the density changing by a factor ≈200 (e.g., [8]). The magnetic field magnitude in the magnetosheath is of the order of 56 nT while that in the outer magnetosphere is of the order of 25 nT, resulting in the magnetic field magnitude changing by a factor of ≈2 at Earth's magnetopause. Component (or guide field) reconnection with magnetic shear angles as small as ~90° may occur at the magnetopause (e.g., [9]).

Reconnection at the heliopause is also asymmetric and may involve a substantial guide field. The outer heliosheath/LISM density at the heliopause is approximately 0.05 cm^{-3} [2] while the inner magnetosphere density is estimated to be approximately 2×10^{-3} cm^{-3} (from Voyager 2 measurements, [10]), resulting in the density changing by a factor of 25. The outer heliosheath magnetic field magnitude, from Voyager 1 measurements is approximately 0.44 nT [3] while the outer magnetosphere magnitude is approximately 0.27 nT, resulting in the magnetic field magnitude changing by a factor of ≈2. At the heliopause, the magnetic shear angle at the Voyager 1 crossing was only the order of 10°. Thus, the guide field (or non-reconnecting component of the magnetic field) may be very large.

A substantial electron diamagnetic drift occurs for asymmetric reconnection in the presence of a strong guide field [11]. If the electron drift speed is greater than the local Alfvén speed at the boundary, then reconnection is suppressed locally. This limiting velocity leads to conditions on the magnetic shear angle and the change in plasma beta across the boundary. Reconnection is suppressed when there is a large change in plasma beta and a small magnetic shear across the boundary [11]. However, although reconnection may be suppressed locally, it may be occurring remotely at a location where the diamagnetic drift velocity is less than the local Alfvén speed.

The suppression of reconnection at the heliopause has been investigated for the Voyager 1 crossing in two studies [11, 12]. The first study [11] was not limited to the location of the Voyager 1 crossing. It used global MHD simulations to determine the plasma properties over nearly the entire heliopause and concluded that reconnection occurs only where the magnetic shear angle is nearly 180°. The second study [12] used local measurements from Voyager 1 and 2, models for the pick up ions in the inner and outer heliosphere, and properties of the plasma depletion layer to determine the plasma properties at the heliopause at the Voyager 1 crossing. This study also concluded that reconnection was probably not occurring locally at the Voyager 1 heliopause location.

The purpose of this paper is to extend the second study [12] to 1) present observations that, by analogy with observations at Earth, suggest that reconnection was occurring at the heliopause remotely from the Voyager 1 location, and 2) determine if reconnection may be suppressed at the projected location of the Voyager 2 crossing of the heliopause. In particular, the effect of the plasma depletion layer (PDL) on predictions for reconnection is investigated because the PDL is predicted to be stronger at the Voyager 2 crossing than at the Voyager 1 crossing.

2. Reconnection at the Voyager 1 heliopause crossing

Table 1 summarizes the plasma parameters at the Voyager 1 crossing of the heliopause from [12]. The plasma density in the depletion layer is extrapolated from the Langmuir wave profile [2]. Over the course of the year following the heliopause crossing, this density increased by more than a factor of two as the spacecraft traversed the plasma depletion layer. A corresponding decrease in the magnetic field magnitude was not observed directly, due to the magnetic field increases caused by several shocks propagating into the interstellar medium [3, 13] but can be discerned if the effects of the first shock is removed [4]. The ion temperature in the outer heliosheath was determined from the properties of the plasma depletion layer and an assumption that pick-up ions are 3% of the total ion density, consistent with pick-up ion models [14,15].

Table 1: Plasma parameters at the heliopause along the Voyager 1 trajectory

Plasma parameter	Value	Origin
Density in the depletion layer adjacent to the heliopause	$N = 0.047\ cm^{-3}$	Extrapolated from the Langmuir wave profile
Magnetic field magnitude in the depletion layer adjacent to the heliopause	$B = 0.44$ nT	Measured by the Voyager 1 magnetometer
Total temperature	$T = 1.5 \times 10^4$ K	Estimated from the plasma depletion layer properties and assuming 0.03% pick up ions
Density in the inner heliosheath adjacent to the heliopause	$N = 2 \times 10^{-3}\ cm^{-3}$	Assumed, consistent with Voyager 2 observations
Magnetic field magnitude in the inner heliosheath at the heliopause	$B = 0.25$ nT	Measured by the Voyager 1 magnetometer
Total temperature in the inner heliosheath at the heliopause	$T = 2 \times 10^6$ K	Estimated from the assumption that there are 16.5% pick up ions
Plasma beta in the depletion layer	0.13	Computed from N, B, and T in the plasma depletion layer
Plasma beta in the inner heliosheath	2.2	Computed from N, B, and T in the inner heliosheath
$\Delta\beta$ across the heliopause = $\lvert\beta_{IHS} - \beta_{PDL}\rvert$	$\Delta\beta = 2.1$	
Magnetic Shear angle across the heliopause at Voyager 1	6°	Measured by the Voyager 1 magnetometer

For the inner heliosheath, [12] used a mixture of Voyager 1 and Voyager 2 observations and some results from modeling. The density in the inner heliosheath at Voyager 2 was used for the inner heliosheath density at Voyager 1. Voyager 1 measured the magnetic field magnitude in the inner heliosheath just before the heliopause crossing [3]. The ion temperature was determined from the assumption that there are 16.5% pick-up ions relative to the total ion population [14, 16] and that pick-up ions dominate the overall plasma temperature.

Table 1 shows the plasma betas on both sides of the heliopause computed from these densities, ion temperatures, and magnetic fields and these betas. Also in Table 1 is the magnetic shear angle at the magnetopause measured by the Voyager 1 magnetometer. Figure 1 shows the test for suppression of reconnection from Swisdak et al. [11] for the Voyager 1 crossing (blue filled circle). For the measured magnetic shear and $\Delta\beta$ from Table 1, Figure 1 shows that reconnection is suppressed locally at the Voyager 1 crossing of the heliopause. The magnetic shear angle at the magnetopause would have had to be of the order of 90°, or about 15 times larger than the shear angle observed by Voyager 1, for reconnection to be possible.

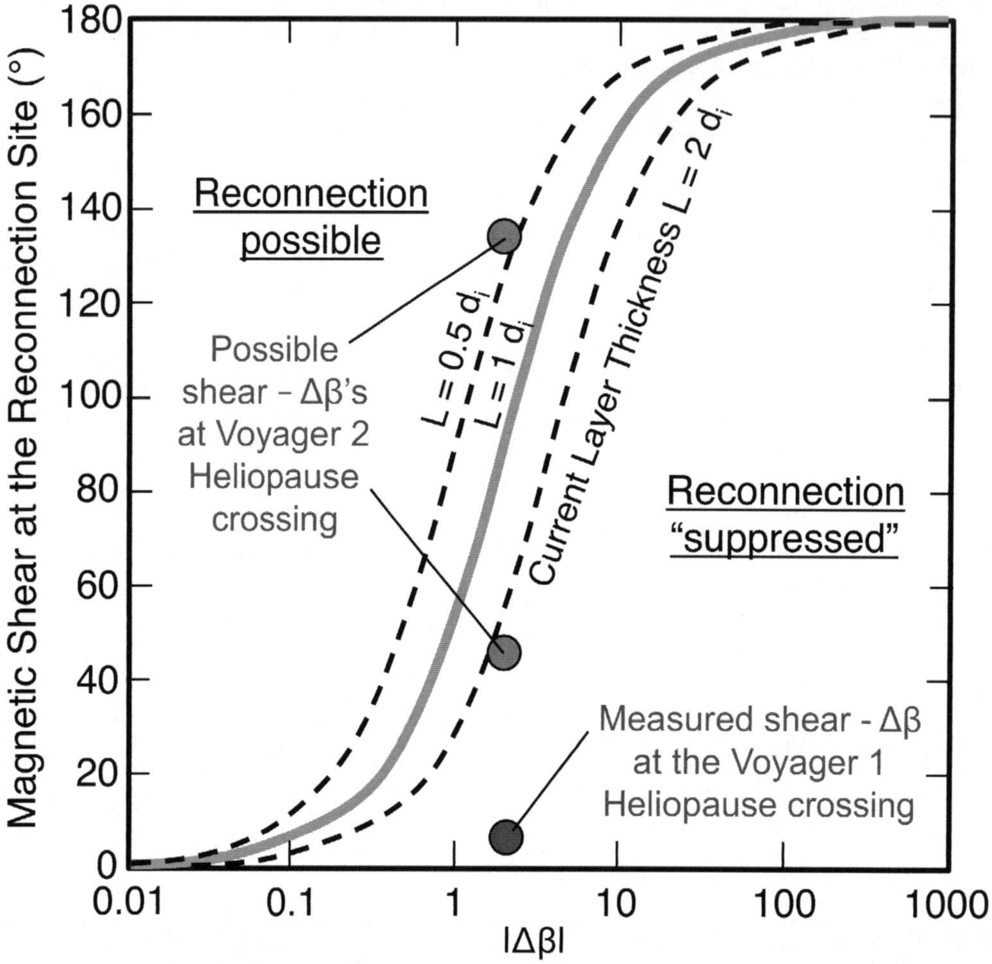

Figure 1. Regions in Δβ-magnetic shear of possible and suppressed reconnection. For the Voyager 1 heliopause crossing, reconnection was probably suppressed locally. For the projected Voyager 2 heliopause crossing, reconnection may not be suppressed if the shear angle is ~135°. However, if the shear angle is ~45°, then reconnection may be suppressed locally.

3. Possible evidence for remote reconnection at the heliopause

At a basic level, the heliopause and magnetopause share many common characteristics. In this section, an analogy is drawn between the magnetopause and the heliopause to help interpret electron observations at the latter boundary.

For some interplanetary magnetic field orientations, there are large regions of the Earth's magnetopause where reconnection is suppressed locally. When the IMF is northward, the region at low latitudes directly in the Earth-Sun line represents one such region. While reconnection is suppressed locally, there are locations at high latitudes where reconnection is not suppressed and is occurring essentially uninterrupted [17]. Often, spacecraft traversing the magnetopause at low latitudes are magnetically connected to reconnection sites at high latitudes. Electron observations during these crossings show two distinct characteristics that indicate that reconnection is occurring remote from the spacecraft location. First, there is a boundary layer inside the magnetosphere, called the low latitude boundary layer, where there is a loss of high-energy (several keV) electrons that are normally present in the magnetosphere. Second, on the sunward side of the magnetopause there is another boundary layer, often called the magnetosheath boundary layer, that contains moderately energized (100's of eV to about 1 keV) electrons streaming along the magnetic field from the reconnection location.

Magnetospheric electrons are lost in the low latitude boundary layer because the magnetic field lines that thread the region are open or were open at one time. Electrons propagate very rapidly along magnetic field lines, so they can exit the magnetosphere along open field lines much faster than the ions. On the sunward side of the magnetopause, the magnetosheath boundary layer is formed by heating of the relatively cold magnetosheath electrons in or near the magnetopause current layer. They stream along reconnected field lines from the direction of the reconnection site and can propagate long distances (more than 60,000 km) from this site.

Voyager 1 did not have electron measurements in the appropriate energy range to observe the heated, streaming electrons in the boundary layer on the outer heliosheath side of the heliopause. However, it did have electron measurements in the appropriate energy range to observe the loss of inner heliosheath electrons in the boundary layer adjacent to the heliopause, equivalent to the low latitude boundary layer at Earth's magnetopause. Figure 2 shows Voyager 1 observations that straddle the heliopause and show possible evidence of reconnection remote from the spacecraft location, as explained next.

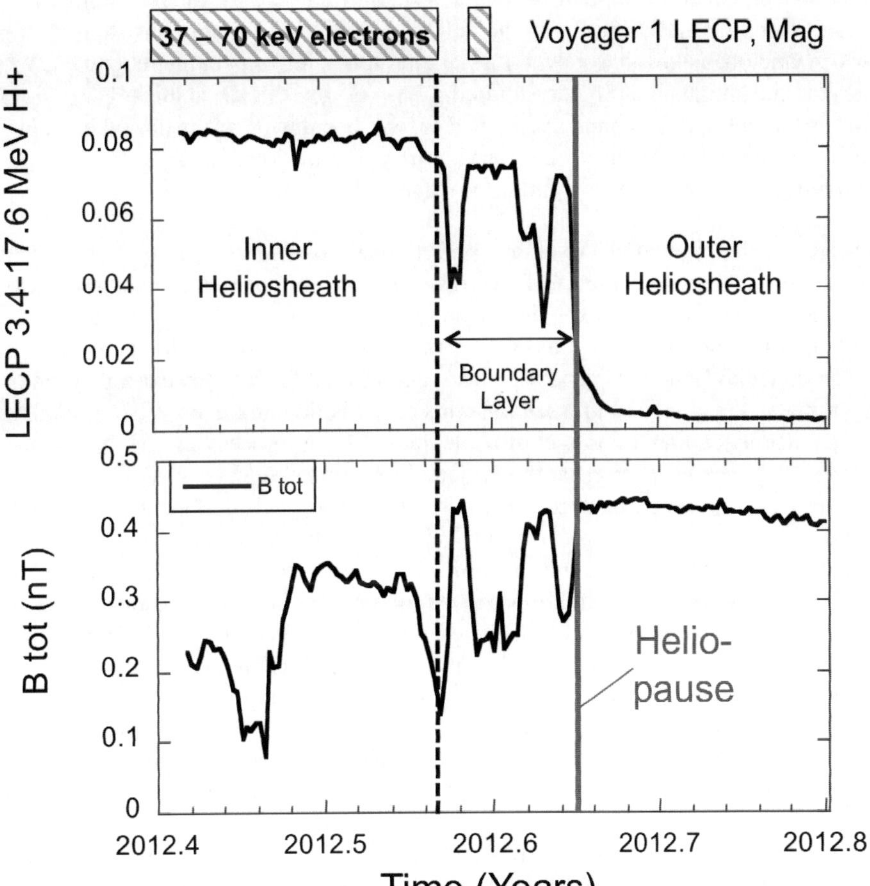

Figure 2: LECP and Magnetometer data surrounding the Voyager 1 heliopause crossing. In the boundary layer, 37 – 70 keV electrons are lost, indicating that there was remote reconnection of the field lines that thread this region.

The heliopause (blue line) is placed at the steep gradient in the cosmic ray flux (not shown) and the equally steep gradient in the 3.4 – 17.6 MeV H^+ in the top panel. As the heliopause is approached from the inner heliosphere side, the 37 – 70 keV electrons are lost prior to the heliopause crossing. This region where the 37 – 70 keV electrons are lost is analogous to the low latitude boundary layer at the magnetopause, where energetic magnetospheric electrons are lost [17]. Also in analogy with the Earth's magnetopause, the 37 – 70 keV electrons are lost prior to the heliopause crossing because those field lines in the inner heliosheath were open to the LISM at one time or remained open throughout the traversal of the boundary layer.

The brief reappearance of the 37 – 70 keV electrons just before 2012.6 may indicate either transient remote reconnection or, more likely, simply the motion of the heliopause first further away from Voyager 1, then closer to the spacecraft. The dips in the 3.4 – 17.6 MeV H^+ fluxes prior to the reappearance of the 37 – 70 keV electrons and after they went away for the last time suggest that the heliopause was in motion the entire time that Voyager 1 was in the boundary layer. The reappearance of energetic magnetospheric electrons because of boundary motion happens very frequently at the Earth's magnetopause [17].

The observations in Figure 2 were also interpreted using global MHD simulations [18,19]. These studies place the heliopause crossing at the dashed line in Figure 2, which is identified here as the transition into the boundary layer. One of the studies [18] identifies the crossing of the outer boundary layer at the solid vertical line in Figure 2, which is identified here as the crossing of the heliopause. The conclusions from the MHD studies on the heliopause location are not consistent with observations at the Earth's magnetopause, which show a loss of energetic electrons in the magnetospheric boundary layer before the magnetopause is encountered. One of the MHD studies [19] did not include reconnection at the heliopause boundary. However, the conclusion from the other study [18] is the same as the conclusion here: the electron observations in Figure 2 are consistent with a heliopause that was undergoing reconnection remote from the Voyager 1 location.

4. Reconnection at the projected Voyager 2 heliopause crossing

The same calculations that were used to investigate the possibility of reconnection at the Voyager 1 crossing of the heliopause are applicable to the projected Voyager 2 crossing of the heliopause. While the predicted date for the crossing is unknown, the conditions at the heliopause at the time of the crossing are predictable. Here, these conditions are determined from 1) the extrapolation of the current Voyager 2 plasma and magnetic field measurements to the heliopause using Voyager 1 observations as a guide, 2) an understanding of the plasma depletion layer process at the heliopause, and 3) a prediction of the magnetic shear angle at the projected crossing.

Table 2 shows the plasma parameters determined from this analysis. For the inner heliosheath, the magnetic field magnitude is assumed to be the same as that for Voyager 1. This magnitude is higher by about a factor of 2 from the magnitude currently measured on Voyager 2. However, the magnitude field magnitude increased as Voyager 1 approached the heliopause. The plasma density is assumed to be approximately equal to the currently measured density at Voyager 2 [10]. The temperature is the same as at Voyager 1 because it is estimated from the same model as that for the Voyager 1 crossing, using the same parameters (see, [12]).

For the outer heliosheath, the basic assumption is that Voyager 2 will cross the heliopause into a much stronger plasma depletion layer than was observed at Voyager 1. The predicted strong plasma depletion layer is consistent with the location of the crossing relative to the IBEX Ribbon and the nose of the heliosphere (see, [4]). The Voyager 1 and 2 measurements of a lower frequency cutoff of the heliospheric radio emissions measured in the 1980's, 1990's, and 2000's (see, [20]) provide a lower bound on the density in the depletion layer adjacent to the heliopause (see [5]). Radio emissions were not observed below approximately 1.8 kHz, which corresponds to a plasma density of 0.04 cm^{-3}. For the magnetic field magnitude in the strong depletion layer, a value of 0.8 nT is assumed. This magnitude is about a factor of 2 larger than the magnetic field magnitude observed by Voyager 1 at the heliopause and is indicative of about a factor of 4 increase in the magnetic field in a strong plasma depletion layer [4].

The temperature anisotropy and temperature at the heliopause are computed using the assumption of an 0.03% pick up ion population and the anisotropy-beta relationship in the strong plasma depletion layer (see, [5]).

Table 2: Plasma parameters at the projected heliopause crossing along the Voyager 2 trajectory

Plasma parameter	Value	Origin
Density in the depletion layer adjacent to the heliopause	$N = 0.04\ cm^{-3}$	Minimum density in depletion layer from 1.8 kHz lower bound of radio wave emissions [Gurnett et al., 2003]
Magnetic field magnitude in the depletion layer adjacent to the heliopause	$B = 0.8\ nT$	Assumed to be a stronger depletion layer than at the Voyager 1 location [Cairns and Fuselier, 2017]
Total temperature	$T = 1.4 \times 10^4\ K$	Estimated from the plasma depletion layer properties and assuming 0.03% pick up ions
Density in the inner heliosheath adjacent to the heliopause	$N = 2 \times 10^{-3}\ cm^{-3}$	Assumed, consistent with Voyager 2 observations [Richardson et al., 2017]
Magnetic field magnitude in the inner heliosheath at the heliopause	$B = 0.25\ nT$	Assumed, measured by the Voyager 1 magnetometer at the heliopause
Total temperature in the inner heliosheath at the heliopause	$T = 2 \times 10^6\ K$	Estimated from the assumption that there are 16.5% pick up ions
Plasma beta in the depletion layer	0.03	Computed from N, B, and T in the plasma depletion layer
Plasma beta in the inner heliosheath	2.2	Computed from N, B, and T in the inner heliosheath
$\Delta\beta$ across the heliopause = $\lvert\beta_{IHS} - \beta_{PDL}\rvert$	$\Delta\beta = 2.2$	
Magnetic Shear angle across the heliopause at Voyager 1	45° or 135°	Estimated for the projected Voyager 2 heliopause crossing

Using the plasma parameters in Table 2, the plasma beta in the inner heliosheath is 2.2 (the same as that for Voyager 1 just before it crossed the heliopause) and the plasma beta in the outer heliosheath/plasma depletion layer is 0.03. Thus the predicted $\Delta\beta$ across the Voyager 2 heliopause location is approximately 2.2 because the plasma beta in the outer heliosheath is very low. This $\Delta\beta$ is very similar to the one at the Voyager 1 crossing (see Table 1) and is dominated by the plasma beta in the inner heliosheath.

The final quantity needed to determine if reconnection is suppressed at the projected Voyager 2 heliopause crossing is the magnetic shear angle. Voyager 2 may be crossing the heliopause at a location where the interstellar magnetic field has nearly the same orientation as the undraped field orientation far from the heliopause (see [4]). In RTN coordinates, the magnitude of the T and N components may be approximately equal, and therefore the angle in the T-N plane would be approximately 45°. In the inner heliosphere, the magnetic field orientation is approximately in either the +T or –T direction, depending on the magnetic sector that the spacecraft is in. Thus, the magnetic shear angle at the heliopause is predicted to be either ~45° or ~135°.

Figure 1 shows $\Delta\beta$ and the two magnetic shear angles from Table 2 for the projected Voyager 2 crossing. If the magnetic shear angle is ~135°, then reconnection may not be suppressed at the Voyager 2 crossing. Even for the smaller magnetic shear angle, the suppression of reconnection may be marginal.

5. Summary and Conclusions

Figure 1 shows that reconnection was likely suppressed locally at the Voyager 1 crossing. However, observations in Figure 2 of 37 to 70 keV electrons measured by LECP near the heliopause, specifically the loss of these electrons in a boundary layer on the inner heliosheath side of the heliopause, are closely analogous to similar observations of energetic electrons lost into the magnetosheath from remote reconnection sites on the Earth's magnetopause. Figure 1's result is in agreement with modeling of the conditions at the crossing [11] and the result inferred in Figure 2 is also not inconsistent with the conclusion of [18] that reconnection may occur at the heliopause away from Voyager 1's location.

For the projected Voyager 2 crossing, Figure 1 shows that the suppression of reconnection will depend on $\Delta\beta$ and the T component of the magnetic field direction at the time of the crossing. $\Delta\beta$ is driven by the plasma beta in the inner heliosheath because the plasma beta in the outer heliosheath and plasma depletion layer is very small. The plasma beta in the inner heliosheath is driven strongly by the ion temperature. This temperature is determined by the percentage of pick up ions in the plasma, which is not directly measured by Voyager 2. Thus, $\Delta\beta$ at the projected Voyager 2 crossing is strongly model dependent.

The magnetic shear angle at the projected Voyager 2 crossing is also strongly model dependent. This angle depends on the shape of the heliopause, which determines how the interstellar magnetic field drapes against the boundary. With a reasonable estimate of the draping (see, [4]), the magnetic shear angle depends on the sign of the T component of the field in the inner heliosheath. The most recently processed magnetic field data available on the CDAWeb site is more than a year out of date. These data indicate that, if Voyager 2 remains in that sector orientation up to the heliopause, then the magnetic shear angle will be approximately 45°. Figure 1 shows that reconnection is likely marginally suppressed for a magnetic shear angle of 45°. However, if Voyager 2 transitions to the opposite sector, then reconnection is possible.

The plasma instrument on Voyager 2 may observe the effects of local reconnection as an increase in the flow velocity and a change in the flow direction of the ions. Moreover, in analogy with observations of reconnection at the Earth's magnetopause, these changes in the ion flow would be accompanied by a loss of the 37 to 70 keV electrons measured by LECP. Even if reconnection is suppressed locally, it is likely that these energetic electrons will be lost in the boundary layer because, like at the Voyager 1 crossing, reconnection is likely to be occurring remotely from the Voyager 2 location.

6. Acknowledgements

Data from the LECP and magnetometer are from the CDAWeb. The authors thank the LECP and Magnetometer PI's of Voyager 1 and 2 for providing these data to the community. Research at Southwest Research Institute is supported by the IBEX grant.

7. References

[1] Krimigis S M, et al. 2013 *Science,* **341,** 144
[2] Gurnett D A, Kurth W S, Burlaga L F, and Ness N F 2013, *ScienceExpress,* 10.1126/science.1241681
[3] Burlaga L F, Ness N F, and Stone E G 2013, *Science,* **341**, 147
[4] Cairns I H, and Fuselier S A 2017, *ApJ*, **834**:197

[5] Fuselier S A, and Cairns I H 2013, *ApJ,* **771**:83
[6] Anderson B J, et al. 1994, *J. Geophys. Res.,* **99,** 5877-5891
[7] Gary S P, et al. 1994, *J. Geophys. Res.,* **99,** 5903-5914
[8] Cassak P A, and Fuselier S A 2016, in: Magnetic Reconnection, edited by W Gonzalez and E Parker, *Astrophys. and Space Sci Library,* **427,** Springer, Switzerland, doi: 10.1007/978-3-319-26432-5, p. 213-276
[9] Vines S K, et al. 2015, *J. Geophys. Res.,* **120,** doi:10.1002/2015JA021464
[10] Richardson J D, et al. 2017, *ApJ,* **834**:190
[11] Swisdak M, Opher M, Drake J F, and Alouani Bibi F 2010, *ApJ,* **710**, 1769-1775
[12] Fuselier S A, and Cairns I H 2015, *Journal of Physics: Conference Series,* **642,** 012010
[13] Gurnett D A, et al. 2015, *ApJ,* **809**, 121
[14] Zank G P, Pauls H L, Williams L L, and Hall D T 1996, *J. Geophys. Res.,* **101**, A10, 21,639
[15] Izmodenov V V , and Alexashov D B 2015, *ApJ Supp. Ser.*, **220**:32
[16] Möbius E, Liu K, Funsten H, Gary S P, and Winske D 2013, *Astrophys. J.,* **766:**129.
[17] Fuselier S A, Trattner K J, Petrinec S M, and Lavraud B 2012, *J. Geophys. Res.,* **117,** A08235, doi:10.1029/2012JA017852
[18] Swisdak M, Drake J F, and Opher M 2013, *ApJ Lett.,* **774,** L8
[19] Washimi H, Zank G P, Hu Q, Tanaka T, and Munakata K 2015 *ApJ,* **809:**16
[20] Gurnett D A, Kurth W S, and Stone E C 2003, *Geophys. Res. Lett.,* **30**, 23, 2209

IOP Conf. Series: Journal of Physics: Conf. Series **900** (2017) 012008 doi:10.1088/1742-6596/900/1/012008

The Acceleration of Thermal Ions at a Strong, Quasi-Parallel Interplanetary Shock: A Hybrid Simulation

Joe Giacalone

Lunar & Planetary Laboratory, University of Arizona, Tucson, AZ 85721

giacalon@lpl.arizona.edu

Abstract. Using a self-consistent hybrid simulation, with kinetic protons and fluid electrons, we investigate the acceleration of thermal protons and minor ions (alphas, $3H_e^{++}$, and C^{5+}) by a quasi-parallel collisionless shock. The results are compared to spacecraft observations of a strong interplanetary shock seen by the Advanced Composition Explorer on DOY 94, 2001, which was associated with significant increases in the flux of > 50 keV/nuc ions. Our simulation uses similar plasma and shock parameters to those observed. The densities of minor ions for two of the species (alphas and C^{5+}) were based on observations at thermal energies for this shock, and we used a nominal value for the density of $3H_e^{++}$, since no observations at thermal energies was available to us. Acceleration of the ions by the shock leads to a high-energy tail in the distribution in the post-shock plasma for all ion species. We find that by extrapolating the simulated tails to the higher energies measured by ACE/EPAM and ACE/ULEIS, the intensity matches well the observations for protons, alphas, and carbon. This suggests that thermal solar wind, accelerated directly at the shock, is a significant source of the observed high-energy protons and these minor ions.

1. Background

The origin and acceleration mechanism involved in creating non-thermal charged-particle distributions in astrophysical, solar, and heliospheric plasmas is an important unsolved problem. Spacecraft observations of particle distributions have noted the existence of a non-thermal tail, even in the absence of an obvious source, like a collisionless shock [4,5,17,9]. It has been reported that non-thermal tails in the distribution often form just behind strong interplanetary shocks [18] suggesting that the formation of the tail is related to plasma heating, particle acceleration, and energy dissipation across the shock. Self-consistent plasma simulations of shocks in which the ions are treated kinetically have also shown this [27,28,14,12]. However, it has also been noted that in the solar wind, particles with energies well above the peak in the thermal distribution have intensities that vary by a few orders of magnitude, which is far more than the variation in the density in the solar wind [25,7]. Moreover, the isotopic composition of "suprathermal" particles can differ considerably from that of the solar wind at 1AU [24,26,6,22]. In some cases of strong interplanetary shocks driven by coronal mass ejections and associated with large intensities of energetic particles the ratio of $3H_e^{++}/4H_e^{++}$ is far greater than the same ratio at thermal energies, based on typical solar wind conditions [24]. This suggests that the source is a pre-existing suprathermal population of particles rather than solar wind.

Collisionless shocks provide an excellent "laboratory" to the formation of non-thermal particle distributions. For one, shocks heat the plasma, and broaden the particle distribution. Moreover, strong interplanetary shocks are known to be associated with large intensities of high-energy particles [12].

Moreover, there is a widely accepted theory for particle acceleration at shocks – diffusive shock acceleration [23,1,2,3]. In this mechanism, particles are accelerated by diffusively moving in the irregular magnetic fields embedded in the plasma flows on both sides of the shock and cross the shock many times gaining energy at the large plasma compression. The theory only addresses acceleration from some initial energy, but does not specifically address what this energy is, or what the value of the distribution is at this energy. Thus, the theory does not address the injection mechanism. Hybrid simulations have shown that shocks naturally accelerate low-energy particles, even those from the thermal distribution [27,28,14,12].

It is the purpose of this paper to address the source of accelerated minor ions at an interplanetary shock. We do this by simulating an interplanetary shock in which multiple ion species are treated kinetically and whose un-shocked, initial distributions are Maxwellian. We then compare the distribution resulting from their interaction with the shock with spacecraft observations of a particular event.

2. Numerical Model

We use the hybrid model described in [15,10] to simulate a quasi-parallel interplanetary shock. The ions are treated kinetically and the electrons as a massless fluid. The bulk flow velocity, number density, electric and magnetic fields are all functions of one spatial coordinate, x. The average magnetic field upstream of the shock, $\mathbf{B}_1$, is in the x direction. The magnetic field consists of this mean plus a fluctuating component derived from a discrete sum of many plane Alfven waves that propagate along the mean field. The amplitudes of each is determined from an assumed Kolmogorov-like power spectrum with variance of $0.3|\mathbf{B}_1|^2$, and an outer scale of 0.01AU, similar to that observed in the interplanetary magnetic field [21].

The shock is created by forcing plasma to reflect off a boundary at one end of the simulation domain. The shock then propagates back into the flow. The calculation is performed in a frame at rest with respect to the reflecting boundary, but results are transformed to the spacecraft frame to compare results with observations. Table 1 lists the input parameters, which are consistent with ACE MAG and SWEPAM observations for the interplanetary shock that crossed ACE at 14:22.27UT on DOY 94, 2001. Averaged over the hour before the shock crossing, ACE observed the number density, field strength, and flow speed to be 4.6cm^{-3}, 7nT, and 490 km/s, respectively. The shock speed was estimated to be 690 km/s. Using these parameters, the proton inertial length, c/ω_p and cyclotron period, Ω_p^{-1}, the characteristic length and time scales of the hybrid simulation, corresponds to 1.1×10^7 cm and 1.6 s, respectively.

We also identify a boundary that co-moves with the shock, a distance Δ_{feb} upstream, which acts as a "free-escape" boundary for any particles that move from the shock back upstream and allowing them to escape. The reason for this boundary is to ensure that the "escape distance" upstream remains constant with time. Note that the shock moves in the simulation frame. For the parameters used, this corresponds to Δ_{feb} = 0.025AU, and the entire simulation domain, x_{max}, is 0.035AU. Such a boundary has been used in diffusive shock calculations previously [8], and can even be made to be a function of energy by setting it to be some number of diffusive skin depths ahead of the shock [31]. The latter requires prior knowledge of the diffusion coefficient's dependence on energy, which we do not have.

Table 1.
Simulation Parameters

M_A'	3
M_A	~4
β_e	0.6
β_p	0.3
ω_p / Ω_p	4500
t_{max}	$12{,}000\Omega_p^{-1}$
x_{max}	$50{,}000c/\omega_p$
Δ_{feb}	$35{,}600c/\omega_p$
$n(4He_e^{++})/n_p$	0.022
$n(3He_e^{++})/n_p$	9.4×10^{-6}
$n(C^{5+})/n_p$	4.3×10^{-5}

The maximum simulation time, t_{max} in Table 1 corresponds to 5.3 hrs of real time. The time step is 0.005 Ω_p^{-1}. The grid spacing is 0.5 c/ω_p. There are 100,000 grid cells in the entire domain.

The initial distributions, and also injected at the inflow boundary, of all ions are drifting Maxwellians. All ions are assumed to have (initially) the same thermal speed. The proton thermal speed is related to the plasma beta, β_p, given in the table. The inflow speed is related to M_A', the Mach number in the simulation frame of reference. A total of 190 particles per grid cell are initially loaded into the simulation domain. Of these, 100 are protons, and 30 each of $4H_e^{++}$, $3H_e^{++}$, and C^{5+}. We also performed a few other simulations with more minor ions per grid cell, doing each species and protons together without including the other species, and found essentially the same results as those presented below. The proton density is $4.6cm^{-3}$, and Table 1 lists the abundance ratios for the other ions that were used in our calculation. The alpha to proton abundance ratio, $n(4H_e^{++})/n_p$, is chosen to match ACE/SWEPAM observations, and the carbon to proton ratio, $n(C^{5+})/n_p$, is also consistent with observations from ACE/SWICS (not shown, but determined from phase-distributions functions provided by S. Lepri). Observations of the $3H_e^{++}$ to proton ratio were not available, so a nominal value was used [16].

As the simulation evolves, the upstream plasma, fields, and ion distributions evolve self-consistently according the fundamental physical equations that govern the hybrid simulation [30].

3. Results

Figure 1 shows some simulated results transformed to the spacecraft frame of reference in real units. The two panels on the left show the differential intensity (upper left) and plasma flow speed (lower left) as functions of the time relative to the shock crossing. This was accomplished by relating the position within the simulation domain, x, at the end of the simulation to "spacecraft time", τ, by $\tau = (x - x_{shock})/V_{sh}$, where $x_{shock} = 38{,}844\ c/\omega_p$ is the position of the shock in the simulation domain at t_{max}. The alternate axis at the top of the upper left panel indicates position relative to the simulation domain. $V_{sh} = 690$ km/s is the observed interplanetary shock speed. The right panel shows the simulated proton differential energy spectrum as a function of kinetic energy as a solid histogram, determined by averaging over the region indicated with the yellow shaded region in the left panels. This is downstream of the shock. The symbols are ACE/EPAM observations averaged over a 15-minute time interval immediately after the shock crossed ACE at 14:22.27UT DOY 94, 2001.

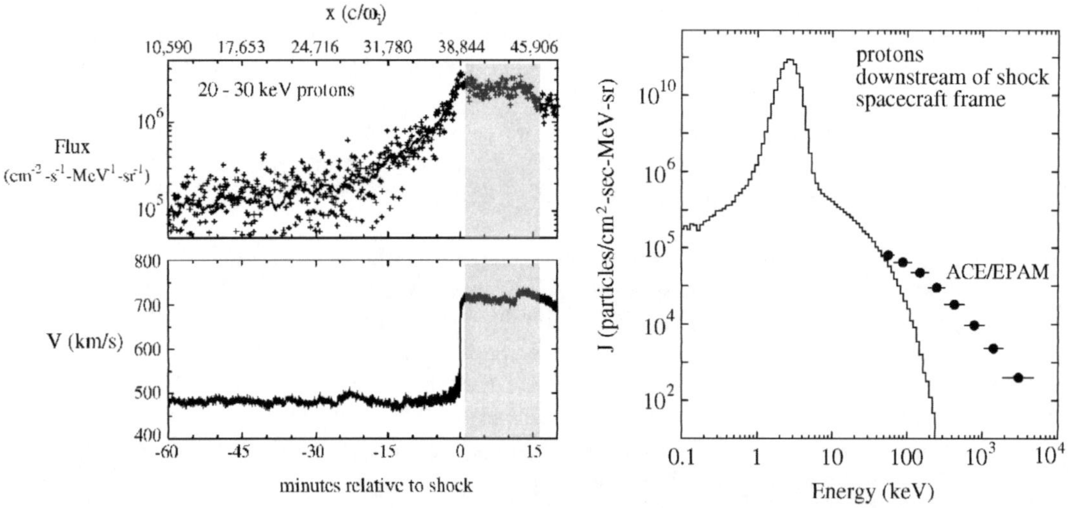

Figure 1. Upper left: Simulated flux of 20-30 keV protons as a function of time relative to the shock crossing. The symbols are averages over $50c/\omega_p$, and the solid curve is a 20-pt running average of this. The upper axis shows the corresponding position within the simulation domain. Lower left: simulated plasma flow speed as a function of time. Right: (Solid histogram) Differential intensity as a function of energy for protons over the region indicated by the yellow-shaded region in the left figures. (Symbols) 15-minute-avarged ACE/EPAM LEMS30 and LEMS120 data at the time of the crossing of the interplanetary shock on DOY 94, 2001.

IOP Conf. Series: Journal of Physics: Conf. Series **900** (2017) 012008 doi:10.1088/1742-6596/900/1/012008

The downstream proton spectrum reveals a high-energy tail. These are shock-accelerated thermal protons [15,10]. The spectrum is approximately a power law from about 8 keV to 30 keV. Above this, the spectrum falls off approximately exponentially, which is the result of particles escaping the free escape boundary (as discussed in the references above). Had we used a larger simulation domain, and run the simulation longer, we expect that the spectrum would have been a power law to larger energies. In fact, it seems by comparing with the ACE/EPAM observations that the simulated high-energy tail would match the observations if it extended to higher energies. The energy where this tail emerges in the distribution is related to the efficiency of the acceleration of thermal protons [15]. In this case, it seems the acceleration efficiency is such that it matches well the observations, suggesting that the source of the high-energy protons observe by ACE/EPAM is accelerated solar wind protons. It is also noteworthy, however, that the acceleration is, in fact, quite slow. The characteristic energy, where the spectrum rolls over from a power law to exponential, is only ~30 keV for a maximum simulation time corresponding to 5.5 hours of real time. This slow acceleration time is because the simulated shock is quasi-parallel, and such shocks are quite slow at accelerating particles. Quasi-perpendicular shocks, in contrast, are much faster [19,20,11,12].

Figure 2 shows downstream energy spectra, averaged over the same region indicated by the yellow-shaded region of the left panels of Figure 1, for the minor ions in our simulation. We only show the spectra up to 30 keV/nuc., and do not show the exponential part of the spectrum, which, as indicated above, is the result of particle escaping our small computational domain. Also shown are ACE/ULEIS observations of the differential flux of the same minor ions. This data was obtained from the ACE science center which provides 1-hour averages. We took the time that most-closely coincided with the passage of the shock, but still downstream of it.

As in the case of the protons, the simulated spectra have a high-energy tail that, when extended to high energies, matches reasonable well with the spacecraft observations for at least two of the minor

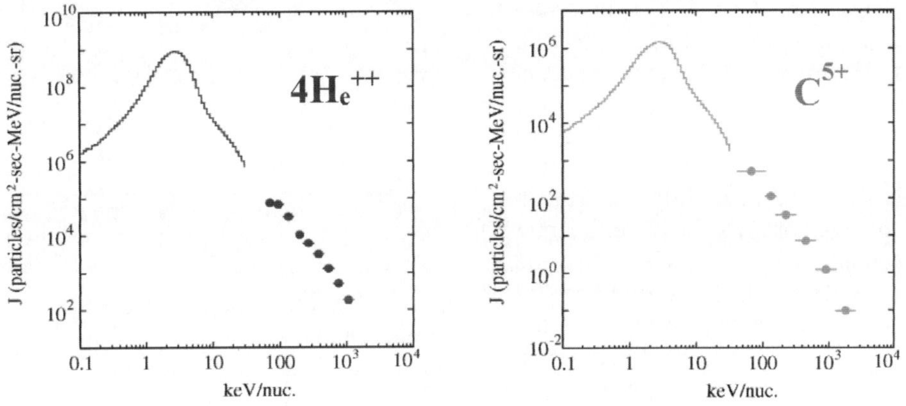

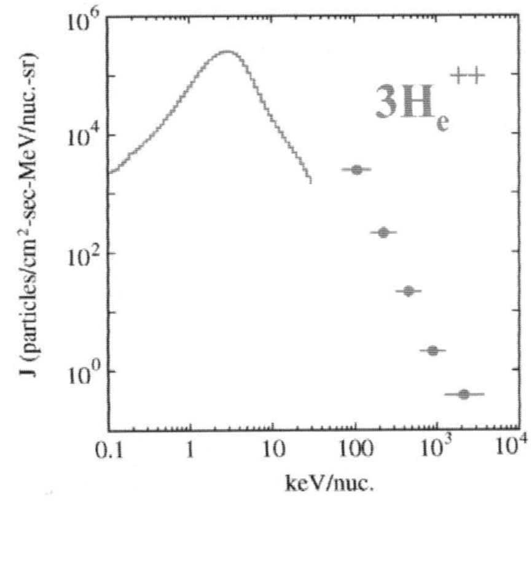

Figure 2. (Solid histograms) Differential intensity spectra, downstream of the sock, as a function of kinetic energy per nucleon for the three minor ions in our simulation, as indicated. (Symbols) 1-hour averaged ACE/ULEIS data at the time of the crossing of the interplanetary shock on DOY 94, 2001. In all of the simulated spectra, only the portion up to 30 keV/nuc. Is shown (see text for details).

IOP Conf. Series: Journal of Physics: Conf. Series **900** (2017) 012008 doi:10.1088/1742-6596/900/1/012008

ion species ($4H_e^{++}$ and C^{5+}). The tail does not match very well the $3H_e^{++}$ data at high energies. It should be noted, however, that this is the only species for which we had no observations of the initial distribution, and, instead had to use a 'nominal value'. For the other two species, which we did have observations at thermal energies upstream of the shock, the simulated energy spectra, if extrapolated to high energies, match quite well with the ACE/ULEIS observations. This suggests that the source of these high-energy ions, as well as the protons, mentioned above, is the solar wind.

Assuming the upstream distribution of solar-wind $3H_e^{++}$ is similar to that which we assumed, then our results would indicate that the source of the high-energy $3H_e^{+}$ would have to be a pre-existing suprathermal population. However, although we do not show it here, inspection of ACE/ULEIS observations of the 160-310 keV/nuc. $3H_e^{++}$ flux over ~36 hour interval of this interplanetary shock (using data obtained from the ACE science center) reveals that the flux increased by almost 2 order of magnitude from far upstream to the peak at the shock. Thus, the high-energy tail would have to be enhanced by this factor across the shock, which is a challenge to understand theoretically. We suggest, alternatively, that it is perhaps more likely that the $3H_e^{++}$ thermal distribution upstream of the shock is larger than we assumed. If this is the case, then our results would also suggest that the source of accelerated $3H_e^{++}$ is solar wind as well. This requires further analysis for verification.

If the source of accelerated $3H_e^{++}$ is indeed thermal solar wind, it would be quite surprising given that previous studies have noted that the abundance ratio of $3H_e^{++}/4H_e^{++}$ at high energies is far greater than the same value at thermal solar wind energies, especially at large solar-energetic particle events [24]. We have found that the acceleration efficiency, which is related to where in the spectrum the high-energy tail emerges, depends on the species, and that $3H_e^{++}$ is the most-efficiently accelerated of the ions. This is shown in Figure 3. This figure shows energy spectra, normalized in such a way to give the same total abundance, or number density – the proton density, in this case. Note that at high energies, the $3H_e^{++}$ is the largest, indicating it is the most efficiently accelerated. Interestingly, the protons are the least efficiently accelerated. A possible interpretation is that the efficiency of the proton acceleration is constrained by the total 'energy budget' since the protons are the dominant species. Total energy must be conserved, and the available energy is that which is swept up by the shock. The shock converts most of this energy into enthalpy flux by heating the protons (and electrons), but a fraction of this is in high-energy protons as shown by [15]. These authors noted that this fraction, related to the acceleration efficiency, does not exceed about 20% in hybrid simulations of strong quasi-parallel shocks. Moreover, a significant fraction of the accelerated particles escape the upstream boundary, representing an energy sink. These authors also noted that a significant energy loss leads to a very large change in the nature of the shock, with a density compression ratio, for example, that is considerably larger than that expected from the standard Rankine Hugoniot jump conditions. The minor ions, however, contribute a negligible amount of energy to the system and, as such, do not have the same restriction as the protons. The minor ions are all accelerated very efficiently, but $3H_e^{++}$, by far is the most efficient. We note that $3H_e^{++}$ is lighter than $4H_e^{++}$, yet $3H_e^{++}$ is accelerated far more efficiently than $4H_e^{++}$, which is not consistent with previous findings showing the acceleration efficiency increasing with mass per charge [8]. The cause of this is not yet determined, but we have found that all the minor ions, especially $3H_e^{++}$, are significantly heated upstream, before they encounter the shock (not shown). We also find

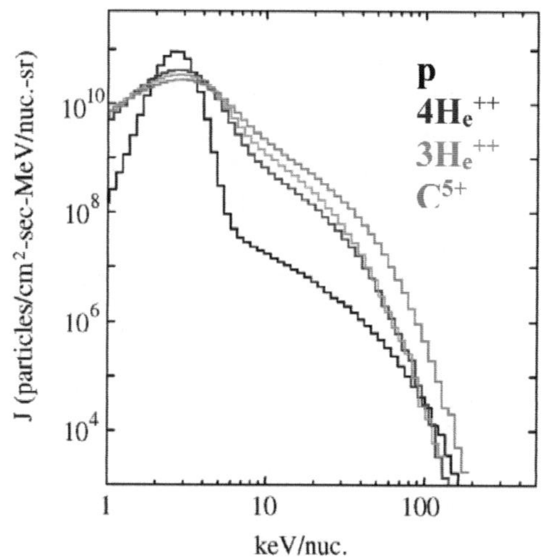

Figure 3. Downstream energy spectra of protons and minor ions, as indicated, normalized so that the density is the same for all species.

that the protons are slightly cooled relative to their injected temperature. Recall that only the distribution at the start of the simulation, and that injected at one boundary of the simulation, have equal thermal speeds for the ions. As the ions move with the flow, before encountering the shock, their distributions evolve in time. The cause of this heating/cooling is not presently understood. It is possibly caused by fluctuations in both the magnetic and electric fields upstream of the shock. These fluctuations are caused, partly, by back-streaming protons accelerated at the shock and evolve non-linearly in time, often forming large magnetic fluctuations such as SLAMS [29]. There also fluctuations in the electrostatic field caused by variations in the electron (equal to the ion) number density. For ions of equal charge, electrostatic fluctuations would preferentially heat lighter particles, which may explain why $3H_e^{++}$ is heated more than $4H_e^{++}$ in our simulations. We are not aware of observations of minor ions heated prior to the passage of interplanetary shocks.

4. Conclusions

We used a one-dimensional hybrid simulation of a shock using input parameters consistent with those observed for an interplanetary shock which crossed the ACE spacecraft at 14:22.27UT on DOY 94, 2001 to investigate the acceleration of thermal protons and minor ions. The simulation treats the ions kinetically, and requires that the ion cyclotron period and thermal ion gyroradius are well resolved. These are very small compared to the typical temporal and spatial scales involved with the acceleration to ions to high energies at a quasi-parallel collisionless shock. We find that acceleration by the shock leads to a high-energy tail in the ion distribution that is approximately a power law up to a characteristic energy where the spectrum becomes exponential. This is consistent with diffusive shock acceleration theory. This change in spectrum is caused by the finite spatial domain of the simulation. Because the shock we studied was only marginally supercritical (M_A~4), and because the simulation domain was finite (although still many thousands of ion inertial lengths) the characteristic energy in our calculations was only about 30 keV/nuc for all ion species, although maximum energies were over 100 keV/nuc. We compared our simulation results to observations. The lowest energy of ACE/EPAM and ACE/ULEIS is higher than the power-law part of our simulated distributions given our computational constraints. Thus, our results could only be compared by extrapolating to higher energy, assuming that a larger simulation domain and longer run time would lead to a power-law part that extended to higher energies and overlap with the observations.

We find that the high-energy tail, caused by acceleration of particles at the shock, emerges from the thermal distribution with an intensity that when extrapolated to higher energies matches well the observed intensities for all ions, except $3H_e^{++}$. The initial distributions in our simulations were assumed to be Maxwellian in the upstream plasma frame. Moreover, we used initial densities and thermal speeds consistent with observed values at thermal energies of all particles (except $3H_e^{++}$). Thus, the results of our simulations suggest strongly that the source of accelerated minor ions at the interplanetary shock seen by ACE on DOY 94, 2001 is thermal solar wind. The source of $3H_e^{++}$ remains a question since we did not have observations of the density and temperature of this ion species. But assuming nominal values, there are fewer high-energy $3H_e^{++}$ in our simulation than that observed by ACE/ULEIS. This suggests the source is pre-existing suprathermal particles.

We also find that the acceleration efficiency depends on the ion species. The protons are the least efficiently accelerated. Their efficiency is constrained by the amount of energy contained in high-energy particles compared to the energy incident on the shock. The minor ions, however, do not have the same constraint because, due to their low density, contribute very little to the total energy. We find that $3H_e^{++}$ is the most efficiently accelerated ion species. We suggest that this may be related to the electrostatic field in the hybrid simulation, which is caused by gradients in the electron density. An electrostatic field effects more significantly particles with a larger charge to mass ratio, and 3He++ has the largest ratio ($(Z/A)_{3He}$ = 2/3) of the three minor ions we simulated ($(Z/A)_{4He}$ = 2/4; $(Z/A)_C$=5/12). In our simulations, the electrostatic field has both a fluctuating part, as well as a steady part, in the shock front, known as the cross-shock electric field.

Our simulations suggest that solar wind is likely a significant source of acceleration particles at strong interplanetary shocks. We also emphasize the importance of measuring the entire distribution function, from thermal energies, to the suprathermal tail, and high energies in order to understand the source of accelerated particles. Future instrumentation with such capabilities will significantly advance our understanding of the origin and physical mechanisms involved in the acceleration of high-energy particles.

Acknowledgements

I thank J. R. Jokipii, J. Kota, and F. Fraschetti, for useful conversations relevant to this study. This study was supported, in part, by NASA under grants NNX15AJ71G and NNX15AJ72G. Numerical simulations were performed with the support of NASA SMD High-End Computing on the Pleiades cluster, under the award SMD-16-7631

References

[1] Axford, W. I., E. Leer, and G. Skadron, Proc. of the 15th International Cosmic Ray Conference, vol. 11, p. 132, 1977.
[2] Bell, A. R., Month. Not. Royal Astr. Soc., vol. 182, p. 147, 1978.
[3] Blandford, R. D., and J. P. Ostriker, Astrophys. J., vol. 221, p. L29, 1978.
[4] Dayeh, M. A., M. I. Desai, J. E. Dwyer, H. K. Rassoul, G. M. Mason, and J. E. Mazur, Astrophys. J., vol. 693, p. 1588, 2009.
[5] Dayeh, M. A., M. I. Desai, G. M. Mason, R. W. Ebert, and A. Farahat, Astrophys. J., vol. 835, article id. 155, 2017.
[6] Desai, M. I., G. M. Mason, R. E. Gold, S. M. Krimigis, C. M. S. Cohen, R. A. Mewaldt, J. E. Mazur, and J. R. Dwyer, vol., 649, p. 470, 2006.
[7] Desai, M. I., and J. Giacalone, Living Reviews in Solar Physics, vol. 13, article id. 3, 2016.
[8] Ellison, D. C, J. Geophys. Res., 90, 29, 1985.
[9] Fisk, L. A., and G. Gloeckler, Astrophys. J., vol. 640, p. L79, 2006.
[10] Giacalone, J., Astrophys. J., vol. 609, p. 452, 2004.
[11] Giacalone, J., Astrophys. J., vol. 624, p. 765, 2005a.
[12] Giacalone, J., Astrophys. J., vol. 628, p. L37, 2005b.
[13] Giacalone, J., Astrophys. J., vol. 761, article id. 28, 2012.
[14] Giacalone, J. D. Burgess, S. J. Schwartz, and D. C. Ellison, Geophys. Res. Lett., vol. 19, p. 433, 1992.
[15] Giacalone, J. D. Burgess, S. J. Schwartz, D. C. Ellison and L. Bennett, J. Geophys. Res., vol. 102, p. 19,789, 1997.
[16] Gloeckler, G., and J. Geiss, Space Sci. Rev., vol. 84, p. 275, 1998.
[17] Gloeckler, G., L. A. Fisk, T. H. Zurbuchen, and N. A. Schwadron, AIP Conf. Proceedings, vol. 528, p. 221, 2000.
[18] Gosling, J. T., J. R., Asbridge, S. J. Bame, W. C. Feldman, R. D. Zwickl, G. Paschmann, G., N. Sckopke, N. and R. J. Hynds, J. Geophys. Res., vol. 86, p. 547, 1981.
[19] Jokipii, J. R., Astrophys. J., vol. 255, p. 716, 1982.
[20] Jokipii, J. R., Astrophys. J., vol. 313, p. 842, 1987.
[21] Jokipii, J. R. and P. J. Coleman, J. Geophys. Res., vol. 73, p. 5495, 1968.
[22] Kahler, S. W., A. J. Tylka, and D. V. Reames, Astrophys. J., 701, p. 561, 2009.
[23] Krymsky, G. F., Soviet Physics – Doklady, vol. 22, p/ 327, 1977.
[24] Mason G. M., J. E. Mazur, and J. R. Dwyer, Astrophys. J., vol. 525, L133, 1999.
[25] Mason, G. M., M. I. Desai, J. E. Mazur, and J. R. Dwyer, AIP Conf. Proc., vol. 781, p. 219, 2005.
[26] Mewaldt, R. A., C. M. S. Cohen. R. A. Leske, E. R. Christian, A. C. Cummings, E. C. Stone, T. T. von Rosenvinge, and M. E. Wiedenbeck, Adv. Space Res., vol. 30, p. 79, 2002.

[27] Quest, K. B., J. Geophys. Res., vol. 93, p. 9649, 1988.
[28] Scholer, M. Geophys. Res. Lett., vol. 17, p. 1821, 1990.
[29] Schwartz, S. J., D. Burgess, W. P. Wilkinson, R. L. Kessel, M. Dunlop, and H. Luehr, J. Geophys. Res., vol. 97, p. 4209, 1992.
[30] Winske, D., and N. Omidi, J. Geophys. Res., vol. 101, p. 17,287, 1996.
[31] Zank, G. P, W. K. M. Rice, and C. C. Wu, J. Geophys. Res., 105, 25079, 2000.

IOP Conf. Series: Journal of Physics: Conf. Series **900** (2017) 012009 doi:10.1088/1742-6596/900/1/01200

A Hierarchical Relationship between the Fluence Spectra and CME Kinematics in Large Solar Energetic Particle Events: A Radio Perspective

N Gopalswamy[1], P Mäkelä[1,2], S Yashiro[1,2], N Thakur[1,2], S Akiyama and H Xie[1,2]

[1]Code 671, NASA Goddard Space Flight Center, Greenbelt, MD 20771, USA
[2]The Catholic University of America, Washington DC 20064, USA

Abstract. We report on further evidence that solar energetic particles are organized by the kinematic properties of coronal mass ejections (CMEs)[1]. In particular, we focus on the starting frequency of type II bursts, which is related to the distance from the Sun where the radio emission starts. We find that the three groups of solar energetic particle (SEP) events known to have distinct values of CME initial acceleration, also have distinct average starting frequencies of the associated type II bursts. SEP events with ground level enhancement (GLE) have the highest starting frequency (107 MHz), while those associated with filament eruption (FE) in quiescent regions have the lowest starting frequency (22 MHz); regular SEP events have intermediate starting frequency (81 MHz). Taking the onset time of type II bursts as the time of shock formation, we determine the shock formation heights measured from the Sun center. We find that the shocks form on average closest to the Sun (1.51 Rs) in GLE events, farthest from the Sun in FE SEP events (5.38 Rs), and at intermediate distances in regular SEP events (1.72 Rs). Finally, we present the results of a case study of a CME with high initial acceleration (~3 km s^{-2}) and a type II radio burst with high starting frequency (~200 MHz) but associated with a minor SEP event. We find that the relation between the fluence spectral index and CME initial acceleration continues to hold even for this minor SEP event.

1. Introduction

The hierarchical relationship recently found between the fluence spectra of large solar energetic particle (SEP) events and the kinematics of the associated coronal mass ejections (CMEs) provides a strong evidence that energetic particles in these events are accelerated by CME-driven shocks [1], most likely by the diffusive shock acceleration process [2]. Shocks can form at any distance from the Sun depending on the speed of CMEs relative to the Alfven speed of the ambient medium. One of the earliest signatures of CME-driven shocks is type II radio bursts, which are due to the plasma emission process involving nonthermal electrons accelerated by the shocks [3]. The plasma emission occurs at the fundamental and second harmonic of the local plasma frequency in the vicinity of the shock, so the starting frequency of a type II burst indicative of the local plasma density and the distance from the Sun where such a density is prevalent. The starting frequency and the frequency range of the type II bursts thus provide important information on where the shocks form and how long they survive in the interplanetary (IP) medium [4]. The existence of a shock is a necessary condition for both type II radio bursts and SEP events because the same shock accelerates both ions and electrons, although the details of the acceleration process may differ for the two species [5]. The high starting frequencies of type II bursts are thus indicative of shock formation close to the Sun. The maximum energy of accelerated particles depends on the magnetic field strength, age of the shock, and the shock speed [2]. Near the Sun, the magnetic field strength is high, so particles can be accelerated to high energies. This

is certainly the case in ground level enhancement (GLE) events, which are hard-spectrum events in which the shock forms at a heliocentric distance of ~1.5 Rs [6]. On the other hand, in large SEP events associated with CMEs involving filament eruption (FE) outside of active regions, the shock forms at large distances from the Sun as indicated by the lack of high-frequency (metric) type II bursts and the presence of IP type II bursts [7]. At large distances, the ambient magnetic field strength is low, so the acceleration efficiency proportional to the magnetic strength is also low, resulting in a lower maximum energy attained by the particles and hence a soft spectrum. A comparison of the shock formation heights between GLE and non-GLE events showed that shocks form closer to the Sun in the case of GLE events [8]. One can see that the high initial speed and high type II starting frequency imply high ambient magnetic field, which is a key factor in accelerating particles to high energies [9].
CMEs attain high speeds early on when the initial acceleration is high, become super-Alfvenic, and drive fast-mode shocks followed by the energization of electrons (type II radio bursts) and ions (SEP events). Such CMEs will have type II bursts with high starting frequencies. However, there are type II bursts starting at high frequencies, but lack high-energy SEP events. Accurate determination of shock formation heights became possible [10] during the early part of solar cycle 24 when data from the Solar Dynamics Observatory (SDO, [11]) became available along with data from the Solar Terrestrial Relations Observatory (STEREO, [12]). For a set of 32 CMEs, the shock formation height was found to be in the range 1.2 to 1.93 solar radii (Rs), with an average of 1.43 Rs [10]. Only 7 of the 32 CMEs had shock formation heights >1.5 Rs. None of the 32 CMEs was associated with a GLE event or a large SEP event. Only a handful of them were associated with minor SEP events with intensity <2 pfu (particle flux unit; 1 pfu = 1 particle per (cm^2 s sr)). The first event in [10] is the 2010 June 12 CME that had a >10 MeV intensity of <1 pfu.
One of the characteristics of the events similar to the 2010 June 12 eruption is the short duration of the associated soft X-ray flares and the rapid decline in CME speed similar to the X-ray time profile after attaining a high speed around the peak time of the soft X-ray flare [13]. Such events provide important constraints on the validity of the hierarchical relationship between the SEP spectrum and CME kinematics.
In this paper, we elaborate on the idea that the initial speed of a CME is a proxy to the high initial acceleration by describing the CME height-time history of a GLE event and an FE SEP event. We then illustrate that the grouping of SEP events into GLE, regular, and FE SEP events is also consistent with the type II starting frequencies and shock formation heights. Finally we study the CME kinematics, SEP association, type II burst association, and the fluence spectrum of an impulsively accelerated CME, which occurred on 2010 June 12 to illustrate that this event behaves like a regular SEP event.

2. CME Initial Speed and Acceleration

The reason for using the initial speed of the CMEs as a proxy is the lack of height-time measurements close to the Sun from the Large Angle and Spectrometric coronagraph (LASCO, [14]) on board the Solar and Heliospheric Observatory (SOHO, [15]) mission. The innermost coronagraph LASCO/C1 was capable of observing the acceleration profile of CMEs [16], but it operated for only less than 3 years during the rise phase of solar cycle 23. Most of the large SEP events occurred after LASCO/C1 ceased operations, so we have to rely on the outer coronagraphs C2 and C3. Here we illustrate how we can use the first two height-time data points from LASCO/C2 to assess the initial acceleration of CMEs. For this purpose, we use two events from cycle 24 that had both SOHO and STEREO observations to confirm the analysis.

2.1. The 2012 May 17 GLE Event

Figure 1 shows the height and speed variation of the 2012 May 17 CME, which produced the first GLE event in cycle 23 [17]. The CME first appeared in the LASCO/C2 field of view (FOV) at 01:48 UT at a height of 3.61 Rs. In the next frame taken ~12 minutes later, the CME leading edge was at 5.43 Rs, indicating a speed of ~ 1757 km s^{-1}. This must be close to the peak speed attained by the CME, but it started slowing down as is clear from the speed – time plot obtained by a second order fit to the height-time measurements. The last two data points are (24.27 Rs, 04:18 UT) and (25.80 Rs, 04:30 UT). The CME moved 1.53 Rs in 12 minutes corresponding to a speed of 1479 km s^{-1}. Clearly, the CME slowed down in the LASCO FOV as confirmed by the second order fit shown in Fig. 1b. Taking into account of the source location at N11W76, the deprojected initial and final speeds become 1810 and 1524 km s^{-1}, respectively. Fortunately, this event was observed by both SOHO and STEREO, so we were able to obtain the three-dimensional speed of the CME. The inner coronagraph COR1 and the Extreme Ultra Violet Imager (EUVI) in STEREO's Sun Earth Connection Coronal and Heliospheric Investigation (SECCHI, [18]) observed closer to the Sun providing the acceleration profile of the CME. Using multiview observations, it was found that the CME attained its maximum speed of 1977 km s^{-1} at 02:00 UT and a peak acceleration of 1.77 km s^{-2} at 01:36 UT. The initial speed from the first two LASCO/C2 data points is only slightly smaller than the peak speed obtained from the multiview observations. If we do not have CME observations near the Sun, we cannot measure the acceleration. However, we can get the average acceleration the CME undergoes from the initial CME speed (1810 km s^{-1}) and the flare rise time. The flare rise time in this event was 22 min, so we get an acceleration of 1.37 km s^{-2}, consistent with the peak acceleration. This simple analysis shows that the initial speed is indicative of the initial acceleration of CMEs.

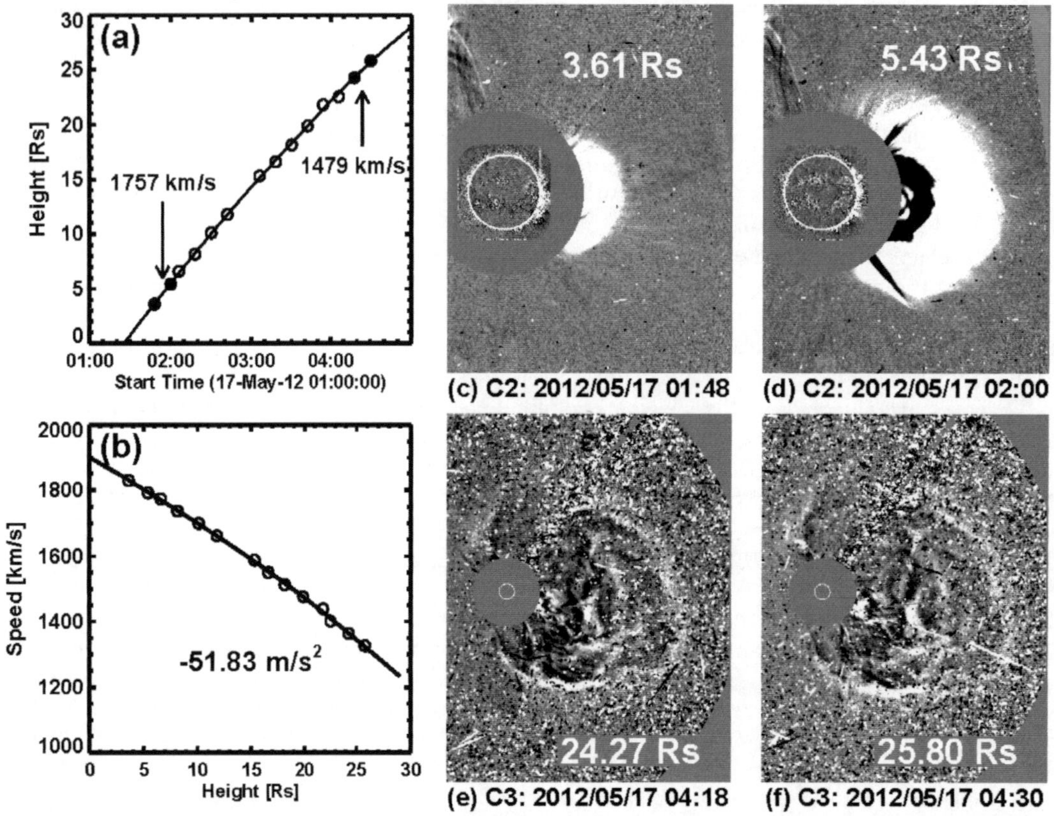

Figure 1. The height-time history of the 2012 May 17 GLE event (a) and the speed variation with height obtained using a second order fit to the height-time measurements (b). The speed obtained using the first two data points is ~1757 km s^{-1}, while the speed using the last two data points is ~1479 km s^{-1}, indicating a clear deceleration in the coronagraph FOV. The deceleration in (b) is the residual acceleration. LASCO difference images corresponding to the first two data points (c, d) and last two data points (e, f) are also shown. We have superposed difference images from SDO's Atmospheric Imaging Assembly (AIA, [19]) at 193 Å on LASCO/C2 images to indicate the eruption location on the Sun. The heliocentric distances of the CME leading edges are also given. The coronal images in (e) and (f) appear corrupted because of the energetic particles hitting the LASCO/C3 detectors.

The eruption was associated with a metric type II burst starting at 01:32 UT and the starting frequency was high (~200 MHz) for the harmonic component. Thus the plasma level at shock formation corresponds to a fundamental plasma frequency of ~100 MHz, indicating a shock formation close to the Sun (~1.38 Rs as derived from the CME height-time plot – see [10]). The CME speed was high enough to be super-Alfvenic and hence form a fast-mode shock. The solar particle release (SPR) was at 01:40 UT, about 8 minutes after the shock formation and the CME had reached a height of 2.32 Rs. Thus the shock travelled ~1 Rs after formation before particles were released.

2.2. The 2011 November 26 FE SEP event

Figure 2 shows the 2011 November 26 FE SEP event that had height-time and speed-time plots different from those of the GLE event in Fig. 1. The CME initial speed from the first two data points is 697 km s^{-1}, while the final speed in the LASCO FOV is 1402 km s^{-1}. The higher final speed is indicative of the continued acceleration of the CME. The centroid of the eruption region was N27W49, which can be used to get the deprojected initial and final speeds as 924 km s^{-1} and 1858 km s^{-1}, respectively. A second order fit to all the height-time data points confirms the continued acceleration of the CME with an average acceleration of ~8.97 m s^{-2}. Observations from STEREO/COR1 showed that the initial acceleration of the CME was ~0.3 km s^{-2}. An extremely faint and narrow band type II burst started around 9 MHz in the Wind/WAVES [20] dynamic spectrum at 07:15 UT, corresponding to a shock formation height of ~3.5 Rs. The type II burst became intense only when the CME reached a height of ~13.5 Rs at 9:15 UT. Thus, the small initial speed is consistent with the low initial acceleration and accordingly the CME became super-Alfvenic only beyond the outer corona [7].

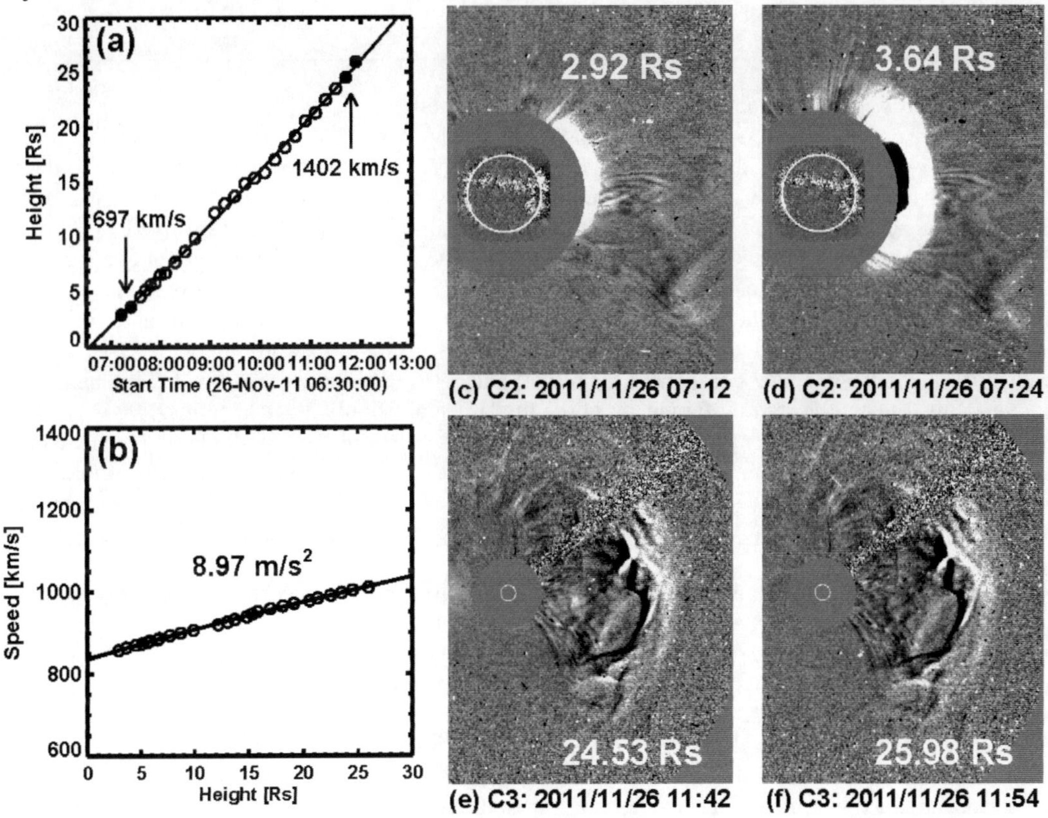

Figure 2. Same as Fig. 1, but for the FE SEP event of 2011 November 26. The initial speed using the first two data points is 697 km s^{-1}, while the speed from the last two data points is 1402 km s^{-1}, indicating that the CME continued to accelerate through the coronagraph FOV. This is confirmed by the positive acceleration in the LASCO FOV (8.97 m s^{-2}) obtained by a second order fit to the height-time measurements.

IOP Conf. Series: Journal of Physics: Conf. Series **900** (2017) 012009 doi:10.1088/1742-6596/900/1/012009

2.3. Summary of comparison

The above discussion provides a clear contrast between a GLE event and an FE SEP event, which differ in their initial acceleration, initial speed, and shock formation height (or starting frequency of the type II radio burst) as summarized in Table 1. Most parameters of the third category – the regular SEP events – fall between those of GLE and FE SEP events as was demonstrated in [1]. With this background we examine the starting frequencies of type II radio bursts and the derived shock formation heights for all SEP events of solar cycles 23 and 24 to understand the initial speed – initial acceleration relationship.

Table 1. Kinematics, shock formation height and fluence spectral indices of a GLE and an FE SEP events.

Property	2012 May 17 GLE	2011 Nov 26 FE SEP
Initial speed (first 2 points) km s^{-1}	1757	697
Final speed (last 2 points) km s^{-1}	1479	1402
Average speed km s^{-1}	1582	933
3-D speed km s^{-1}	1997	1187
Initial acceleration km s^{-2}	1.77	0.32
Starting frequency of type II MHz	100	9
Shock formation height Rs	1.38	3.5
Fluence Spectral Index	2.48	5.46

3. Type II Radio Bursts Associated with Large SEP Events

One of the main results of [1] is that the high initial CME speed is a good proxy to the initial acceleration of CMEs. High initial speed guarantees the formation of shocks near the solar surface. Since type II radio bursts are good indicators of CME-driven shocks in the corona, we can use the starting frequency of type II bursts as an indicator of the height at which the shocks formed.

We examined the dynamic spectra of type II bursts associated with large SEP events of cycle 23 and 24 and identified the type II starting frequencies. For determining the shock formation height, we need the starting frequency of the fundamental component. When both fundamental and harmonic components are observed in the dynamic spectrum, it is straight forward to obtain the starting frequency. If the fundamental component is present, but not well observed, we divided the harmonic frequency by 2 to get the starting frequency of the fundamental. If only a single component is present, we dropped the event. Another parameter we need is the onset time of the type II burst, which we take as the time of shock formation [10]. We then determined the leading edge of the CME at the onset time of the type II burst as the shock formation height. We assumed that the radio emission originates from the leading edge of the CME and hence ignore the possibility of radio emission coming from the shock flanks.

Figure 3a shows the distribution of the starting frequencies of type II radio bursts associated with the three types of large SEP events (GLE events, regular SEP events, and FE SEP events). Although there is significant overlap between the three groups of SEP events, we see that the GLE events have the highest average starting frequency, 107 MHz. Type II bursts associated with FE SEP events have the lowest starting frequency, 22 MHz, while those associated with regular SEP events have an intermediate starting frequency, 81 MHz. The shock formation height obtained by matching the onset time of the type II bursts with the CME leading edge height-time history is shown in Fig. 3b. Clearly the shock formation heights in GLE events are at the bottom of the plot (average 1.51 Rs). The shock formation heights of the regular SEP events overlap heavily with the GLE events and extend to larger heights. The average shock formation height is only slightly larger (1.72 Rs). The shock formation heights of the FE SEP events are generally large and have no overlap with those of GLE events. They also did not have overlap with the regular SEP events except for one event. The average shock formation height of FE SEP events is 5.38 Rs, much greater than those of the other two types.

According to the empirical relationship between the starting frequency of type II bursts and the heliocentric distance of the shock obtained in [10], the shocks should form at a heliocentric distance of 1.34 Rs for a starting frequency of 107 MHz in GLE events. Similarly, the starting frequencies of 81 MHz (regular SEP events) and 22 MHz (FE SEP events) correspond to heliocentric distances of 1.44 and 2.1 Rs, respectively. The actual measurements of CME leading edge heights are generally greater than these numbers as seen in Fig. 3b, although the hierarchical relationship is clearly preserved. It must be noted that the empirical relationship in [10] was derived from metric type II bursts that corresponded to shock heights <2 Rs. Here we have included type II bursts with starting at frequencies below 14 MHz. Even though the average starting frequencies of the three SEP populations are distinct, we see that there is considerable overlap between regular SEP events and GLE events. In particular, there are many high-starting frequency type II bursts in SEP events that are not GLEs. A preliminary look at these events shows that the shock speeds are not as high as that in typical GLE events. In some cases, the connectivity is poor either in latitude or longitude [21].

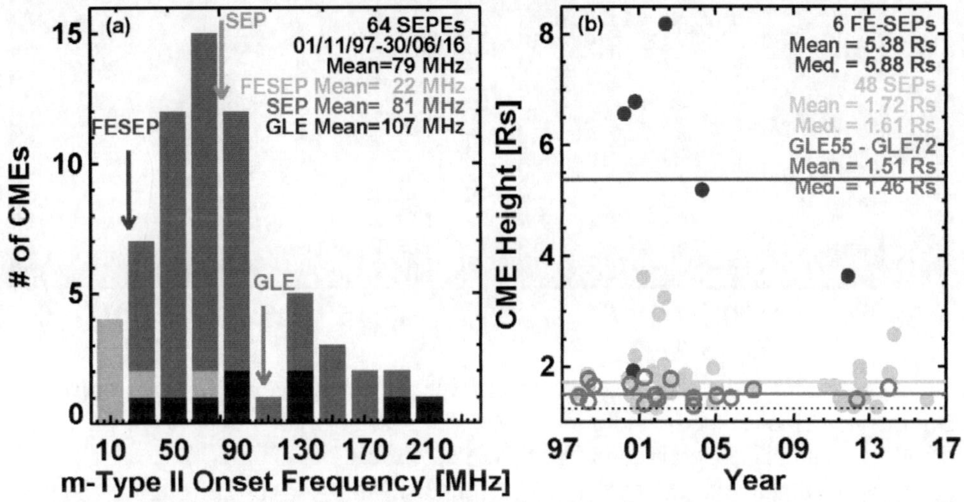

Figure 3. (a) Distribution of starting frequencies of type II bursts associated with large SEP events. The starting frequency is the lowest for FESEP events (average ~22 MHz), highest for GLE events (average ~107 MHz) and intermediate for regular SEP events (average ~81 MHz). The entire population of SEP events has a mean staring frequency of 79 MHz. (b) The shock formation height for the three types of SEP events: 5.38 Rs (FESEP events), 1.72 Rs (regular SEP events), and 1.51 Rs (GLE events). The shock formation height is consistent with the dispersion in starting frequencies among the three types of SEP events.

4. The 2010 June 12 Event

We consider the 2010 June 12 CME that was associated with a metric type II burst and a minor SEP event. The flare was of very short duration, lasting only for ~8 minutes between 00:54 and 01:02 UT. The type II burst started at 00:56 UT, just two minutes after the flare onset. Figure 4 shows the GOES soft X-ray flare profile and the type II radio burst. The type II burst has multiple lanes in the dynamic spectrum. We identify the highest frequency (~400 MHz) as harmonic emission. Therefore, the plasma level where the type II burst starts is at ~200 MHz. Previous estimate of the starting plasma frequency as 107 MHz was based on dynamic spectra with limited frequency extent [10]. Examining the extended dynamic spectrum from the Hiraiso Radio Spectrograph (HiRAS) [22], we found that the harmonic starts around 400 MHz, so we revised the fundamental plasma frequency to be ~200 MHz. The type III burst started at a frequency of ~2000 MHz, suggesting that the eruption occurred deep in the corona.

The CME associated with the eruption was observed as a slow but wide CME in the LASCO FOV. Following the analysis shown in Figs. 1 and 2, the first two LASCO data points give the initial speed is 476 km s^{-1} and the last two data points give a speed of 412 km s^{-1}. These speeds are consistent with the average speed in the LASCO FOV (486 km s^{-1}) obtained by a linear fit to the height-time data points and an average deceleration of -5.2 m s^{-2}. Using the solar source location of this eruption (N22W57), we can deproject the speeds to 568 km s^{-1} (initial) and 491 km s^{-1} (final). The CME first appeared in the LASCO FOV at 01:31 UT, which is well after the end of the type II burst. This suggests that LASCO missed the initial phase of the CME that was driving a shock.

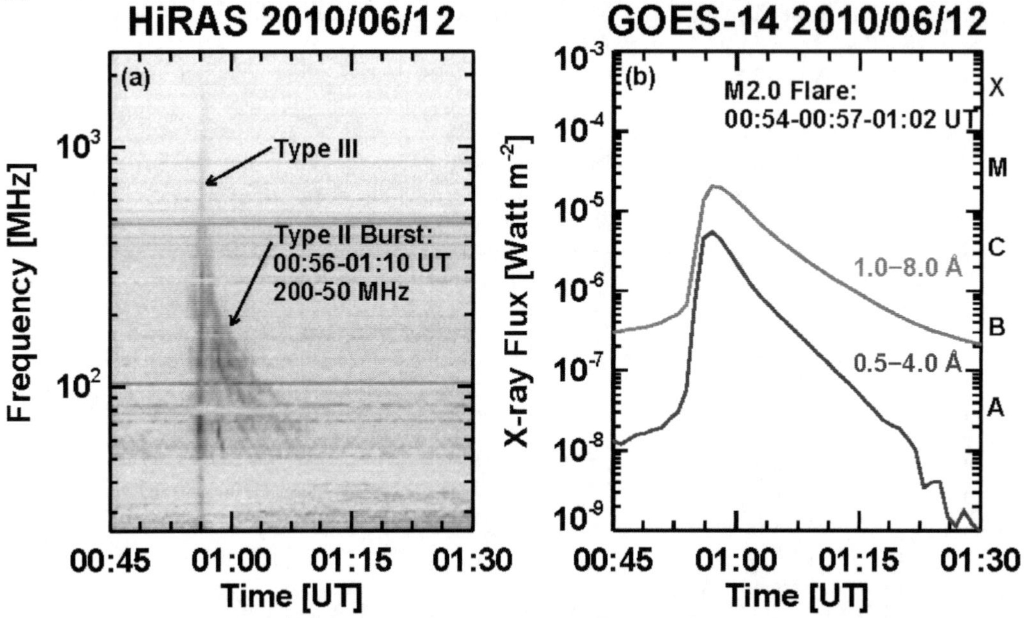

Figure 4. (a) Radio dynamic spectrum from the Hiraiso Radio Spectrograph (HiRAS) showing the type II and type III radio bursts. (b) The soft X-ray light curves of the M2.0 flare in two energy channels as observed by the GOES satellite.

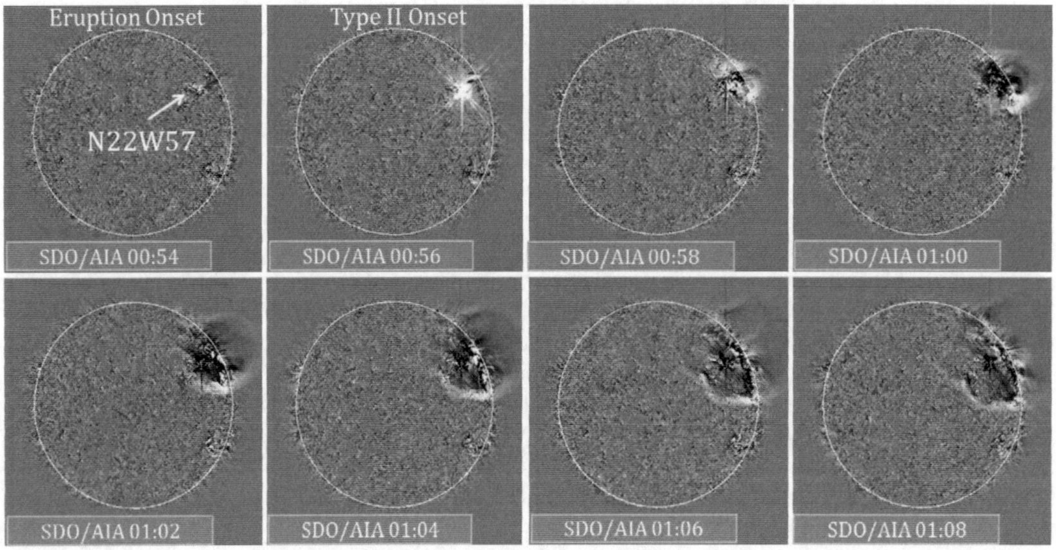

Figure 5. A series of SDO/AIA 193 Å difference images of the Sun on 2010 June 12 showing the EUV wave originating from the eruption site at N22W57. The snapshots are at every 2 minutes between 00:54 and 01:08 UT. The metric type II radio burst started when the leading edge of the wave was at the limb in projection, which corresponds to an actual height of 1.20 Rs from the Sun center. The type II burst ended around 01:10 UT.

Figure 5 shows the evolution of the eruption between 00:54 and 01:08 UT, roughly over the life time of the type II burst. A hint of the CME can be already seen in the 00:56 UT image extending to the limb in projection. Given the eruption location at N22W57, the CME leading edge can be estimated to be ~1.20 Rs, consistent with the high starting frequency. The empirical formula between the starting frequency (f in MHz) of type II burst and the shock formation height (r in Rs): $f = 308.17r^{-3.78} - 0.14$ [10], gives a shock formation height of ~1.13 Rs for $f = 200$ MHz in agreement with the measurements.
The bubble-like structure can be interpreted as the shock, although the flux rope is not seen distinctly. The shock was also observed as a well-defined circular front in STEREO-A EUVI close to the disk center (not shown). The CME was a limb event in STEREO-B FOV. Combining the SDO and STEREO-B/COR1 images, we were able to track the CME during its accelerating phase.

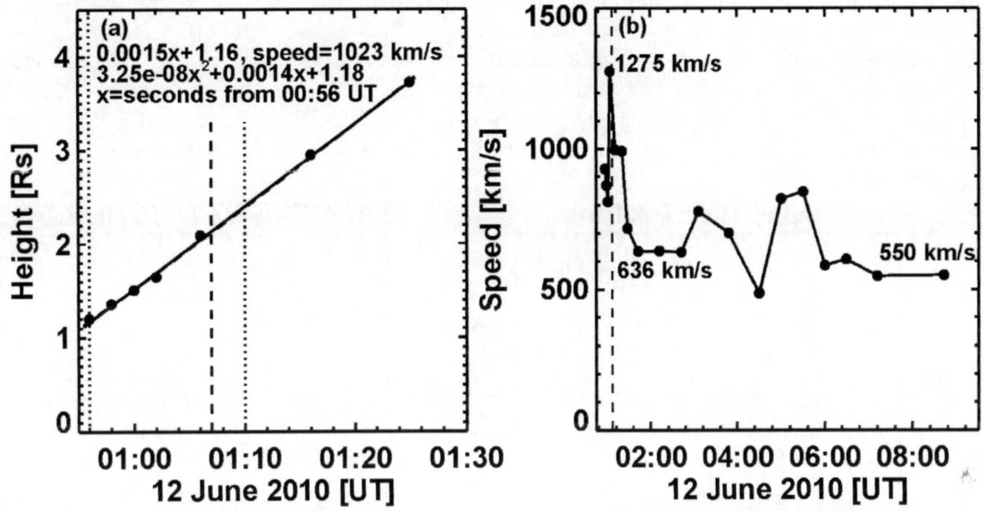

Figure 6. (a) The CME leading-edge height as a function of time obtained by combining SDO and STEREO/COR1 images. The solid vertical lines mark the onset and end of the type II burst. The dashed vertical line marks the time of solar particle release. (b) The CME leading-edge speed as a function of time obtained by combining SDO, STEREO, and SOHO images over an extended period of time. The vertical line at 01:07 UT marks the approximate solar particle release time, which is about 9 minutes after the onset of the type II burst.

4.1. CME height-time history

Figure 6 shows the CME height-time plot and the evolution of the leading edge speed. During the life time of the type II burst, the CME travelled ~1 Rs, from 1.2 to 2.2 Rs. The CME speed was obtained from successive height-time measurements and plotted as a function of time. When the type II burst started, the CME leading edge had already reached a speed of ~900 km s^{-1} and further increased to a maximum value of ~1275 km s^{-1}. The speed dropped back to ~1000 km s^{-1} when the type II ended. A similar event on 2007 June 3 with an impulsive soft X-ray flare and CME speed during the life time of the type II burst was reported in [13]. The main reason for the end of the type II in that event was an increase in the Alfven speed after the minimum around 1.5 Rs [23]. The evolution of the 2010 June 12 shock is also consistent with such a changing Alfven speed. The impulsive nature of the type II burst is indicative of the short-lived propelling force. The CME speed settled down around 600 km s^{-1} through the SOHO/LASCO FOV and there was no IP type II burst. This suggests that either the shock died or it was too weak to accelerate significant number of electrons to produce type II radio emission.

4.2. Energetic particles

The 2010 June 12 eruption did produce an SEP event with a peak intensity of 0.3 particles per (cm^2 s sr MeV) in the 13.8 – 16.9 MeV energy channel of SOHO's Energetic and Relativistic Nuclei and Electron (ERNE) instrument [24]. The time profile of the SEP intensity and the fluence spectrum are shown in Fig. 7. The event lasted for about three days. The SEP event started at about 01:30 UT at Earth, which is about 34 minutes after the onset of the type II burst. Assuming a Parker spiral length of 1.2 AU, we estimated that 50 MeV protons would take ~31 minutes to reach Earth. Thus the particle release time at the Sun is 00:59 UT. Normalizing the travel time to the electromagnetic emissions (X-rays, type II radio emission), the solar particle release time becomes 01:07 UT, which is just a few minutes before the end of the type II burst (see Fig. 4). Thus the acceleration time for the 50 MeV protons is ~10 minutes, similar to other large SEP events.

The fluence spectrum of the 2010 June 12 SEP event is remarkably similar to those of regular SEP events [1]. The spectral index (3.14) lies between the 2001 April 12 regular SEP event and the 2004 April 11 FE SEP event. In fact, the spectral index of the 2010 June 12 SEP event is smaller than the average spectral index of all regular SEP events (3.83) reported in [1]. The plot between CME acceleration (a) and the fluence spectral index γ for large SEP events reported in [1] (their Fig. 8f) yields a regression line, $\gamma = -0.29a + 4.2$. Substituting $\gamma = 3.14$, we get $a = 3.6$ km s^{-2}. We can see that the initial acceleration of the CME is in this range because the CME speed increased from ~900 km s^{-1} to 1275 km s^{-1} in 2 min, which corresponds to $a = 3$ km s^{-2}. Thus the 2010 June 12 SEP event, despite being a minor event, is consistent with the hierarchical relationship between CME acceleration and the fluence spectral index [1].

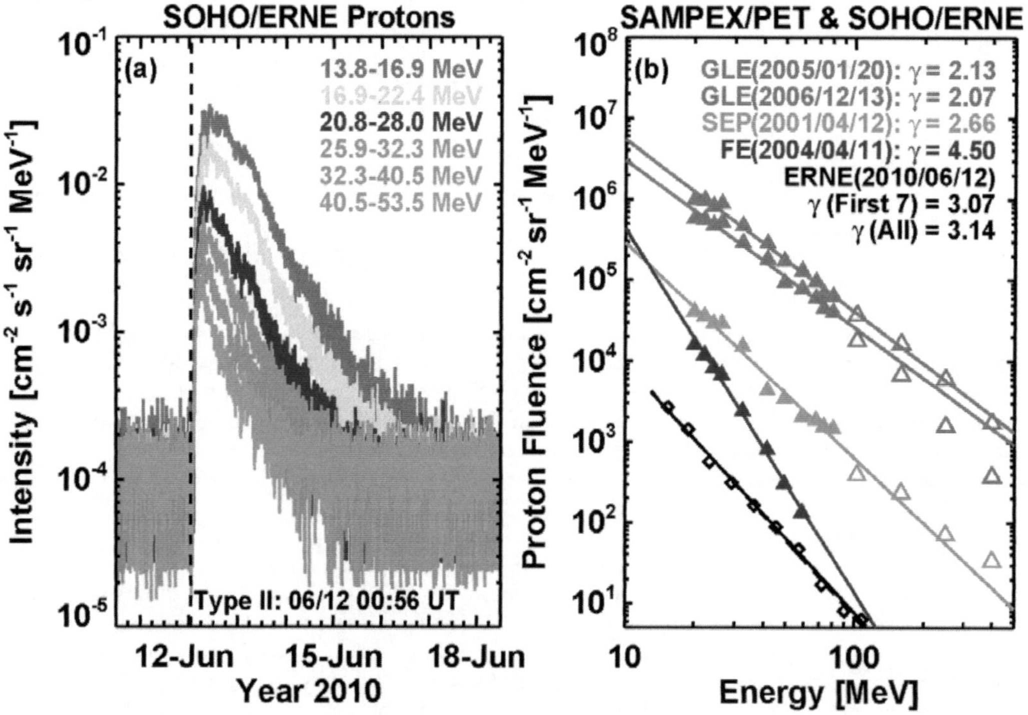

Figure 7. (a) Proton intensity of the 2010 June 12 event from SOHO/ERNE in several energy channels noted on the plot. The onset time of the type II burst is marked by the vertical dashed line. The onset time of the SEP event was around 1:30 UT. (b) Fluence spectrum of the 2010 June 12 event compared with a FE SEP, GLE, and regular SEP event from [1].

5. Summary and Conclusions

It has been recently found that CMEs with high initial acceleration are associated with SEP events with the hardest fluence spectra, while those with lowest initial acceleration have the softest SEP

fluence spectra. Consistent with this trend, CMEs with intermediate initial acceleration have moderately hard SEP fluence spectra. Physically speaking, impulsive acceleration leading to high initial CME speeds close to the Sun results in shock formation close to the Sun, where the ambient magnetic field and density are high and the particles are energized more efficiently. On the other hand, slowly accelerating CMEs can form shocks only at several solar radii from the Sun, where the magnetic field and density have fallen off significantly, reducing the efficiency of shock acceleration. The GLE SEP events represent the high initial acceleration class, while the FE SEP events represent the low initial acceleration class. We illustrated this hierarchical relationship using the starting frequency of type II radio bursts associated with large SEP events in solar cycles 23 and 24. We found that the average starting frequency of type II bursts progressively decreases as one goes from GLE events (107 MHz), to regular SEP events (81 MHz) and to FE SEP events (22 MHz). Since the fundamental component of the type II burst is emitted at the local plasma frequency, and the local plasma frequency decreases with heliocentric distance, the frequency variation reflects the height of shock formation measured from the Sun center. We confirmed this by measuring the height of the CME leading edge at the time of type II burst onset assuming that the shock formation coincides with the onset of type II burst. We found that the shock formation distance increases as one goes from GLE events (1.51 Rs), to regular SEP events (1.6 Rs), and to FESEP events (5.38 Rs). This finding strongly supports the idea that large SEP events are primarily due to CME-driven shocks and reflect the changing ambient magnetic field strength influencing the efficiency of particle acceleration [9].
We also performed a case study of the 2010 June 12 CME, which had high initial acceleration, but the associated SEP event did not fall into the class of large SEP events because the intensity was <1 pfu. Interestingly, the relation between the initial CME speed and acceleration continued to hold. Despite the low SEP intensity, we found the fluence spectral index to be similar to that of regular SEP events. As in large SEP events, the energetic particles were released several minutes after the shock formation. However, the CME speed dropped abruptly, similar to the spiky soft X-ray profile, suggesting a short-lived propelling force. The sudden drop in the CME speed weakened the shock and the injection of the energetic particles into the heliosphere ended up being a delta function. A detailed analysis of such small SEP events and the associated type II bursts and CMEs will be reported elsewhere.

6. Acknowledgments

This work was supported by NASA's Heliophysics Guest Investigator program. PM was partially supported by NASA grant NNX15AB77G and NSF grant AGS-1358274. NT was partially supported by NSF grant AGS-1622377. HX was partially supported by NASA LWS TR&T grant NNX15AB70G.

7. References

[1] Gopalswamy N, Yashiro S, Thakur N, Mäkelä P, Xie H, and Akiyama S 2016 The 2012 July 23 Backside Eruption: An Extreme Energetic Particle Event? *Astrophys. J.* **833** 216
[2] Zank, G P, Rice, W K M., Wu, C C 2000 Particle acceleration and coronal mass ejection driven shocks: A theoretical model. *J. Geophys Res.* **105** 25079
[3] Melrose, D B 1987 Plasma emission - A review. *Solar Phys.* **111** 89
[4] Gopalswamy N, Xie H, Mäkelä P, Akiyama S, Yashiro S, Kaiser M L, Howard R A, Bougeret J-L 2010b Interplanetary Shocks Lacking Type II Radio Bursts. *Astrophys. J.* **710** 1111; Gopalswamy, N, Aguilar-Rodriguez, E, Yashiro, S, Nunes, S, Kaiser, M L Howard, R A 2005 Type II radio bursts and energetic solar eruptions. *J. Geophys Res.* **110**, A12S07
[5] Masters, A et al. 2013 Electron acceleration to relativistic energies at a strong quasi-parallel shock wave. *Nature Physics* **9** 164; Amano, T and Hoshino, M 2010 A critical Mach number for electron injection in collisionless shocks. *Phys. Rev. Lett.* **104** 181102; Lembege, B, Giacalone, J, Scholer, M, Hada, T, Hoshino, M Krasnoselskikh, V, Kucharek, H, Savoini, P, and Terasawa, T 2004 Selected Problems in Collisionless-Shock Physics. *Space Sci. Rev.* **110** 161

[6] Cliver E W, Nitta NV, Thompson B J, Zhang J 2004 Coronal Shocks of November 1997 Revisited: The Cme Type II Timing Problem *Solar Phys.* **225** 105; Reames D V 2009 Solar Energetic-Particle Release Times in Historic Ground-Level Events *Astrophys. J.* **706** 844; Gopalswamy N, Xie H, Yashiro S, Akiyama S, Mäkelä P, Usoskin I G 2012 Properties of Ground Level Enhancement Events and the Associated Solar Eruptions During Solar Cycle 23. *Space Sci. Rev.* **171** 23; Desai, M and Giacalone, J 2016 Large gradual solar energetic particle events. *Living Rev. in Solar Phys* **13** 3

[7] Gopalswamy N, Mäkelä P, Akiyama S, Yashiro S, Xie H, Thakur N, Kahler S W 2015 Large Solar Energetic Particle Events Associated with Filament Eruptions Outside of Active Regions *Astrophys. J.* **806** 8

[8] Mäkelä P, Gopalswamy N, Akiyama S, Xie H, Yashiro S 2015 Estimating the Height of CMEs Associated with a Major SEP Event at the Onset of the Metric Type II Radio Burst during Solar Cycles 23 and 24 *Astrophys. J.* **806** 13

[9] Zank, G P, Li, G, Florinski, V, Hu, Q, Lario, D, Smith, C W 2006 Particle acceleration at perpendicular shock waves: Model and observations. *J. Geophys Res.* **110** 6108; Li, G, Shalchi, A, Ao, X, Zank, G, Verkhoglyadova, O P 2012 Particle acceleration and transport at an oblique CME-driven shock. *Adv. Space Res.* **49** 1067

[10] Gopalswamy et al. 2013 Height of shock formation in the solar corona inferred from observations of type II radio bursts and coronal mass ejections *Adv. Space Res.* **51** 1981

[11] Schwer K, Lilly R B, Thompson B J, and Brewer D A 2002, The SDO Mission *AGU Fall Meeting Abstracts*, SH21C-01

[12] Kaiser M L, Kucera T A, Davila J M St. Cyr O C Guhathakurta M, Christian E 2008 The STEREO Mission: An Introduction *Space Sci. Rev.* **136** 5

[13] Gopalswamy N, Thompson W T, Davila J M, Kaiser M L, Yashiro S, Mäkelä P, Michalek G, Bougeret J-L, Howard R A 2009 Relation between type II bursts and CMEs inferred from STEREO observations. *Solar Phys.* **259** 227

[14] Brueckner G E, Howard R A, Koomen M J, Korendyke C M, Michels D J, Moses J D 1995, The large angle spectroscopic coronagraph (LASCO), *Solar Phys.* **162** 357

[15] Domingo V, Fleck B, and Poland A I 1995 SOHO: The Solar and Heliospheric Observatory *Space Sci. Rev.* **72** 81

[16] Gopalswamy N and Thompson B J 2000 Early life of coronal mass ejections *JASTP* **62** 1457; Zhang J, Dere K P, Howard R A, Kundu M R, White S M 2001 On the Temporal Relationship between Coronal Mass Ejections and Flares *Astrophys. J.* **559** 452

[17] Gopalswamy N, Xie H, Akiyama S, Mäkelä P, Yashiro S, Usoskin, I G, Davila, J M 2013 The first ground level enhancement event of solar cycle 24: direct observation of shock formation and particle release heights. *Astrophys. J.* **765** L30

[18] Howard RA et al. 2008 Sun Earth Connection Coronal and Heliospheric Investigation (SECCHI). *Space Sci. Rev.* **136** 67

[19] Lemen, J et al. 2012 The Atmospheric Imaging Assembly (AIA) on the Solar Dynamics Observatory (SDO) *Solar Phys.* **275** 17

[20] Bougeret J-L et al. 1995 Waves: The Radio and Plasma Wave Investigation on the Wind Spacecraft *Space Sci. Rev.* **71** 231

[21] Gopalswamy N, Xie H, Akiyama S, Mäkelä P, Yashiro S, Major solar eruptions and high-energy particle events during solar cycle 24 *Earth, Planets and Space* **66** 104

[22] Kondo T, Isobe T, Igi S, Watari S, Tokumaru M 1995 The Hiraiso Radio Spectrograph (HiRAS) for Monitoring Solar Radio Bursts *J. Comm. Res. Lab.* **42** 111

[23] Gopalswamy N, Lara A, Kaiser ML, Bougeret, J-L 2001 Near-Sun and near-Earth manifestations of solar eruptive events. *J. Geophys Res.* **106**, 25261

[24] Torsti et al. 1995 Energetic Particle Experiment ERNE *Solar Phys.* **162** 505

Recent Voyager Evidence for Rapid Transport of Flare-Generated Disturbances by Polar Coronal Hole Streams

D S Intriligator[1], W D Miller[1], J Intriligator[1,2], W Webber[3], W Sun[4], T Detman[1], M Dryer[1], and C Deehr[4]

[1]Carmel Research Center, Inc., Santa Monica, CA
[2]Tufts University, Medford, MA
[3]New Mexico State University, Las Cruces, NM
[4]University of Alaska, Fairbanks, AK

E-mail: devriei@aol.com

Abstract. Disturbances observed by Voyagers 1 and 2 during the past five years or more may have been transported by plasma emitted from polar coronal holes, thereby having travelled much faster from the Sun to the termination shock than previously recognized. Estimating the average speed to the shock as 750 km/s has produced consistently good associations between solar flares, or groups of them, and dynamic pressure increases at Voyager 2 and plasma wave events at Voyager 1. Furthermore, magnetograph observations confirm that polar coronal holes were present around the times of the flares to which the events at the Voyagers have been attributed. These calculations also provide revised estimates of the transport of heliospheric current sheet fluctuations. We discuss the possibilities that extrapolations from past observations and simulations based on them may provide insight into currently challenging issues and possible future developments.

1. Introduction

Spacecraft observations obtained over the past fifteen or twenty years from regions of the heliosphere that formerly were solely the domains of guesswork and theoretical calculations have given a more definite view of the vast plasma bubble around the Sun and the planets than previously was ever possible. Nevertheless, these observations extend over such widely separated locations and long time intervals that many questions arise that can only be answered by comparing results from many spacecraft. These observations need to be combined with model calculations, and any other ground-based or inner solar system observations that can be applied.

In particular, the importance of understanding the propagation and interactions of the interplanetary disturbances generated by major solar flares has long been understood. However, it has been given additional urgency by Voyager 1 (V1) observations in recent years indicating that the strongest of these disturbances propagate all the way through the heliosheath and out into the local interstellar medium (LISM). Thus, the results we present here go substantially beyond our previously reported studies connecting solar events with observations in the heliosheath and LISM. We describe evidence that disturbances generated by solar flares that occurred from the summer of 2011 to early 2015 may have been convected to Voyager 2 (V2) and V1 in high-speed solar wind plasma emitted from polar coronal holes, producing phenomena observed at the Voyagers from the middle of 2012 to late 2016.

As we did last year, we concentrate on combining various in-situ observations from remote spacecraft with ground-based observations of the Sun, the heliosphere, the heliosheath, and the LISM.

2. A Consistency Question and a Transport Hypothesis

Our analyses in Intriligator et al. [1] reported the period of major solar flares in the first seven months of 2012 may be associated with phenomena observed at V2 in the middle months of 2013. In particular, a collection of powerful flares in early March, 2012, is associated with particle and plasma phenomena at V2 peaking around April 18, 2013. The even more powerful flares of July, 2012, are associated with particle and plasma phenomena peaking around August 19, 2013. These V2 observations were put into a larger context by Richardson, et al. in [2] as parts of a period of enhanced plasma dynamic pressure that they identify from magnetometer observations as a merged interaction region (MIR). It is the fourth of seven such pressure pulses (designated as pulses A through G as shown in Figure 1, which we adapted from Figure 4 of [2]) occurring in the V2 plasma observations from the middle of 2011 to late 2016. The last five of them (i.e., pulses C through G) are identified as MIRs. These regions are caused when shock waves resulting from closely spaced flares merge in propagating over the 100 or more AU from the Sun to the outer heliosphere. Thus, an obvious question is: whether the associations in [1] could be extended to associate the earlier and later MIRs with respectively earlier and later flares?

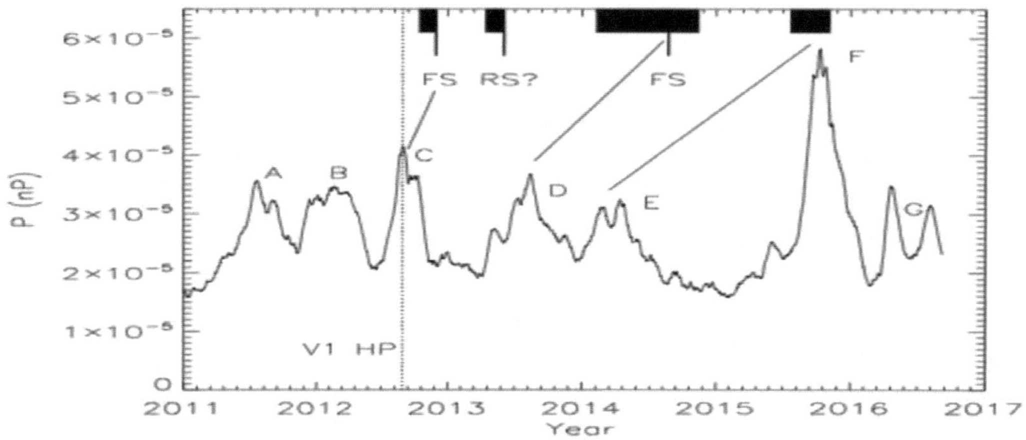

Figure 1. Smoothed history of plasma dynamic pressures at V2 in nanopascals. The Forward Shock (FS), Reverse Shock (RS), and the V1 Heliopause (HP) are indicated. The black bars are the times of the Plasma Wave System (PWS) events at V1. The onsets of the PWS events are shown in Figure 6.

However, the effort to do this raised a question of mathematical consistency. As we noted in [3], the period of great solar flares around Halloween, 2003, produced an MIR that reached V2 in late April, 2004, in which the peak convection speed (as of the middle of May, 2004) was about 560 km/s, and during most of the MIR the speed was around 500 km/s or less. Also, Richardson, et al. report in [2] that the convection speed of the heliosheath plasma was between about 90 and 120 km/s during MIR F, and in many other publications of the V2 PLS team they have reported heliosheath speeds of between 100 and 150 km/s. Assuming a speed of 500 km/s between the Sun and the termination shock (TS) at 84 AU (in 2007), and a speed of 150 km/s from the TS to V2, at 101.4 AU gives a predicted travel time of 493 days. This was three months longer than the 404 days from March 10, 2012 to April 18, 2013 shown in [1] in the table of delays. An even bigger discrepancy arises for the phenomena observed by V2 in August, 2013.

Richardson et al. may have recognized that difficulties are raised if the speed for the MIRs is only modestly above the observed plasma speeds, because they hypothesize a 320 km/s shock speed in the heliosheath which is more than twice as high as the presumed bulk plasma speed. This would bring the predicted delay times for the 2012 flares into much closer agreement with the actual delays at V2 observed in 2013.

An alternative hypothesis is provided by recalling that by the second half of 2012, when MIR C was observed, V2 was at an HGI latitude of 30 degrees south. In the model calculations of Usmanov, et al. [4] the solar wind at latitudes of 30 degrees or more is dominated by high-speed plasma flowing from the polar coronal holes. This raises the question: what results would be obtained from assuming that during the period of MIR observations from late 2012 to late 2016 V2 was in heliosheath plasma that the TS had decelerated from a higher speed than 500 km/s?

Simply assuming that flare disturbances travel 750 km/s from the Sun to a TS at 84 AU yields the travel time to the TS as a constant 194 days. Then a heliosheath speed of 150 km/s gives an additional time in days of (r_{V2}-84)*1,000,000/86,400, where r_{V2} is the V2 distance, for a total time of 194+11.574(r_{V2}-84) days, as shown in Table 1.

Table 1. V2 associations with events in Figure 1

MIR	Peak day	V2 Distance (AU)	Estimated delay (days)	Flare date(s)	Delay(s) (days)
C	Aug. 28, 2012	99.4	373	Aug. 9, 2011 Sept. 6, 2011	385 357
D (early)	Apr. 18, 2013	101.42	396	Mar. 10, 2012	404
D (late)	Aug. 19, 2013	102.48	408	Jul. 12, 2012 Jul. 23, 2012	403 392
E	Feb. 10, 2014	104.00	426	Dec. 14, 2012	423
F	Oct. 17, 2015	109.32	488	Jun. 10, 2014	494
G(early)	Apr. 30, 2016	111.02	507	Dec. 13, 2014 Dec. 17, 2014	504 500
G (late)	Jul. 20, 2016	111.72	515	Feb. 21, 2015	515

3. Solar Cycle Effects and Actual Coronal Hole Observations

We obtained excellent agreements in the arrival times in Table 1 with the simple assumption that solar flare disturbances had been transported to the TS in fast polar coronal hole plasma (specifically, from the south polar coronal hole for the pressure pulses at V2). Thus, the next step was to try to determine whether the agreements were merely fortuitous or whether a south polar coronal hole was actually present and large enough to be a plausible source of high-speed plasma around latitude 30 south. Actually, such a southern (high-latitude) polar coronal hole was directly observed by Ulysses during two solar minima and one solar maximum. This can be seen in Figure 2 (in which each top panel shows the spacecraft's distance from the Sun in AU, the middle of its heliographic inertial (HGI) latitude, and the bottom the solar wind speed in km/s). Ulysses found that during the Solar Cycle (SC) 23 maximum there was very little high-speed solar wind from the south polar region and only a small region of fast plasma at the highest latitudes in the north polar region.

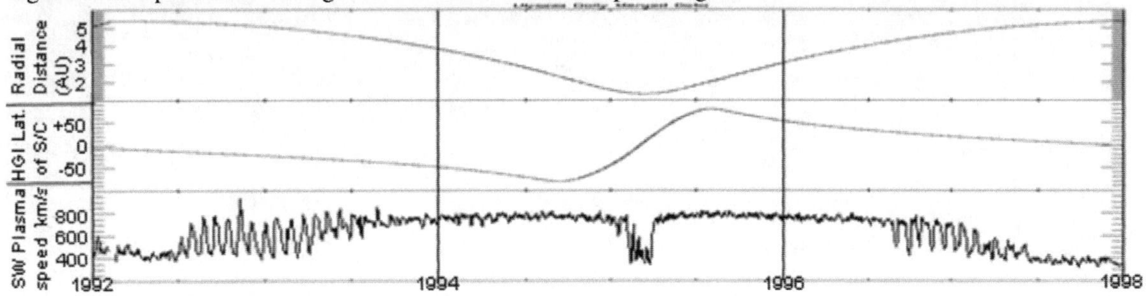

Figure 2a. Ulysses HGI distance and latitude and solar wind speed from 19920101 to 19971231

IOP Conf. Series: Journal of Physics: Conf. Series **900** (2017) 012010 doi:10.1088/1742-6596/900/1/012010

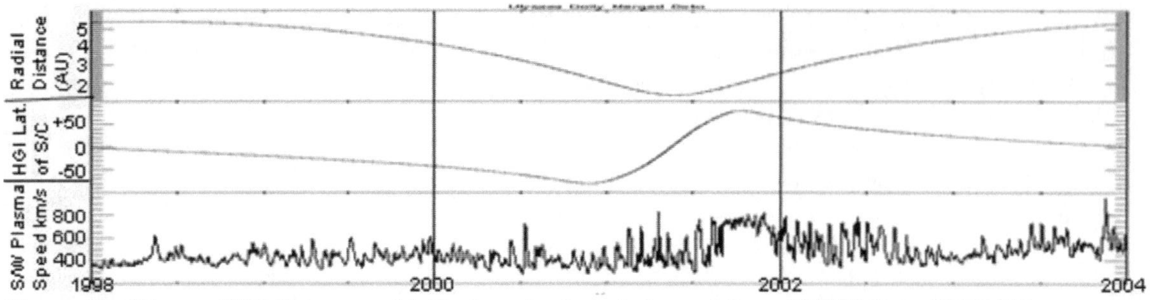

Figure 2b. Ulysses HGI distance and latitude and solar wind speed from 19980101 to 20031231

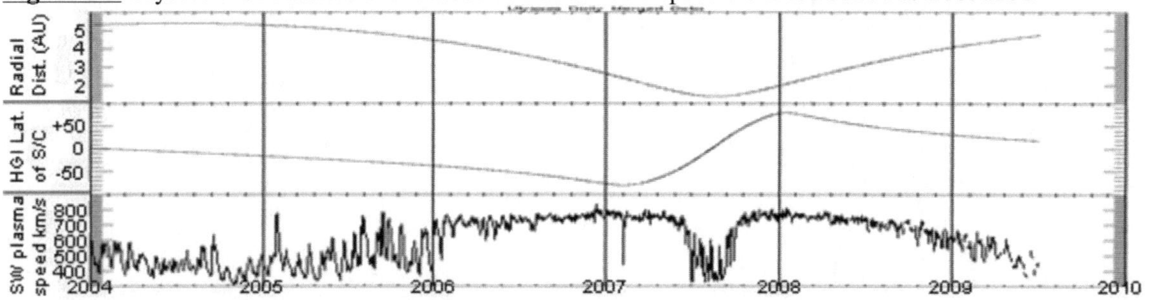

Figure 2c. Ulysses HGI distance and latitude and solar wind speed from 20040101 to 20090630

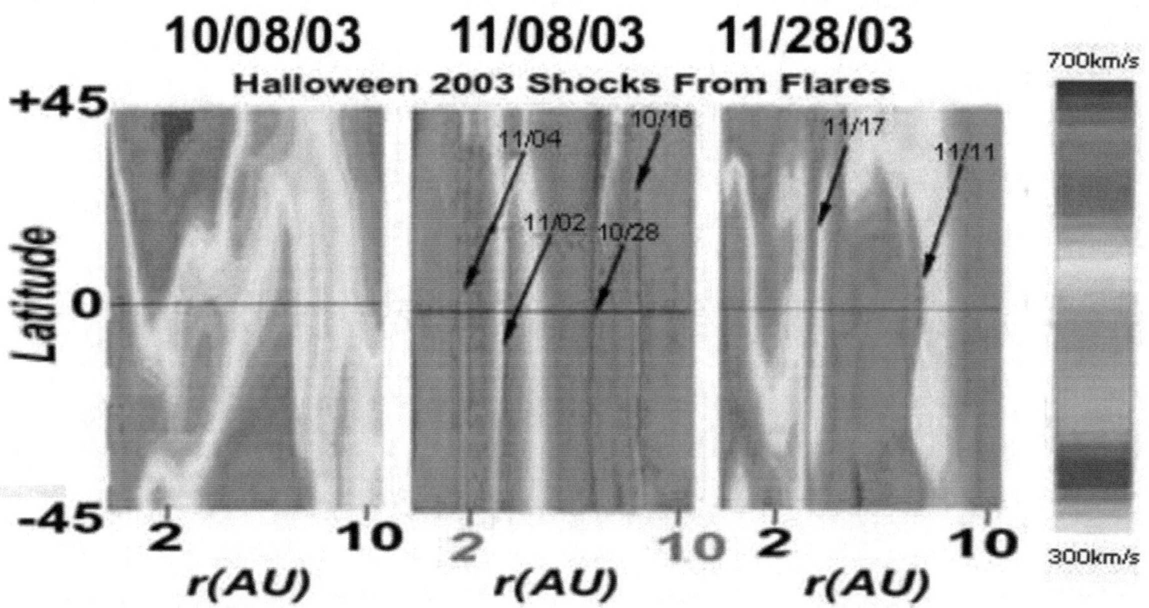

Figure 3. HHMS simulations of solar wind speed on the Sun-Earth line for latitudes N45 to S45 [3]

Correspondingly, as shown in Figure 3, for early October and late November, 2003 (after Ulysses had completed its polar pass), our HHMS program used inputs from the Wang-Sheely-Arge model (based on magnetograms extending up to high solar latitudes) to calculate that there were only small regions of solar wind with speeds as high as 700 km/s, as indicated by the pink and purple blobs in the left and right rectangular regions of the figure. These are indicative of the recurring high speed stream structures before and after the flares were launched.

On the other hand, since the 2012-2014 period of peak sunspot activity of SC 24 was notably weaker than the SC 23 peak observed by Ulysses in 2000-2003, this raised the possibility that the polar

IOP Conf. Series: Journal of Physics: Conf. Series **900** (2017) 012010 doi:10.1088/1742-6596/900/1/012010

coronal holes in SC 24 might have been bigger, even at the peak of the SC, than the polar coronal holes during SC 23.

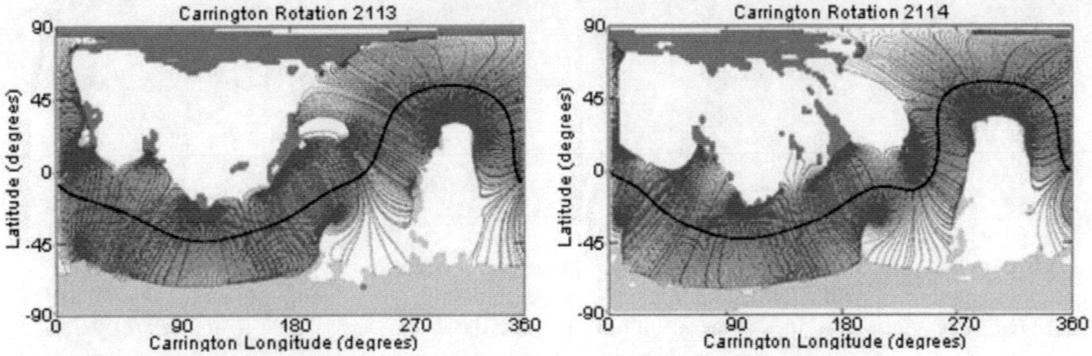

Figure 4a. Coronal holes in August and September, 2011 (CR 2113 and 2114) [5]

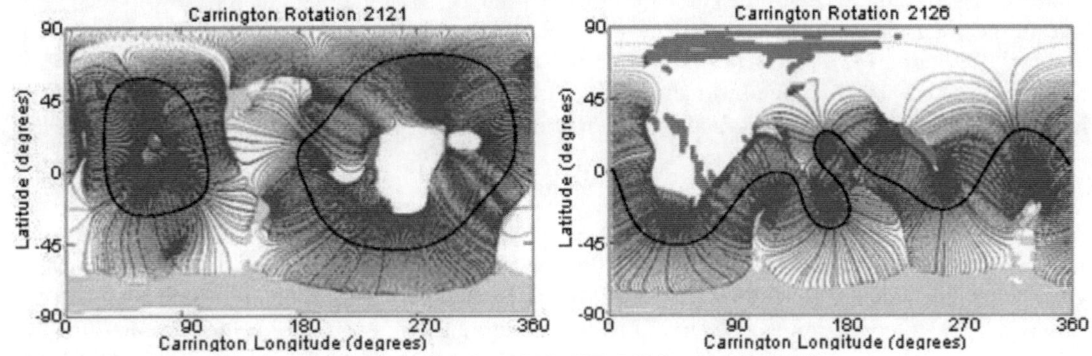

Figure 4b. Coronal holes in March and July, 2012 (CR 2121 and 2126) [5]

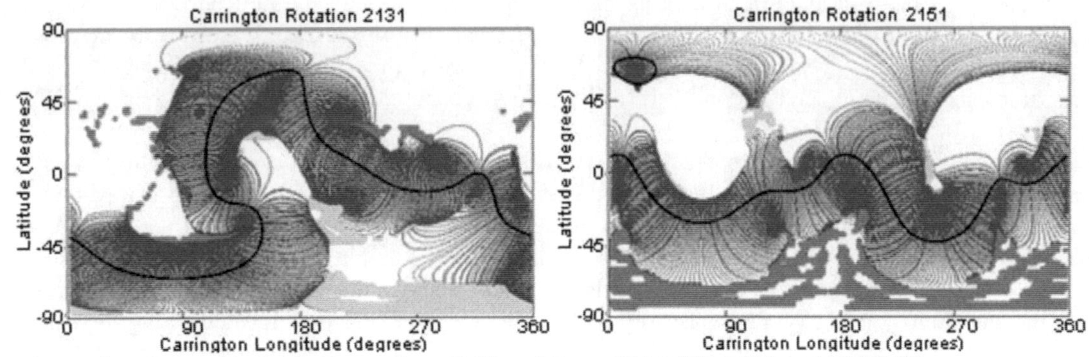

Figure 4c. Coronal holes in December, 2012 and June, 2014 (CR 2131 and 2151) [5]

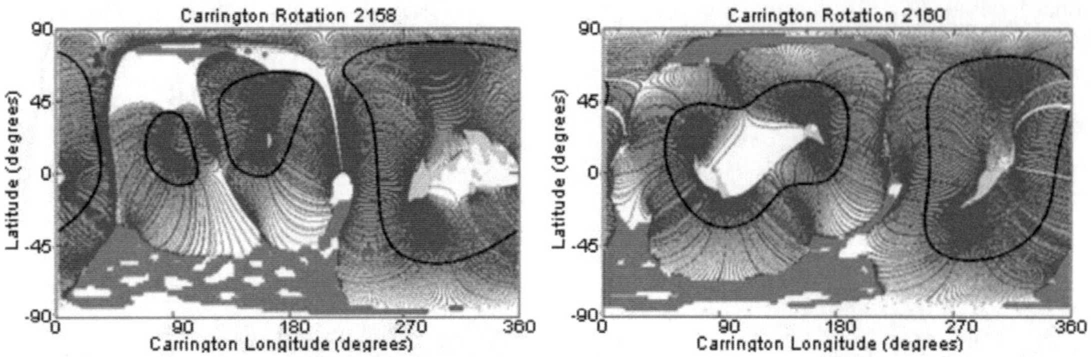

Figure 4d. Coronal holes in December, 2014 and February, 2015 (CR 2158 and 2160) [5]

As shown in Figures 4a - 4d, it was possible to resolve these questions using synoptic coronal hole maps prepared by the National Solar Observatory Integrated Synoptic Program (NISP) [5]. They show that large southern coronal holes were present during all the Carrington rotations (CR) during which the flares listed in Table 1 occurred. The lack of events during 2013 may be connected with the rapid change in the solar magnetic field at the end of 2012 and during the first months of 2013, so that in Carrington rotations 2133 - 2137 the polar coronal holes were almost completely closed. In these images the blue lines represent closed magnetic field lines, while the red regions are coronal holes in which the polarity is toward the Sun, and the green regions are holes with polarity away.

4. Comparisons with Voyager 1 Plasma Wave System Events

As V1 is now in the LISM, some 24.1 AU (as of early April, 2017) farther from the Sun than V2, and more than 60 degrees of heliographic latitude north of V2, so that they are more than 100 AU apart, they are approximately as far from each other as either is from the Sun. Thus, there are obvious questions of how to compare their observations and environments, and some differences are already known. For example, V1 crossed the TS at about 94 AU from the Sun, and emerged into the LISM after traveling about 27.6 AU through the heliosheath. By contrast, V2 crossed the TS at a little less than 84 AU from the Sun and after traveling more than 30 AU through the heliosheath has not yet emerged. If Roelof, et al (2012) and Krimigis et al. (2016) [6] are correct, V2 probably may not emerge for several more years at the earliest. This is because they combine the Voyager LECP observations and energetic neutral atom data from Cassini to conclude that the heliosheath along V2's trajectory may be 40 - 70 AU thick.

However, MIRs are expected to spread out very widely both in longitude, as suggested in Figure 5 from Intriligator et al., (2014) [7], and in latitude from the locations of the flares that produced them. Thus, an obvious question is whether the PWS events observed at V1 (see Figure 1) are associated with the pressure pulses observed at V2 and hence with the flares listed in Table 1?

Since V1 is a few degrees farther from the heliographic equator than V2 is, at about 34 degrees north, the success of the high-latitude assumption for the V2 calculations suggested trying the same approach for V1. Specifically, the assumption is a speed of 750 km/s from the Sun to the TS at 94 AU, then 150 km/s for 27.6 AU through the heliosheath, and finally a speed of around 40 km/s through the LISM to V1. The times predicted by this approach were compared with the onset dates of the 3 kHz (green) PWS events as shown in Figure 6. We obtained these values from the Univ. of Iowa [8]. (Note that the readings in the 1000 Hz (blue) and 1780 Hz (red) channels visible in the plots for May, 2014, and to a lesser extent in the plots for September, 2015, are caused by spacecraft roll maneuvers to calibrate the magnetometer (or "magrols") that happened to be in progress at the times these PWS events started, Thus, unlike the 3 kHz signals, these lower-frequency readings do not represent detections of physical phenomena external to the spacecraft. The nearly silent purple channel is 5620 Hz.)

The results in Table 2 of the calculations for the three PWS events in 2013, 2014, and 2015 and the forward shock in August, 2014, show fairly good agreement with predictions made by assuming that the shock speed through the LISM is 43 km/s with respect to the heliosphere. Since the agreement is very close for the onsets of the PWS events, but associating the forward shock with either of the listed July, 2012 flares implies that it arrived significantly sooner than the prediction, this is consistent with the expectation that the shock would propagate through the plasma faster than the bulk convection speed of the CME plasma that formed the MIRs.

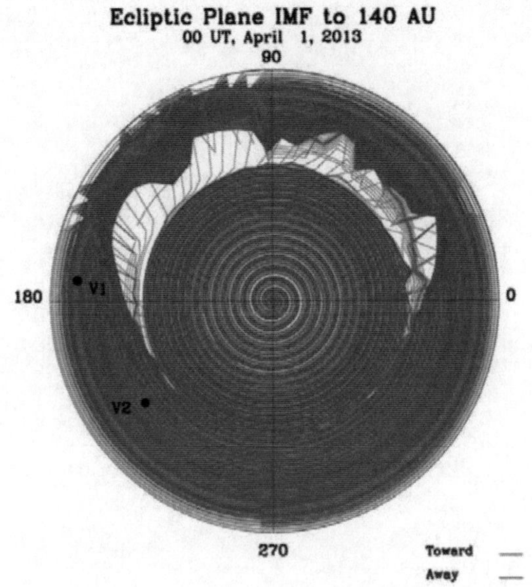

Figure 5a. Our HAFSS ecliptic plane plot [7] of the B field from the Sun to 140 AU.

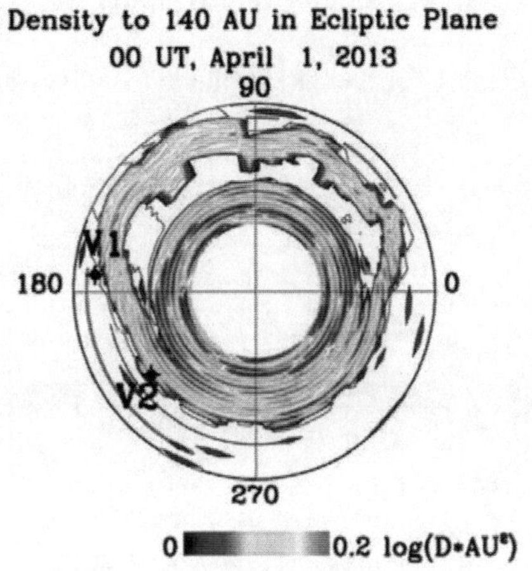

Figure 5b. Our HAFSS ecliptic plane simulation plot of the plasma density [7] from the Sun to 140 AU.

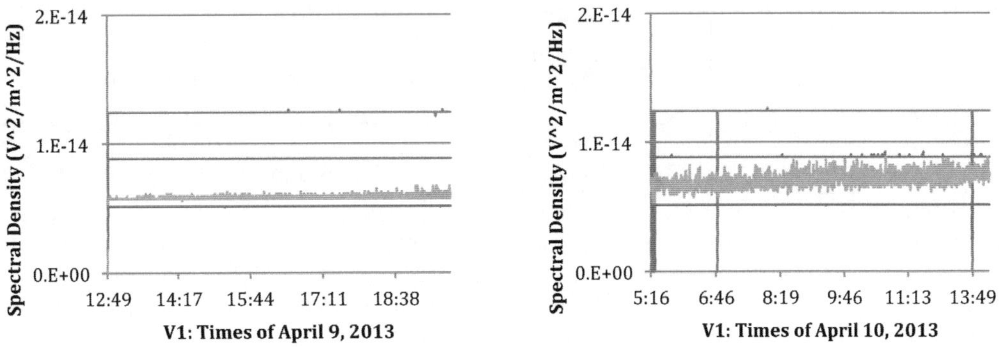

Figure 6a. V1 for the 3 kHz plasma wave system event onset April 9-10, 2013

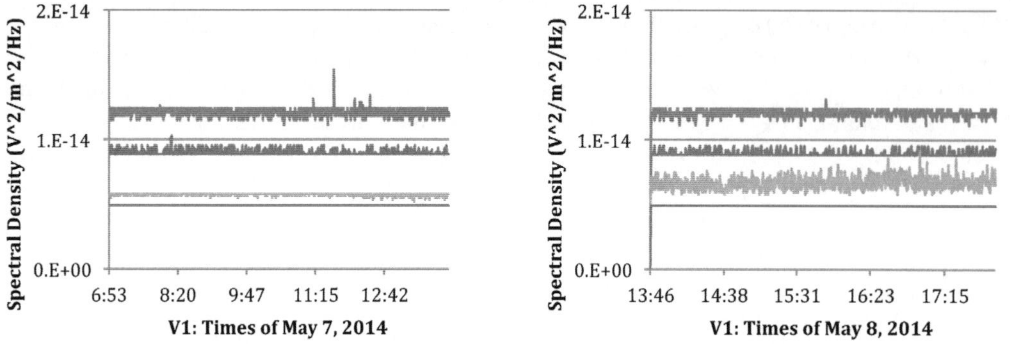

Figure 6b. V1 for the 3 kHz plasma wave system event onset May 7-8, 2014

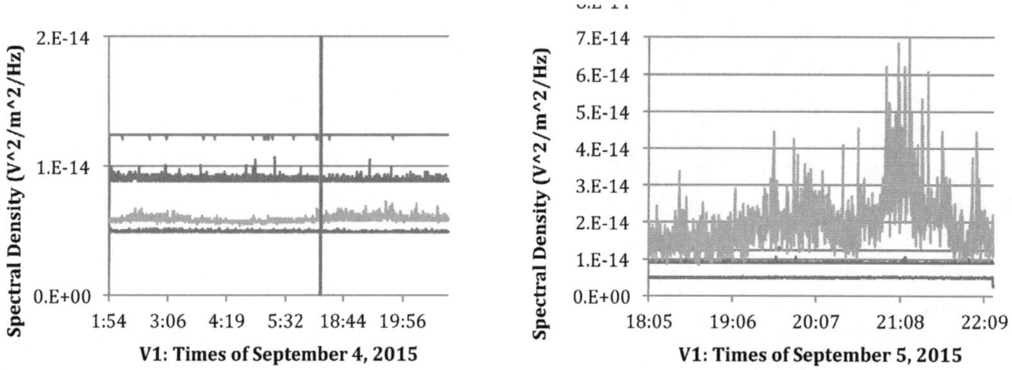

Figure 6c. V1 for the 3 kHz plasma wave system event onset September 4-5, 2015

Table 2. V1 associations with events in Figure 1 and Table 1

MIR?	V1 PW event onset	V1 Dist. (AU)	Estimated delay (days)	Flare date(s)	Delay(s) (days)
C	April 10, 2013	123.82	627	Aug. 9, 2011 Sept. 6, 2011	610 582
D (early)	May 8, 2014	127.66	783	Mar. 10, 2012	789
D (late)	Aug. 22, 2014 (FS)	128.70	824	Jul. 12, 2012 Jul. 23, 2012	771 760
E	Sept. 4, 2015	132.40	974	Dec. 14, 2012	994
F	(pred.) Apr. 1, 2018	(pred.) 140.87	(pred.) 1314	Jun. 10, 2014	(pred.) 1314

The last line of Table 2 shows that extrapolating the calculation to MIR F, which was unusually powerful at V2, implies that it could reach V1 and set off another PWS event in January of 2018. This may be a little sooner than the "early 2018" prediction of Richardson, et al. [2], since they assume a propagation speed through the LISM of 40 km/s with respect to the heliosphere. (A matter for future consideration is that since the IBEX and Ulysses observations have been reconciled to conclude that the heliosphere is moving 26 km/s with respect to the rest frame of the LISM [9], shock or MIR speeds of 40 and 43 km/s with respect to the heliosphere correspond to speeds of 66 and 69 km/s through the LISM.)

These associations in Table 2 suggest that the apparently unusual length of the 2014 PWS event (from early May to early November, much longer than the other two) may be attributable, at least in part, to the prolonged period of large flares in 2012 [1, 7]. It began with large flares in late January, continued with several larger ones in early March, and ended with still more in July, culminating in the great flare of July 23. Correspondingly, MIR D at V2 is broader with respect to its height (as depicted in Figure 1) than most of the other pressure pulses.

5. Additional Flare Contributions to the MIRs and Plasma Wave System Events

As each of MIRs C through G is likely to have formed as CMEs from several flares merged on their way outward from the Sun, Table 3 lists additional information [10] about the more important flares that may have contributed to each MIR, including many that were not listed in Table 1.

Table 3a. Possible Flares for MIR C

Date, 2011	Time (UT)	Lat.	Earth Lon.	HGI Lon.	X-ray Imp.	CME (km/s)
August 4	0415	N19	W36	274	M9.3	1315
August 8	0820	N17	W69	312	X6.9	1610
September 6	0200	N14	W07	277	M5.3	782
September 6	2230	N14	W18	288	X2.1	575

Table 3b. Possible Flares for MIR D (early)

Date, 2012	Time (UT)	Lat.	Earth Lon.	HGI Lon.	X-ray Imp.	CME (km/s)
January 23	0400	N28	W21	67	M8.7	2175
January 27	1830	N27	W71	121	X1.7	2508
March 5	0400	N17	E52	36	X1.1	1531
March 7	0100	N17	E27	63	X5.4	2684
March 10	1755	N15	W24	117	M8.4	1296

Table 3c. Possible Flares for MIR D (late)

Date, 2012	Time (UT)	Lat.	Earth Lon.	HGI Lon.	X-ray Imp.	CME (km/s)
July 6	2310	S13	W59	268	X1.1	1828
July 12	1645	S15	W01	216	X1.4	> 885
July 19	0530	S13	W88	316	M7.4	1631
July 23	0230	S17	W132	1	? on far side	> 2003

The coronagraph underestimated the CME speeds for both the July 12 flare, because it was at the solar centerline, and the July 23 flare, because it was on the far side.

Table 3d. Possible Flare for MIR E

Date, 2012	Time (UT)	Lat.	Earth Lon.	HGI Lon.	X-ray Imp.	CME (km/s)
Dec. 14	0100	N07	W120 - 140	126 - 146	? on far side	> 1000

No Type II radio burst is recorded for this flare, but it was identified from the GSFC daily plot of CMEs for December 14, 2012 [11], which also shows that it was associated with a modest rise at the Earth in the count rate for particles in the energy range of 10 - 50 MeV. The COHO website hourly average data for the STEREO spacecraft [12] also shows that substantial particle counts from this event were observed at STEREO A, about 110 degrees ahead of the Earth, but not at STEREO B, about 120 degrees behind. This implies (consistent with evidence from the GOES x-ray data that an active region rotated out of view a day or two earlier) that this flare occurred beyond the west limb of the Sun as seen from the Earth.

Table 3e. Possible Flares for MIR F

Date, 2014	Time (UT)	Lat.	Earth Lon.	HGI Lon.	X-ray Imp.	CME (km/s)
May 8	0321	S09	W108	258	? on far side	> 847
May 9	0240	S11	W122	273	? on far side	> 1099
May 10	0432	S11	W136	288	? on far side	> 1086
June 10	1258	S17	E82	100	X1.1	1469

Table 3f. Possible Flares for MIR G (early)

Date, 2014	Time (UT)	Lat.	Earth Lon.	HGI Lon.	X-ray Imp.	CME (km/s)
Dec. 13	1427	?	W90	94	? at limb	2222
Dec. 17	0409	S11	E33	335	M1.1	869
Dec. 17	0500	S20	E09	359	M8.7	587
Dec. 18	2231	S11	E15	354	M6.9	1195

Table 3g. Possible Flare for MIR G (late)

Date, 2015	Time (UT)	Lat.	Earth Lon.	HGI Lon.	X-ray Imp.	CME (km/s)
February 21	0430	S20	W87	162	C1.5	> 1000

No Type II radio burst is recorded for this flare, but it was identified from the GSFC daily plot of CMEs for February 21, 2015 [11], which also shows that it was associated with a modest rise at the Earth in the count rate for particles in the energy range of 10 - 50 MeV.

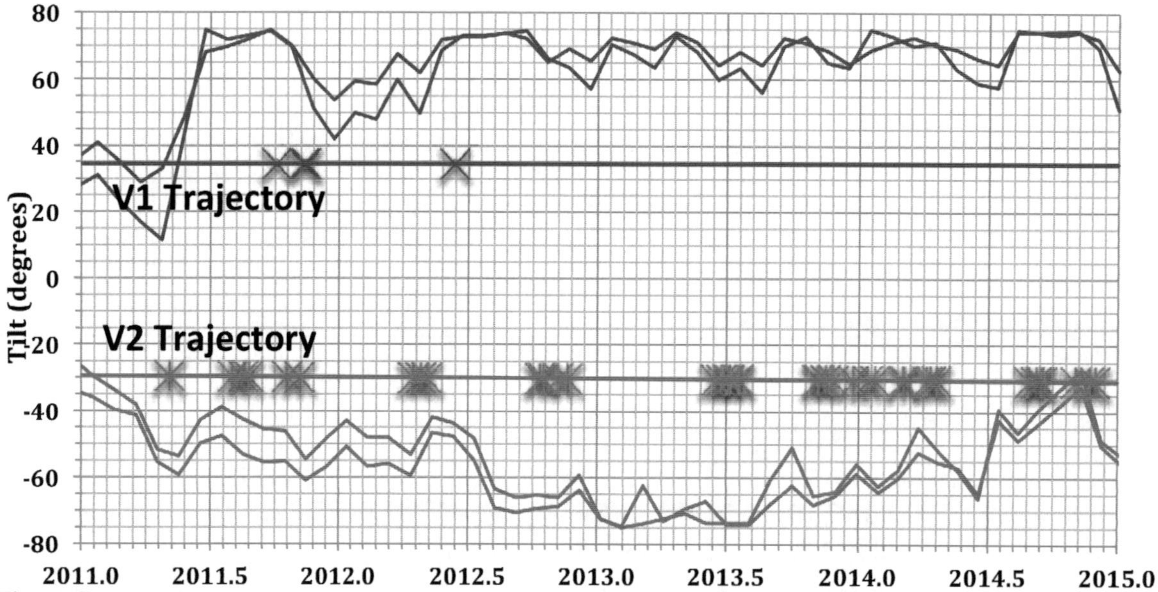

Figure 7. WSO limit estimates [1] for maximum HCS latitude, adjusted for delays to spacecraft. Stars are at times of sector boundaries observed at each spacecraft (V1 and V2).

6. Solar Cycle Variations in the Extent of the Sectored Heliosheath

Finally, we consider possible implications that these observations of temporally dependent latitudinal variations in the solar wind environment may have for the larger-scale structure of the heliosphere. Figure 7 provides a more comprehensive depiction of the indications in the plots in Figure 4 that during 2011 through 2014 (the period for which V2 magnetometer data are available to allow identifying sector boundaries) the heliospheric current sheet (HCS, the footprints of which are shown as the thick black lines in the Figure 4 plots) extended to high solar latitudes. This figure shows estimates of the northern and southern limits of the extent of the HCS made at the Wilcox Solar Observatory (WSO), with the times from the WSO dataset [13] adjusted for the delays to the spacecraft, using the formulas that made the predictions in Tables 1 and 2.

This approach to calculating the delays is similar to one used in Hill, et al [14], a paper written to propose that the regions of unipolar heliosheath in polar latitudes and their boundaries with the region of sectored heliosheath could explain the disparity between the nearly constant count rates of energetic particles in many energy bands observed at V1 and the much more variable count rates in the same energies observed at V2. Thus, it may be significant that in the period shown in Figure 7 (which is mostly after the period covered by the data in [14]) the regions of unipolar heliosheath shrank nearly to vanishing and the southern boundary of the sectored heliosheath was far from V2.

The authors of [14] note that this paper was written before V1 emerged into the LISM (it was first submitted to the *Astrophysical Journal* at the end of June, 2012). It also does not address the arguments in [6] (which records information presented at a conference in March, 2010) that the heliosheath is likely to be thicker along the trajectory of V2 than for V1. However, since it now appears that in [6] Roelof, Krimigis, et al. may have been at least partially correct (though the total thickness along the V2 trajectory is still unknown), a more realistic depiction of the heliosheath than what appears in the figures in [14] would look more like a geological diagram. It would have a north-south asymmetry and thicker and thinner regions of sectored plasma from respectively more and less active phases of the SC, flowing at speeds of 100 - 150 km/s. Furthermore, as the IBEX experimenters estimate that the tail of the heliosphere extends for hundreds of AU downstream, this would imply that the downstream portion of the heliosheath contains fluctuations dating from a number of years or even more than one SC in the past. (However, the only foreseeable way to test this hypothesis would be if evidence could be found in additional IBEX observations or in the more comprehensive data expected from the planned IMAP mission, for which near the end of [15] the IBEX experimenters include a recommendation for a timely launch.)

As the Sun is now in a period of declining activity and the latitudinal extent of the HCS has substantially decreased from what it was during the maximum of SC 24, this raises a question of whether the unipolar heliosheath region will expand enough to once again come close to or even to reach V2. In their discussion of the magnetic field the authors of [14] state that they do not expect this to happen because of the deflection of the solar wind flow observed by the spacecraft. However, we note that the transmission of the dynamic pressure of the interstellar plasma through the heliosheath to cause the deceleration at the TS and the gradually increasing deflection of the plasma flow that V2 has been observing since entering the heliosheath [16] is likely to be occurring by interactions of the solar wind flux tubes with the interstellar magnetic field and with each other, since the densities of these plasmas are far below those of laboratory plasmas that are collisionless in terms of the interactions of their particles. Thus, the tangential orientation of the field, and hence of the flux tubes, is likely to account for at least part of the observation in [16] that the deflection of the solar wind during V2's first several years in the heliosheath has been much more in the radial-tangential (RT) plane than in the radial-normal (RN) plane. As long as the deflection is more tangential than normal, this increases the chance that a more nearly equatorial HCS will bring the unipolar region close to the spacecraft again.

7. Concluding Remarks

In seeking to compare the discussion in this paper with the extensive explicit computations of Washimi et al. [17] and references therein about the heliospheric boundary and termination shock, we note that their computations start with interstellar medium conditions and extend to 65 AU with generic latitudinal backgrounds. The previous sections have described our fundamentally different approach of estimating how solar flare coronal mass ejections form merged interaction regions affecting the Voyager observations of the boundary and shock. Combining these approaches is beyond the scope of this work, but might be useful in the future. The contrast between the transport speed estimates in this paper and the results of the simulations in [17] is not a direct contradiction because those simulations were for 2001-2009, while the flares and spacecraft observations here are from late 2011 through 2016, during weak SC 24.

Observations by the Voyagers and various other spacecraft are gradually providing a comprehensive understanding of the effects of solar activity on phenomena in the front half of the heliosphere during the past decade or more. In particular, we have evidence that from the summer of 2011 to early 2015 solar flares produced disturbances that may have been convected to V2 and V1 in high-speed solar wind emitted from polar coronal holes, producing phenomena observed at the Voyagers from the middle of 2012 to late 2016. The time may now be right for more detailed efforts of modeling and analysis, as in [17], to see how MIRs from these events have migrated out to the TS, how they have affected the heliosheath and the LISM, and what is likely to happen as solar activity declines. Furthermore, our description of this whole region may also evolve as we gain more understanding of the basic physics involved.

8. References

[1] Intriligator D S, et al., 2016, *15th AIAC, "The Science of Ed Stone Celebrating his 80th Birthday:" Journal of Physics Conference Series,* **761**, 012013
[2] Richardson J D, et al., 2017, *Astrophys. J.*, **834**, 190
[3] Intriligator D S, et al., AGU Fall Presentation, 2015
[4] Usmanov AV, et al., 2014, *Astrophys. J.*, **788**, 43
[5] NSO Integrated Synoptic Program (NISP) Integral PFSS model synoptic coronal hole plot http://gong2.nso.edu/archive/patch.pl?menutype=z
[6] Roelof E C, et al., 2012, *AIP Conf. Proc.*, **1436**, 239. updated by Krimigis S M, et al., AGU Fall Presentation, **SH23A-04,** 2016
[7] Intriligator D S, et al., 2014, 13th AIAC, *Journal of Physics Conference Series,* **577**, 012013
[8] PWS Gurnett and Kurth, http://www-pw.physics.uiowa.edu/das/voyager-pws-sa-list
[9] Moebius E, et al., AGU Fall Presentation, **SH11C-4060,** 2014
[10] Wind/WAVES type II bursts and CMEs list https://cdaw.gsfc.nasa.gov/CME_list/radio/waves_type2.html
[11] GSFC daily CME list: https://cdaw.gsfc.nasa.gov/CME_list/daily_plots/sephtx/2015_02/sephtx_20150221.png
[12] COHO websites for STEREO data: https://omniweb.gsfc.nasa.gov/coho/form/stereoa.html and https://omniweb.gsfc.nasa.gov/coho/form/stereob.html
[13] Wilcox Solar Observatory (WSO) http://wso.stanford.edu courtesy of J T Hoeksema
[14] Hill M E, et al., 2014 *Astrophys. J.*, **781**, 94
[15] McComas D H, et al., 2014, *Astrophys. J. Sup.* **213**, 20
[16] Richardson J D, 2015, 14th AIAC, *Journal of Physics Conference Series,* **642**, 012022
[17] Washimi H, et al., 2011, *Mon. Not. R. Astron. Soc.*, **416**, 1475

Acknowledgments
We gratefully thank the Voyager projects for the vast information we have received. We acknowledge the support of Carmel Research Center, Inc. The V1 PWS data were gratefully obtained from Drs. Gurnett and Kurth, and the University of Iowa website.

Flare magnetic reconnection fluxes as possible signatures of flare contributions to gradual SEP events

S.W.Kahler[1], M.Kazachenko[2], B.J.Lynch[2], and B.T.Welsch[3]

[1] Air Force Research Laboratory, Kirtland AFB, NM 87117, USA
[2] Space Sciences Laboratory, University of California, Berkeley, CA 94720, USA
[3] University of Wisconsin, Green Bay, WI 54311, USA

E-mail: stephen.kahler@us.af.mil

Abstract. The primary sources of solar energetic (E > 20 MeV) particle (SEP) events are flares and CME-driven shocks. Some studies claim that even up to GeV energies solar flares are major contributors to SEP events. There are several candidate flare processes for producing SEPs, but acceleration in magnetic reconnection regions is probably the most efficient. Previous studies have relied on flare radiation signatures to determine the times and locations of SEP injections. An alternative approach is to use the amount of magnetic flux that gets reconnected during solar flares. The photospheric magnetic flux swept out by flare ribbons is thought to be directly related to the amount of magnetic reconnection in the corona and is therefore a key diagnostic tool for understanding the physical processes in flares and CMEs. We use the database of flare magnetic reconnection fluxes to compare these parameters with peak intensities of SEP events. We find that while sizes of 15 ~25-MeV SEP events in the western hemisphere correlate with both CME speeds and reconnection fluxes, there are many cases of large reconnection fluxes with no observed SEP events. The occurrence of large reconnection fluxes accompanied by slow CMEs but no SEP events suggests that the CME shocks are the primary, if not the only, sources of high energy (E > 100 MeV) SEP events.

1. Introduction

It has long been established that there are two classes of SEP events [1-4]. Impulsive SEP events have small peak intensities, durations of hours, and substantially enhanced abundances of high-Z elements. Their solar associations are narrow white-light jets from magnetic reconnection regions, fast-drift type III radio bursts from electron beams, small impulsive flares, and well connected solar source longitudes of W20° to W80°. These SEPs are generally measured in the 2 to 5 MeV/nuc range where the intensities are highest and least susceptible to energy spectral variations [2]. The gradual SEP events are defined by their relatively high peak intensities, durations of at least a day, and abundances representative of the ambient corona, although in some events residual ions from previous impulsive events are also accelerated by the shocks. Their solar event associations are CMEs of widths W > 60° and speeds of Vcme > 900 km/s [5], slow-drift type II radio bursts, assumed to be produced in CME-driven shocks, and relatively long duration soft X-ray and Hα flares. The solar source longitudes can extend to beyond the east and west limbs as a result of the broad extent of the CME-driven shocks. The gradual SEP

IOP Conf. Series: Journal of Physics: Conf. Series **900** (2017) 012011 doi:10.1088/1742-6596/900/1/012011

event energy spectra can extend to several hundreds of MeV/nuc and are typically fitted by a double power law spectrum, expected from the theory of shock acceleration [6].

The division of all SEP events into two separate groups has never been completely accepted as a definitive paradigm by some SEP investigators. No one argues for CME-driven shocks as the sources of impulsive SEP events, but a flare contribution has often been invoked as a significant factor in some gradual events, particularly for those reaching GeV energies, known as ground level events (GLEs) because they can be detected with ground-based neutron monitors. All gradual SEP events are associated with fast CMEs, but both CMEs and flares arise in approximate spatial and temporal coincidence in solar energetic eruptive events, so parsing a flare contribution to an SEP event from the solar and SEP observations, taken separately or together, has been contentious. If a flare SEP contribution is present in a gradual event, one expects two signatures of that component. First, the elemental abundance ratios, such as Fe/O, should reflect the enhanced heavy ion abundances characteristic of the impulsive SEP event abundances. Second, the times of solar particle release (SPR), associated with the flare impulsive phase, should precede those of the shock-accelerated components, which arise in the subsequent development phase of the CME shock and type II burst.

2. Arguments for and against flare contributions to gradual SEP events

2.1. Flare contributions to ground-level events (GLEs)

The GeV protons of the GLE events travel from the Sun to 1 AU with minimal interplanetary scattering, so their solar injection times are readily determined from the 1 AU intensity-time profiles. The onsets of SEP events have been studied by Reames for 16 cycle-23 GLEs [7] and then for 30 well observed GLEs dating back to 1973 [8]. The GLEs were analyzed with plots of 1-AU arrival times versus c/v for particles with a wide range of speeds v to get the SEP injection times. The initial SPR times followed the onsets of type II bursts in all cases, as expected if all particles came from the type II shock and not from the preceding flare phases. Those initial SPR times were confirmed in a separate study of [9]. In addition, it was found that the SPR coronal heights increased with longitudinal distance away from the source, again as expected if only expanding coronal shocks were the sources of GLE SEPs.

The question of flare contributions to GLEs was directly confronted by Aschwanden [10] who compared the SPR profiles of Reames [8] with inferred flare impulsive-phase hard X-ray profiles for 12 non-occulted GLE sources. He found that 6 of the 12 SPR profiles overlapped in time with flare impulsive phases, which he took to be consistent with prompt GLE components from flares, as previously inferred by Vashenyuk et al.[11] for 35 large GLEs. Those authors found 29 out of 35 GLEs to have prompt components (PCs) defined by impulsive profiles, strong anisotropies, and exponential energy spectra. PCs were followed by delayed components (DCs) characterized by gradual profiles, moderate anisotropies, and power-law energy spectra. Other authors have found PCs in relatively fewer GLEs. Moraal and McCracken [12] reported double-pulse profiles in only three of 16 GLEs of cycle 23, but McCracken et al.[13] found high-energy impulsive enhancements (HEIs) averaging 21 minutes in length in 10 of 19 large GLES that they studied. All ten HEIs were associated with well connected ($\geq$ W24°) flare sources. Other investigators looking at selected GLEs [14-18] have also found two or more components, consistent with flare contributions.

2.2. Flare contributions to Fe/O ratios

Impulsive SEP events are distinguished by their enhanced Fe/O ratios, usually measured in the 2 to 5 MeV/nuc energy range [19]. Their relatively low intensities and durations of less than a day [20] would seem to preclude any direct contributions of these events to large gradual SEP events at energies of $\sim$ 25 MeV/nuc. However, remnant low-energy ions from previous impulsive SEP events could form a suprathermal seed population that would be favored by

IOP Conf. Series: Journal of Physics: Conf. Series **900** (2017) 012011 doi:10.1088/1742-6596/900/1/012011

quasi-perpendicular shocks [21], with resulting enhanced Fe/O ratios at $\sim$ 25 MeV/nuc. The alternative interpretation favored by many investigators is that the impulsive SEP event occurs along with the CME eruption and directly contributes to the resulting mixed SEP event. Here again there are two possibilities for a direct flare contribution. First, the impulsive SEPs could interact with the CME-driven shock and be further accelerated to higher energies, particularly in quasi-perpendicular shocks. In the Particle Acceleration and Transport in the Heliosphere (PATH) model [22, 23] a direct flare SEP component extending to 1 GeV is assumed. The shock is transparent to the high-energy particles, but the lower-energy particles are further accelerated at the shock in addition to suprathermals of the solar wind. The preferred fraction of flare seed particles is $\sim$ 25% [21, 23]. The second possibility is that the flare and shock make independent contributions to the mixed SEP event. In that case the flare SEPs are assumed to be more energetic than shock SEPs and a good magnetic connection is needed to observe such events.

Tylka et al.[24] found that only about one third of the event-integrated Fe-enhanced gradual events of their study had sources in a well connected longitude range. On the contrary, Cane et al.[25] found that the majority of the Fe-rich events of their 80 gradual SEP events occurred in the western hemisphere, which they took as evidence for flare contributions. As in their earlier work [26], they also found enhanced Fe/O ratios to occur more often with prompt SEP intensity profiles, reflecting an early flare injected component. However, the early occurrence of enhanced Fe/O ratios has been explained as an expected rigidity-dependent transport effect from self-amplified waves at shocks [3, 27, 28]. Mason et al.[29, 30] showed that for most western hemisphere SEP events, the Fe/O and O/He ratios were approximately constant in time-intensity profiles when plotted at appropriately scaled different energies, reflecting similar injection profiles near the Sun but different subsequent rigidity-dependent propagation characteristics. For observers not well connected to SEP source regions Dalla et al.[31] showed that the larger gradient and curvature drifts of Fe compared to O result in initially enhanced Fe/O ratios, which then decay in time. This drift effect would add to the transport effect from waves at shocks.

Further arguments against flare components are based on the fluence spectra of 16 GLEs compiled by Mewaldt et al.[32]. Those energy spectra were well fit by a double power-law shape expected from diffusive shock theory [33], and in no case was there a flattening at the high ($\sim$ 1 GeV) range that might indicate the presence of a flare component [34]. The slightly enhanced enrichments and highly-ionized charge states of Fe in the GLE events were consistent with shock acceleration beginning at heights of $\sim$1.3 to 1.6 Rs and involving seed particles of remnant suprathermal ions of prior ^{3}He-rich flares.

3. Resolving the flare versus shock SEP debate with flare and CME signatures

It is clear from the preceding section that the GLE time profiles and Fe/O compositions of large SEP events have not settled the question of flare contributions to SEP events, and there remain two camps of investigators. A more simplified approach is to compare peak intensities of SEP events with associated flare X-ray peaks and CME speeds. An early comparison of logs of 2-MeV and 20-MeV peak proton intensities with associated CME speeds [35] yielded correlation coefficients (CCs) of 0.6-0.7 and were taken as evidence of shock acceleration of the SEPs. However, subsequent comparisons of peak SEP intensities with both CME speeds and flare X-ray peaks [36-39] have generally yielded comparable CCs of 0.4 to 0.7 for the CMEs and 0.3 to 0.6 for flares (Figure 1). While some authors [39] have inferred from these comparisons evidence of flare contributions to SEP events, the correlation between the CME speeds and associated flare X-ray peaks [40, 41] is acknowledged to be a confounding factor [39] in deducing whether or how much flares might contribute to SEP events.

Grechnev et al.[42] compared logs of E $>$ 100 MeV proton fluences, corrected for source longitude, with associated logs of 35-GHz burst fluences observed with the Nobeyama Radio Polarimeter. By excluding from a "main sequence" of events four "proton-abundant" outliers, which they considered as likely shock-source events, they derived a CC = 0.67 for 28 SEP events versus only 0.001 for the similar comparison with CME speeds, which they took as evidence of flare sources for those events. Cliver [43], however, pointed out that inclusion of the four proton-abundant events in the analysis reversed the results, now giving a CC = 0.09 for the 35-GHz burst fluences and 0.33 for the CME speeds. As a second line of argument, he compared the e/p ratios of 0.5-MeV electron peak intensities to those of E > 10 Mev protons for 22 well connected events of the [42] study with similar ratios of established impulsive-flare and shock-dominated SEP events to show that the [42] study events were consistent with the lower e/p ratios of the shock-dominated SEP events.

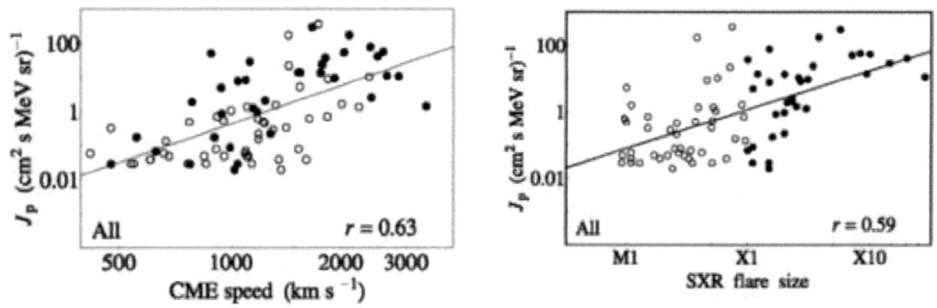

Figure 1. Comparison of peak GOES 15-40 MeV proton intensities of 81 SEP events versus CME speeds (left) and X-ray peak fluxes (right) showing the similarity of CCs for the two groups. Solid and open circles are M and X-class flare associations. From [36].

Fast CMEs seem to be a requirement for all gradual SEP events, but flares are not. Energetic (E > 50 MeV) SEP events have been observed following filament eruptions (FEs) associated with only C-class X-ray enhancements, apparently arising from late-forming CME shocks [44]. Do we find any events that might arise primarily or solely from flare injections only? On the assumption that the magnetic reconnection regions following solar eruptive events might be the flare sources of 25-Mev proton events, Kahler et al.[45] carried out a search for SEP events arising from well connected sources of soft X-ray arcades or metric noise storms. Of the 35 arcade/storm cases they investigated none was associated with a credible SEP event. More indirect evidence for at least a dominance of the shock contribution comes from a survey of major ($\geq$ M5) eruptive flares in cycle 24 [5]. The E $\geq$ 80-MeV proton events of that study were associated with a broad range of flare sizes, but the difference in average CME speeds Vcme of those proton events with and without large $\geq$ M5 flares was not large (2329 km/s versus 1724 km/s). Furthermore, large flares without SEP events had much lower average Vcme (1093 km/s), again consistent with a shock dominance for the sources of the SEP events.

4. Flare reconnecting magnetic fluxes as a better diagnostic for SEP event studies

Flare X-ray peak fluxes and fluences have been the tools of choice for comparative signatures in studying both the role of flares as sources of SEPs, as discussed above, and for forecasting SEP events (e.g., [46]). The X-ray observations (GOES full-Sun observations) are available in real time and readily quantified with a log-based C-M-X flux system, but they are a measure of thermal flare heating, and not of the energy release available for SEP acceleration. The active region (AR) magnetic flux is the energy source of solar eruptive events, so it provides a starting point for the quest of a better measure of flare energy. Nitta et al.[47] considered the AR unsigned magnetic fluxes associated with GLEs and other large SEP events, but found no AR property to be a discriminating factor for GLEs. However, proxies for the available active

IOP Conf. Series: Journal of Physics: Conf. Series **900** (2017) 012011 doi:10.1088/1742-6596/900/1/012011

region (AR) free energy can be deduced from algorithms using AR magnetic field maps, which are now available as SHARPs (Space-Weather active region patches) [48] based on magnetic images from the SDO/HMI instrument. Success in forecasting flares [49] and CMEs [50] based on AR magnetic flux algorithms suggests that we look not simply at the AR magnetic fluxes, but rather at the reconnecting fluxes in flares, the presumed source of any flare SEP production.

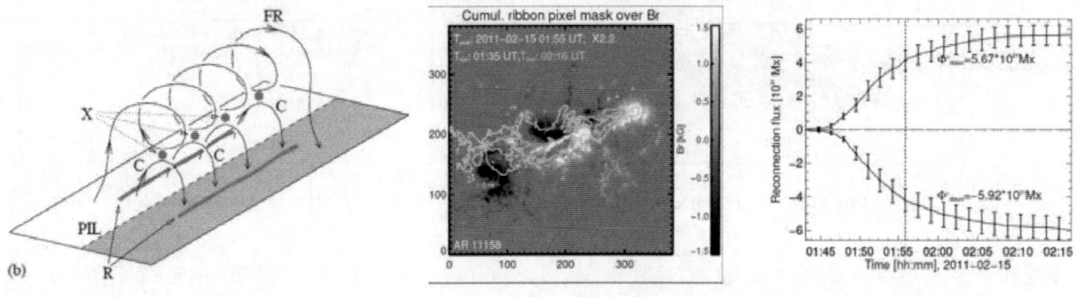

Figure 2. Basic concepts of the reconnecting flare magnetic fluxes. **Left**: Cartoon showing magnetic fields of erupting flux rope (FR) with reconnection X points and reconnection ribbons R lying along the polarity inversion line (PIL). From [54]. **Center**: Flare ribbon contours (green) overlay the SDO/HMI grey-scale radial component of the magnetic field of a 2011 flare example. **Right**: Plot of accumulated positive (top) and negative (bottom) reconnected flux for an example flare. The vertical dotted line marks the GOES X-ray peak time. Center and right figures are from [53].

Flare ribbons are emission structures that are frequently observed during flares in transition-region and chromospheric radiation. They typically straddle polarity inversion lines (PILs) of the radial magnetic field at the photosphere and move apart as the flare progresses. The ribbon flux, the amount of unsigned photospheric magnetic flux swept out by flare ribbons, is thought to be related to the amount of coronal magnetic reconnection. Previous measurements of the magnetic flux swept out by flare ribbons required time-consuming co-alignment between magnetograph and intensity data from different instruments, explaining why those studies only analyzed, at most, a few events. The launch of the Helioseismic and Magnetic Imager (HMI, [51]) and the Atmospheric Imaging Assembly (AIA, [52]), both aboard the Solar Dynamics Observatory (SDO), presented a new opportunity to compile a large sample of flare-ribbon events. Kazachenko et al.[53] have compiled a catalog of of flare-ribbon positions and reconnection fluxes, based on vector magnetograms from HMI and 1600 Å UV images from AIA (Figure 2). The catalog consists of 3137 > C1.0-class X-ray flares within 45° of disk center observed by the SDO from June 2010 to April 2016. We will use a subset of 406 unsigned magnetic reconnection fluxes from > C8.0-class flares to look for signatures of flare contributions to SEP events.

How well do the commonly used GOES flare X-ray peak fluxes and fluences compare with their associated AR unsigned reconnection fluxes? We compare the X-ray peak fluxes and fluences with the reconnection fluxes of the catalog on log-log plots in Figure 3. The linear correlation coefficients (CC) are 0.66 for the 406 flux peak events and 0.71 for 382 available fluence events, but there is a considerable scatter between the X-ray and magnetic flux values, indicating an uncertainly of at least 0.5 in the log values between the matched parameters. Kazachenko et al.[53] did a comparable fit for X-ray peak fluxes over their data base of all 3137 > C1 flares using the rank Spearman correlation r_s and found $r_s = 0.64 \pm 0.01$.

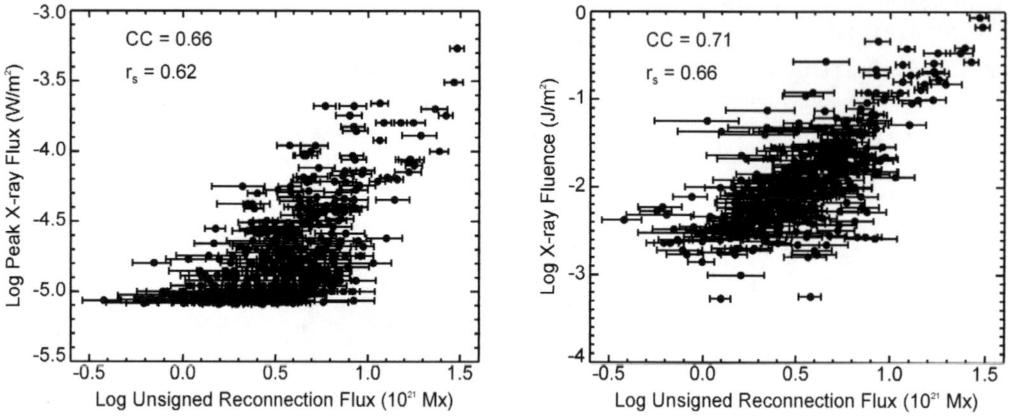

Figure 3. Log-log plots of the GOES flare peak X-ray fluxes (left) and fluences (right) versus the flare unsigned magnetic reconnection fluxes. Only flares of peak intensity > C8 (log > –5.1) are included in the catalog. The Pearson CC and Spearman r_s are indicated.

5. Data Analysis

The goal is to select flares at relatively well connected solar longitudes to search for associated SEP events that result mainly or solely from flare injections rather than from CME-driven shocks. Since the optimum well connected longitude range is ~ W30° to W80° and the catalog extends only to W44°, we selected flares occurring in the range W20° to W44°, a total of 128 flares. We also look for CMEs associated with these flares, so a further preference is to avoid flare locations near disk center where the CME identifications and projected speeds are most prone to error [55]. We use an updated list of E > 25-MeV gradual SEP events observed at 1 AU [56]. There were 15 SEP events associated with the 128 selected catalog flares.

Each flare is characterized by separate positive and negative reconnection fluxes, which ideally should have the same magnitude. Here we use the total unsigned reconnection fluxes. For each flare we looked for associated CMEs in the LASCO CDAW catalog [57]. In two cases there were catalog data gaps, and in 59 cases there were no suitable CME associations, as expected from previous comparisons of flares with CMEs [58, 59]. Two of the authors did independent CME associations and resolved differences in four cases. Figure 4 shows the plots of the 126 reconnection fluxes versus the CME speeds, where we arbitrarily assign speeds of 40 km/sec to the cases of no CMEs to include those cases in the comparison. The dashed lines delimit the upper left quadrant where we look for candidate cases of high reconnection fluxes, indicating maximum flare reconnection energy, and low CME speeds, suggesting unlikely SEP production from shock formation. We not only find no SEP events in this quadrant, but also that all the 15 SEP events (red crosses) lie in the regions of highest CME speeds and are broadly distributed over the dynamic range of the flare reconnection fluxes.

We have examined the daily low-frequency (20 kHz to 13.8 MHz) dynamic spectral plots of the Wind/WAVES radio instrument [60] for each of the 59 flares without associated CMEs. With only 2 or 3 possible exceptions, there were no associated type III radio bursts with those flares, consistent with previous results for large non-eruptive flares [58]. We note that Reames et al.[19] found associated type III bursts for 95 of their 111 impulsive SEP events, so in the absence of a CME an associated type III burst would be a reasonable requirement in our search for candidate flare injections of 25-MeV protons. This reduces the optimum flare-SEP candidates in the upper left quadrant of Figure 4 to only 8 or 10 flares. For 11 of the 15 CMEs with SEP events we

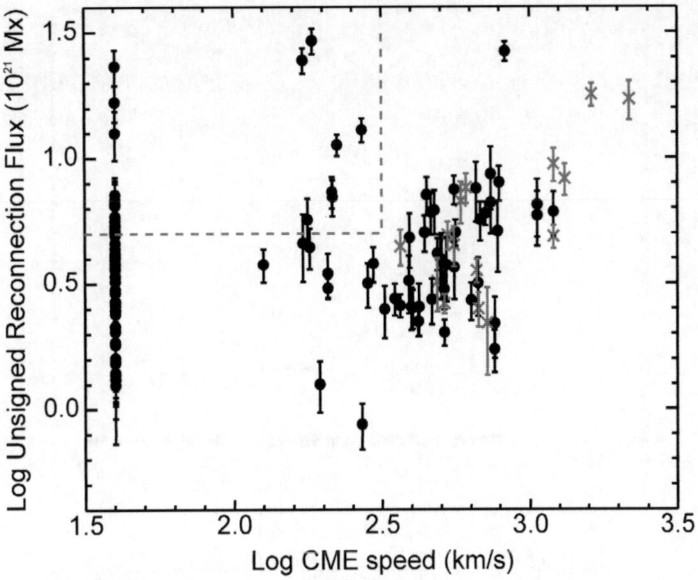

Figure 4. Log-log lots of the flare unsigned reconnection fluxes versus the CME speeds for the 126 flares. Black dots are flares with no associated SEP events, and red crosses are the 15 flares with SEP events. The flares with log CME speed = 1.6 are cases of no associated CMEs, for which Vcme was taken as 40 km/sec. The best candidates for flare SEP events are those in the upper left quadrant indicated by dashed lines. Plot units of the reconnection fluxes and CME speeds are 10^{21} Maxwells and km/sec, respectively.

found possible type II burst listings in the Wind/WAVES data. There were only metric type II bursts for three other CMEs, indicating likely shock acceleration for 14 of the 15 SEP events of the study.

The remaining SEP event, with no type II burst, on 30 September 2015, might be our best candidate for a flare source event. There were two candidate associated CMEs, the more likely of which had only a modest CME speed of 586 km/sec, but also an associated type IV radio burst, which often accompanies type II bursts. From SOHO/COSPIN and GOES observations, we found a peak 0.5 MeV/(> 10 MeV) e/p ratio of 200, right at the overlapping range of the high end of shock events and the low end of flare events [43]. The associated flare is a well connected (W42° S18°) M1.3 X-ray event with a typical reconnection flux of 6.73×10^{21} Mx (log = 0.83, Figure 3). However, the 0.5 MeV electron onset in the COSPIN observations is ~ 1600 UT and the 25 MeV proton onset at 1900 UT, both more than 5 hours after the 1049 UT flare X-ray peak, so our assumed solar association is tenuous. Thus, we do not find in this data set a good candidate flare-source SEP event.

Finally, we compare the SEP peak 25-MeV proton intensities of each SEP event with their corresponding flare reconnection fluxes in Figure 5. When no associated SEP event was observed, we use the approximate instrument background value of 1×10^{-4} p/(cm^2 s sr MeV). There is a weak trend for larger SEP peak intensities to scale with the reconnection fluxes, but in each plot range the cases of no SEP events are dominant.

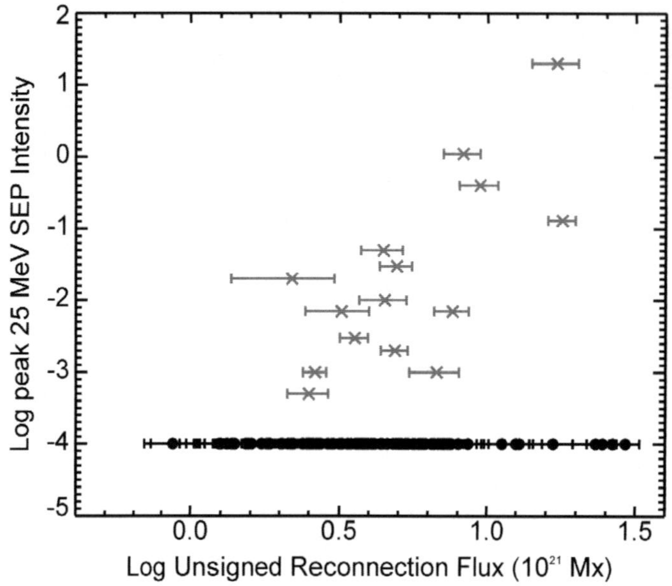

Figure 5. Log-log lots of the peak 25-MeV SEP intensities in protons/(MeV cm^2 sr sec) versus unsigned reconnection fluxes. Cases of no observed SEP events are shown with black dots at instrumental background values.

6. Discussion

The goal of this study has been to use a new quantitative measure of flare magnetic-reconnection energy to search for $\geq$ 25-MeV gradual SEP events that might have a sole or dominant flare contribution rather than only the shock contribution assumed in the two-class SEP event paradigm. The flare X-ray peak or fluence is the product of an AR emission measure distribution in temperature that reflects the thermal flare heating and may not be relevant to the nonthermal process(es) involved in SEP acceleration. We have seen in Figure 1 that the X-ray signatures do not differ substantially from CME speeds in their correlations with SEP event peak intensities.

We found good correlations between logs of the X-ray flare peaks and fluences and logs of the flare reconnection fluxes. However, if the conjectured flare SEP injections arise from magnetic flux reconnection, then the reconnection fluxes may be the best available signatures of that acceleration process. We have compared the flare reconnection fluxes with associated CME speeds to attempt to isolate SEP events with strong flare components (Figure 4), but except for one case the gradual SEP events are characterized by fast CMEs with associated type II bursts expected from the shock-source scenario for those events. A questionable flare/CME association characterizes the one remaining SEP event.

Our result constitutes weak evidence against the flare-SEP concept. The flare longitudes of the study (W20° – W45°) are optimized for magnetic field measurements and are not well suited for good magnetic connections to Earth or for CME detections and speed measurements. The event numbers (15 SEP events and 126 flares) are small, and nearly half of the flares have no observed associated CMEs or type III bursts. Figure 4 could be construed as indicating a flare reconnection contribution to SEP events, but most flares with similar reconnection fluxes are not associated with SEP events. An inherent limitation of this kind of study is that the reconnection fluxes correlate with flare peak X-ray fluxes (Figure 3), which correlate with both

CME speeds and SEP intensities (Figure 1).

The question of flare contributions to gradual SEP events may be better addressed by simply requiring the demonstration of one good example of an SEP event with no associated fast CME and shock. When have we ever seen such an event? All the arguments reviewed in Section 2 claim the existence of a flare contribution in addition to another, usually shock-based, contribution. The enhanced Fe/O ratios early in the SEP events can be explained by rigidity-dependent propagation, and the claimed GLE PCs have exactly the properties of initially harder energy spectra and higher anisotropies that one expects from the propagation effects of single shock-accelerated injections. Why should the alleged flare-based PC occur only in conjunction with the shock-based DC in GLEs?

Finally, we should demand more theoretical insights into the mechanism(s) and observational consequences of the presumed flare SEP acceleration process. Is there necessarily an elemental abundance enhancement, as is seen in impulsive SEP events? Does the acceleration process occur in coincidence with the hard X-ray impulsive phase of the flare? Where in the corona, in what kind of magnetic configuration or topology, and with what timescale does the flare SEP acceleration and injection to space occur? In view of the considerable recent progress in detailed modeling of type II emission [61, 62] and SEP injection from coronal shocks [63], the vague attributions of a SEP flare acceleration process remain at best primitive.

Acknowledgments

Acknowledgments We acknowledge use of the LASCO CME catalog. This CME catalog is generated and maintained at the CDAW Data Center by NASA and The Catholic University of America in cooperation with the Naval Research Laboratory. SOHO is a project of international cooperation between ESA and NASA. We acknowledge use of the Wind/WAVES dynamic plots and thank I. Richardson for sharing his list of > 25-MeV proton events and A. Ling for the figure plots. The work was supported at AFRL by AFOSR task 2301RDZ4.

References

[1] Reames D V 2013 *Space Sci. Rev.* **175** 53
[2] Reames D V 2016a *Solar Phys.* **291** 911
[3] Reames D V 2016b *Solar Phys.* **291** 2099
[4] Reames D V 2017 *Solar Energetic Particles: A Modern Primer on Understanding Sources, Acceleration and Propagation* (New York: Springer)
[5] Gopalswamy N, Xie H, Akiyama S, Makela P A, and Yashiro S 2014 *Earth Planets Space* **66** 104
[6] Desai M and Giacalone, J 2016 *Liv. Rev. Solar Phys.* **13** 3
[7] Reames D V 2009a *Astrophys. J.* **693** 812
[8] Reames D V 2009b *Astrophys. J.* **706** 844
[9] Gopalswamy N, Xie H, Yashiro S, Akiyama S, Makela P, and Usoskin I G 2012 *Space Sci. Rev.* **171** 23
[10] Aschwanden M J 2012 *Space Sci. Rev.* **171** 3
[11] Vashenyuk E V, Balabin Yu V, and Gvozdevsky B B 2011 *Astrophys. Space Sci. Trans.* **7** 459
[12] Moraal H, and McCracken K G 2012 *Space Sci. Rev.* **171** 85
[13] McCracken K G, Moraal H, and Shea M A 2012 *Astrophys. J.* **761** 101
[14] Perez-Peraza J, Vashenyuk E V, Miroshnichenko L I, Balabin Yu V, and Gallegos-Cruz A 2009 *Astrophys. J.* **695** 865
[15] Masson S, Klein K-L, Butikofer R, Fluckiger E, et al. 2009 *Solar Phys.* **257** 305
[16] Kocharov L, Klassen A, Valtonen E, Usoskin I, and Ryan J M 2015 *Astrophys. J. Lett.* **811** L9
[17] Klein K-L, Agueda N, and Butikofer R 2015 *J. Phys.: 34th Int.Cosmic Ray Conf.* 121
[18] Ding L-G, Jiang Y, and Li G 2016 *Astrophys. J.* **818** 169
[19] Reames D V, Cliver E W, and Kahler S W 2014 *Solar Phys.* **289** 3817
[20] Reames, D V 2014 *Solar Phys.* **289** 977
[21] Tylka A J, and Lee M A 2006 *Astrophys. J.* **646** 1319
[22] Li G, and Zank G P 2005 *Geophys. Res. Lett.* **32** L02101
[23] Verkhoglyadova O P, Li G, Zank G P, Hu Q, et al. 2010 *J. Geophys. Res.* **115** A12103
[24] Tylka A J et al. 2005 *Astrophys. J.* **625** 474

[25] Cane H V, Mewaldt R A, Cohen C M S, and von Rosenvinge T T 2006 *J. Geophys. Res.* **111** A06S90
[26] Cane H V, von Rosenvinge T T, Cohen C M S, and Mewaldt R A 2003 *Geophys. Res. Let.* **30** 8017
[27] Ng C K, Reames D V, and Tylka A J 2012 *AIP Conf. Proc.* **1436** 212
[28] Tylka A J, Malandraki O E, Dorrian G, Ko Y-K, Marsden R G, Ng C K, and Tranquille C 2013 *Solar Phys.* **285** 251
[29] Mason G M, Desai M I, Cohen C M S, Mewaldt R A, Stone E C, and Dwyer R 2006 *Astrophys. J.* **647** L65
[30] Mason G M, Li G, Cohen C M S, Desai M I, Haggerty D K, et al. 2012 *Astrophys. J.* **761** 104
[31] Dalla S, Marsh M S, Zelina P, and Laitinen T 2017 *Astron. Astrophys.* **598** A73
[32] Mewaldt R A, Looper M D, Cohen C M S, Haggerty D K, et al. 2012 *Space Sci. Rev.* **171** 97
[33] Desai, M I, Mason G M, Dayeh M A, Ebert R W, et al. 2016 *Astrophys. J.* **828** 106
[34] Reames D V 2015 *Space Sci. Rev.* **194** 303
[35] Kahler S W 2001 *J. Geophys. Res.* **106** 20947
[36] Miteva R, Klein K-L, Malandraki O, and Dorrian G 2013 *Solar Phys.* **282** 579
[37] Park J, and Moon Y-J 2014 *J. Geophys. Res.: Space Physics* **119** 9456
[38] Dierckxsens M, Tziotziou K, Dalla S, Patsou I, et al. 2015 *Solar Phys.* **290** 841
[39] Trottet G, Samwel S, Klein K-L, Dudok de Wit T, and Miteva R 2015 *Solar Phys.* **290** 819
[40] Yashiro S, and Gopalswamy N 2009 *Proc. IAU Symp. 257, Universal Heliophysical Processes*, eds. N Gopalswamy and D F Webb, 233
[41] Salas-Matamoros C, and Klein K.L. 2015 *Solar Phys.* **290** 1337
[42] Grechnev V V, Kiselev, V I, Meshalkina, N S, and Chertok, I M 2015, *Solar Phys.* **290** 2827
[43] Cliver E W 2016, *Astrophys. J.* **832** 128
[44] Gopalswamy N, Makela P, Akiyama S, Yashiro S, Xie H, Thakur N, and Kahler S W 2015 *Astrophys. J.* **806** 8
[45] Kahler S W, McAllister A H, and Cane H V 2000 *Astrophys. J.* **533** 1063
[46] Balch C C 2008 *Space Weather* **6** S01001
[47] Nitta N V, Liu Y, DeRosa M L, and Nightingale R W 2012 *Space Sci. Rev.* **171** 61
[48] Bobra M G, et al. 2014 *Solar Phys.* **289** 3549
[49] Leka K D, and Barnes G 2007 *Astrophys. J.* **656** 1173
[50] Falconer D A, Moore R L, Barghouty A F, and Khazanov I 2012 *Astrophys. J.* **757** 32
[51] Scherrer P H, et al. 2012 *Solar Phys.* **275** 207
[52] Lemen J R, et al. 2012 *Solar Phys.* **275** 17
[53] Kazachenko M D, Lynch B J, Welsch B T, and Sun, X 2017 http://cdsads.u-strasbg.fr/abs/2017arXiv170405097K
[54] Longcope D, Beveridge C, Qiu J, et al. 2007, *Solar Phys.*, **244**, 45
[55] Yashiro S, Gopalswamy, N, Akiyama, S, Michalek, G, and Howard R A 2005 *J. Geophys. Res.* **110** A12S05
[56] Richardson I G, et al. 2014 *Solar Phys.* **289** 3059
[57] Gopalswamy N, Yashiro S, Michalik G, Vourlidas A, Freeland S, and Howard R 2009b *Earth Moon Planet* **104** 295
[58] Gopalswamy N, Akiyama, S, and Yashiro, S 2009a *Proc. IAU Symp. 257, Universal Heliophysical Processes*, eds. N Gopalswamy and D F Webb, 283
[59] Sun X, Bobra M G, Hoeksema J T, Liu, Y et al. 2015 *Astrophys. J. Lett.* **804** L28
[60] Bougeret J.-L., Kaiser M L, Kellogg P J, et al. 1995 *Space Sci Rev* **71** 5
[61] Kozarev K A, and Schwadron N A 2016 *Astrophys. J.* **831** 120
[62] Schmidt J M, and Cairns I H 2016 *Geophys. Res. Lett.* **43** 50
[63] Bain H M, Mays M L, Luhmann J G, Li Y, Jian L K, and Odstrcil D 2016 *Astrophys. J.* **825** 1

IOP Conf. Series: Journal of Physics: Conf. Series **900** (2017) 012012 doi:10.1088/1742-6596/900/1/012012

Large Energetic Particle Pressures in Solar Cycles 23 and 24

D Lario[1], R B Decker[1], E C Roelof[1], A F Viñas[2], R F Wimmer-Schweingruber[3] and L Berger[3]

[1]The Johns Hopkins University. Applied Physics Laboratory
[2]Goddard Space Flight Center. National Aeronautics and Space Administration
[3]Institut für Experimentelle und Angewandte Physik (IEAP), Christian-Albrechts-Universität zu Kiel

Corresponding author's e-mail: david.lario@jhuapl.edu

Abstract. We study periods of elevated energetic particle intensities observed at the L1 Sun-Earth Lagrangian point when the partial energy density associated with energetic ($\geq$80 keV) particles (P_{EP}) dominates that of the local magnetic field (P_B) and thermal plasma populations (P_{PLS}). These periods are not uncommon and are frequently observed prior to the passage of interplanetary (IP) shocks. Because of the significant decreases in key solar wind parameters observed during solar cycle 24 [e.g., 1], we were motivated to perform a comparative statistical analysis to determine if the occurrence rate of periods when P_{EP} exceeded P_B or P_{PLS}, or both, differed between solar cycles 23 and 24. We find that the general decrease of P_B and P_{PLS} in solar cycle 24 was also accompanied by a general decrease of periods with elevated P_{EP}. The result is that solar cycle 24 showed a lower number of time intervals dominated by P_{EP}. We analyze whether these differences can be related to the properties of the IP shocks observed at L1. Incomplete datasets of shock parameters do not show significant differences between solar cycles 23 and 24 that would allow us to explain the difference in the number of periods with $P_{EP}>P_B$ and $P_{EP}>P_{PLS}$. We analyze then the averaged plasma parameters measured in the upstream region of the shocks and find significantly lower solar wind proton temperatures and magnetic field magnitude upstream of IP shocks in solar cycle 24 compared with those in solar cycle 23. These factors, together with the lower level of solar activity, may explain the lower particle intensities in solar cycle 24 and hence the fewer events with $P_{EP}>P_B$ and $P_{EP}>P_{PLS}$.

1. Introduction

Magnetohydrodynamic studies of solar wind structures generally assume that the pressure exerted by energetic particles is negligible compared to both (1) the hydrostatic magnetic field pressure computed as $P_B = B^2/2\mu_0$ (where B is the magnetic field intensity and μ_0 is the magnetic permeability), and (2) the pressure exerted by the solar wind thermal particles P_{PLS}, usually computed as $P_{PLS} = n_pkT_p + n_ekT_e$ (where k is the Boltzmann constant, n_p the solar wind proton density, n_e is the solar wind electron density, T_p the solar wind proton temperature, T_e the solar wind electron temperature, and where the contribution of alpha particles and heavier ions is assumed negligible). However, there have been a number of reports showing distinct periods and/or regions of interplanetary (IP) space when the partial pressure exerted by energetic particles P_{EP} is locally comparable to or even exceeds both P_B and P_{PLS} [e.g., 2, 3, 4, 5, 6, 7, 8]. In particular, Lario et al. [7] analyzed periods observed by STEREO-A at ~1 AU during the rising and maximum phases of solar cycle 24 when the pressure exerted by ≥83 keV protons was larger than P_B. These periods occur predominantly in association with the passage of IP shocks when depressed magnetic field in the upstream region of the shocks coincides with increases of energetic particle intensities. Occasionally, periods when P_{EP} exceeds P_B may also occur when solar energetic particles (SEPs) are injected into regions of depressed magnetic field, although the occurrence of these periods is much more infrequent than those associated with IP shocks.

The observation of periods with $P_{EP}>P_B$ and $P_{EP}>P_{PLS}$ depends on both the existence of intense sources of energetic particles (such as strong IP shocks) and the properties of the background medium where these particles propagate. Compared to previous solar cycles, the first part of solar cycle 24 showed a significant drop in the solar wind output with significant decreases in solar wind speed V_{sw}, proton density n_p, proton temperature T_p, thermal pressure P_{PLS}, and magnetic field magnitude B [1, 9]. Under the assumption that energetic particle intensities in both solar cycles are similar, the decline observed in the solar wind parameters may translate into more frequent observations of periods with $P_{EP}>P_B$ and $P_{EP}>P_{PLS}$. On the other hand, the medium where IP shocks propagate may also have important implications for the strength of the IP shocks arriving at 1 AU, and hence their capability to accelerate particles. In this article we analyze how the frequency of periods dominated by the pressure exerted by energetic particles has varied over the last two solar cycles and try to find possible explanations in terms of the shock strength and the IP medium where these shocks propagate.

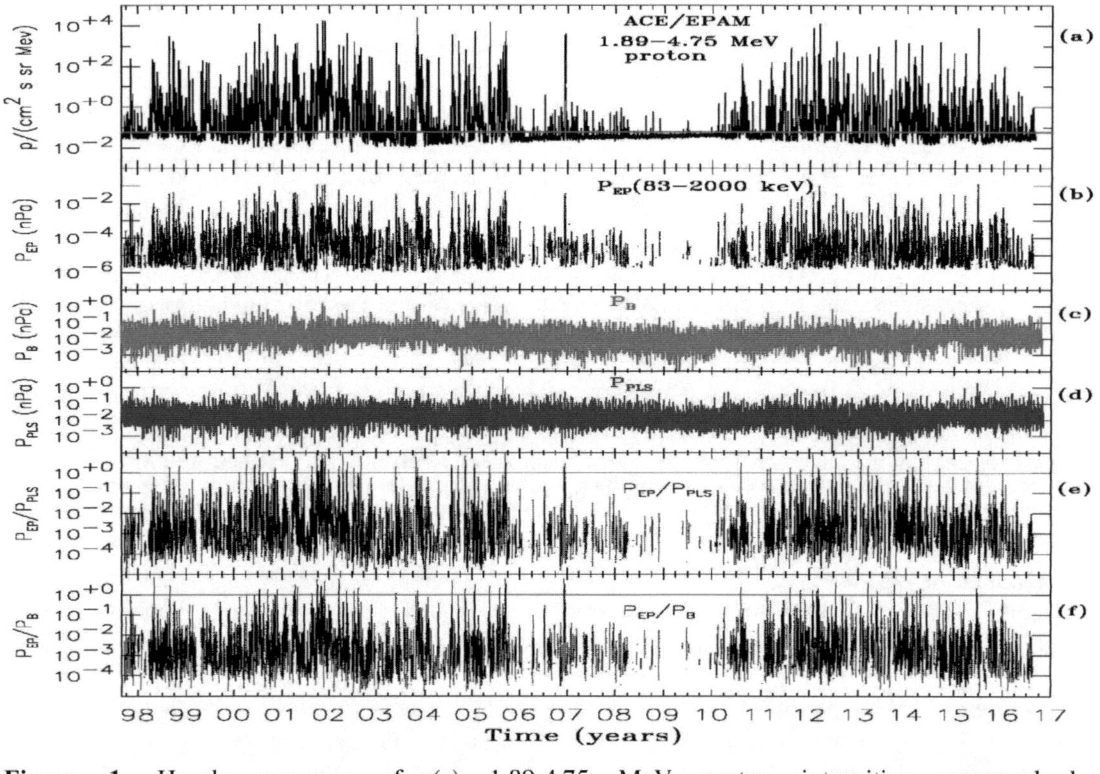

Figure 1. Hourly averages of (a) 1.89-4.75 MeV proton intensities measured by ACE/EPAM/LEMS120, (b) the pressure P_{EP} exerted by protons in the energy range 83-2000 keV, (c) magnetic field pressure P_B, (d) solar wind thermal pressure P_{PLS}, (e) the ratio between P_{EP} and P_{PLS}, and (f) the ratio between P_{EP} and P_B. The horizontal straight line in (a) indicates the differential intensity above which P_{EP} has been computed. The horizontal straight lines in (e) and (f) indicate the values $P_{EP}/P_{PLS}=1$ and $P_{EP}/P_B=1$, respectively.

2. Frequency of events with P_{EP}>P_B and P_{EP}>P_{PLS}

Figure 1 shows from top to bottom and over the time interval from day 242 of 1997 to day 243 of 2016 hourly averages of (a) 1.89–4.75 MeV ion differential intensities measured by the LEMS120 telescope of the EPAM instrument [10] on board the Advanced Composition Explorer (ACE); (b) P_{EP} computed over the energy interval 83–2000 keV; (c) $P_B = B^2/2\mu_0$; (d) $P_{PLS} = n_p k\ (T_p + T_e)$ (computed assuming $T_e = 2\ T_p$, [e.g., 11]); (e) the ratio P_{EP}/P_{PLS}; and (f) the ratio P_{EP}/P_B. Solar wind parameters B, n_p, T_p in Figure 1 come from OMNI data sets at spdf.gsfc.nasa.gov/pub/data/omni/low_res_omni/. Energetic particle pressure P_{EP} in Figure 1 was computed as

$$P_{EP} = \left(\frac{4\pi}{3}\right)\left(2m_p\right)^{1/2}\int_{E_1}^{E_2} E^{1/2}\, j(E)dE$$

where E is the proton kinetic energy, j(E) is the proton differential flux, m_p is the proton mass, and E_1 and E_2 are the limits of the energy range over which P_{EP} is computed (in this case E_1=83 keV and E_2=2000 keV). Since ACE/EPAM/LEMS120 does not distinguish ion species, we have used hourly averages of the differential particle intensities measured by ACE/EPAM/LEMS120 to determine j(E), assuming that heavy ions make a negligible contribution into the measured particle intensities. A functional form described as a power law with an exponential cutoff over the energy range 83-2000 keV has been fitted to j(E) measured in the energy channels of ACE/EPAM/LEMS120. When doing this fit we have considered only those points when the differential intensity of the energy channel 1.89-4.75 MeV was above 0.06 particles (cm^2 s sr MeV)$^{-1}$, as indicated by the horizontal line in Figure 1a. This intensity threshold allowed us to consider only cases where intensities were above the instrumental background and the energy spectra over the differential energy channels of ACE/EPAM/LEMS120 were well fitted by the chosen functional form. The horizontal solid lines in Figures 1e and 1f indicate the values $P_{EP}/P_{PLS} = 1$ and $P_{EP}/P_B = 1$, respectively. Periods with P_{EP}>P_B and P_{EP}>P_{PLS} were frequently observed, and most of them occurred in association with the arrival of shocks accompanied by elevated particle intensities.

In order to estimate the frequency distribution of time intervals with elevated particle intensities, we have computed the number of hourly data points falling within different ranges of particle intensities. This technique (explained in detail elsewhere [e.g., 12]) consists of dividing the particle intensity data points displayed in Figure 1a in equally logarithmic spaced bins and computing the number of data points in each bin. Figure 2a shows the number of hourly averaged data points from the time series plotted in Figure 1a that fall in each of the equally logarithmic-spaced intensity bins for the time interval between day 242 of 1997 to day 4 of 2008 (Fig. 2a) and from day 4 of 2008 to day 243 of 2006 (Fig. 2b). The first time interval covers most of solar cycle 23 whereas the second covers most of solar cycle 24. We have arbitrarily considered that the division between the two solar cycles occurs on day 4 of 2008 (day when the first sunspot with inverted magnetic field polarity after the prolonged solar minimum was observed). Since the time coverage (total number of points) is not the same in each time interval, we have normalized the number of hours spent in each specific bin to the total number of hours of each time interval. The dashed red lines in Figure 2a delimit the particle intensity frequency distribution in solar cycle 23 and are only plotted to guide the eye. The same dashed red lines are reproduced in Figure 2b to show that the periods with elevated particle intensities (i.e., above ~1 (cm^2 sr s MeV)$^{-1}$) were less frequent in solar cycle 24. Similarly, we have applied the same technique for the P_{EP} data points in Figure 1b. The frequency distributions of P_{EP} for solar cycles 23 and 24 are shown in Figures 2c and 2d, respectively. Correspondingly, solar cycle 24 showed a clear deficit of periods with elevated P_{EP} at the higher pressures (i.e. P_{EP}>10^{-2} nPa).

We have repeated the same technique with the values of P_B and P_{PLS} displayed in Figures 1c and 1d. Figure 3 shows the resulting frequency distributions. The values m and σ in each panel indicate the mean and standard deviation of the $\log_{10}(P_B)$ and $\log_{10}(P_{PLS})$ distributions shown in Figures 1c-d, whereas the dashed red lines are only plotted to guide the eye. As found in prior studies [e.g., 1, 9] solar cycle 24 exhibited lower average values of P_B and P_{PLS} as displayed by the leftward shifted distributions in the bottom panels of Figure 3. In other words, mean elevated values of P_B and P_{PLS}

were more frequent in solar cycle 23 than in solar cycle 24.

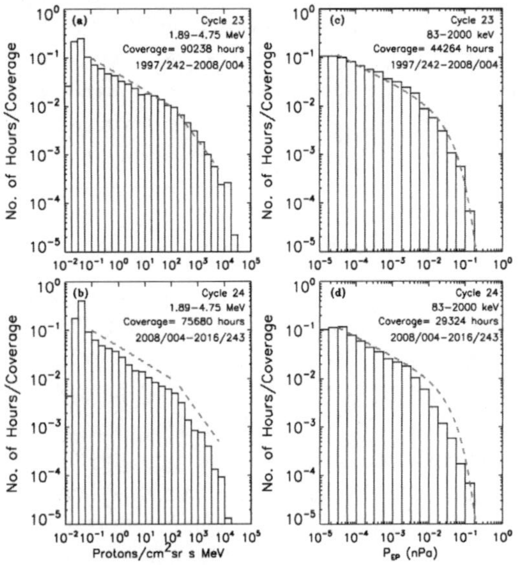

Figure 2. (Left) Distribution of hourly averaged 1.89-4.75 MeV proton intensity data points in intensity bins for solar cycles (a) 23 and (b) 24. (Right) Distribution of hourly values of P_{EP} in pressure bins for solar cycles (c) 23 and (d) 24. The dashed lines indicate the solar cycle 23 distributions and are reproduced in the bottom panels.

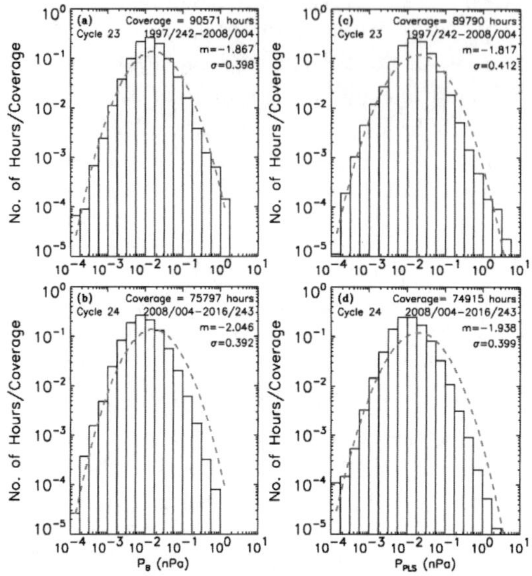

Figure 3. Distribution of hourly values of (left) P_B and (right) P_{PLS} for solar cycles (a and c) 23 and (b and d) 24. The dashed lines indicate the shape of the solar cycle 23 distributions are just plotted to guide the eye (i.e., are not Gaussian fits to the distributions) and are reproduced in the bottom panels to show the differences between distributions of both solar cycles.

Finally, Figure 4 shows the frequency distributions of the ratios P_{EP}/P_B (left) and P_{EP}/P_{PLS} (right) computed from the data points shown in Figures 1e and 1f for solar cycles 23 (top) and 24 (bottom). The vertical dashed lines indicate the values $P_{EP}/P_B=1$ and $P_{EP}/P_{PLS}=1$, whereas the filled bars of the histograms indicate the occurrence of periods with $P_{EP}>P_B$ and $P_{EP}>P_{PLS}$. Clearly, periods with $P_{EP}>P_B$ and $P_{EP}>P_{PLS}$ were more frequent in solar cycle 23 than in solar cycle 24. Over the whole time interval considered in Figure 4, the probability of having $P_{EP}>P_B$ was 0.284% in solar cycle 23 in contrast to 0.155% in solar cycle 24, whereas for $P_{EP}>P_{PLS}$ the percentage was 0.845% in solar cycle 23 and 0.327% in solar cycle 24. Of course, these percentages may seem negligible when considering the whole duration of a solar cycle, but when considering only those time intervals associated with strong shocks the percentage occurrence of periods with $P_{EP}>P_B$ and $P_{EP}>P_{PLS}$ is not negligible.

3. Shock parameter statistics

Since most of the periods with $P_{EP}>P_B$ and $P_{EP}>P_{PLS}$ are usually observed in the upstream region of IP shocks [7, 8] we proceed here to analyze the properties of the shocks observed at 1 AU. There exist several catalogs of shocks observed at 1 AU (e.g., ipshocks.fi; www.cfa.harvard.edu/shocks; www.ssg.sr.unh.edu/mag/ace/ACElists/obs_list.html) that use diverse methods to first select the shocks and then compute the shock parameters. Selection criteria and the lack of plasma measurements for the most intense events (i.e., IP shocks accompanied by high energetic particle intensities that usually saturate plasma instruments) are limiting factors that prevent us from having a complete and exhaustive list of shocks throughout the solar cycles.

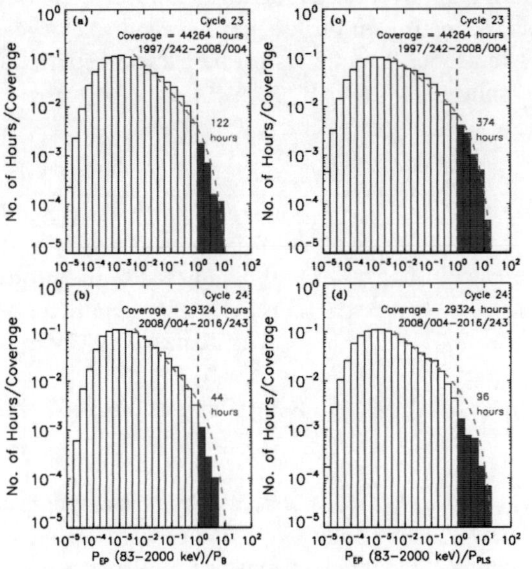

Figure 4. Distribution of hourly values of (left) P_{EP}/P_B and (right) P_{EP}/P_{PLS} for solar cycles (a and c) 23 and (b and d) 24. The vertical dashed lines indicate the value (left) P_{EP}/P_B =1 and (right) P_{EP}/P_{PLS} =1. The blue bars indicate the number of hours when (left) P_{EP}/P_B >1 and (right) P_{EP}/P_{PLS} >1.

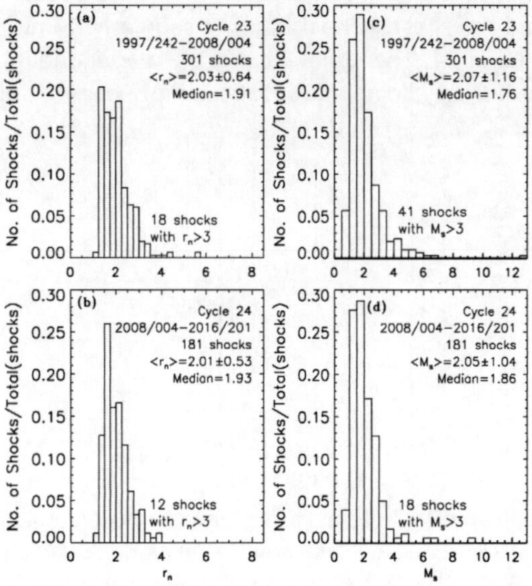

Figure 5. Frequency distribution of (left) compression ratios r_n and (right) magnetosonic Mach numbers M_s for the fast forward shocks cataloged in ipshocks.fi observed by the Wind spacecraft during solar cycles (a and c) 23 and (b and d) 24.

Another limiting factor when using the shock catalogs is the accuracy of the shock parameters for the most intense events since none of the methods used to compute the shock parameters includes the pressure exerted by the energetic particles when solving the Rankine-Hugoniot conservation equations. The contribution of P_{EP} in the momentum and energy conservation equations may not be negligible when P_{EP} is of the same order or even exceeds P_B and P_{PLS}. Therefore, the statistical study presented in this section is just provided for completeness of the possible differences between the two solar cycles, but it does not allow us to draw firm conclusions of the differences in the shock parameters for those events where $P_{EP} > P_B$ and $P_{EP} > P_{PLS}$.

Here we use the list of fast forward shocks observed by the Wind spacecraft cataloged in the heliospheric shock database maintained at the University of Helsinki by Kilpua et al. at ipshocks.fi [13]. At the time of writing this report, the Wind data set was the most complete of all the different data sets considered in the catalog, covering from 15 Nov 1994 up to 19 July 2016 (day of year 201). Although not coming from the same spacecraft as the energetic particle data, the Wind data set do enable us to compare on a statistical basis the shock properties between the two solar cycles (comparison of the shock parameters with the particle intensities on an event by event basis is beyond the scope of this article).

Figures 5a and 5b show the density compression ratio r_n of the fast forward shocks observed by the Wind spacecraft as listed in ipshocks.fi. From day 242 of 1997 to day 4 of 2008 (Fig. 5a) a total of 301 fast forward shocks were analyzed in ipshocks.fi, whereas form day 4 of 2008 to day 201 of 2016 (Fig 5b) only 181 fast forward shocks were analyzed in ipshocks.fi. The distribution of shock compression ratios r_n during both solar cycles had similar average and median values. Similarly, Figures 5c and 5d show the distribution of magnetosonic Mach numbers M_s as reported in ipshocks.fi for solar cycles 23 and 24, respectively. The distributions of M_s for both solar cycles were quite similar, with the only small difference of a more extended tail in solar cycle 23. Distributions of other shock parameters such as magnetic compression ratio r_b, shock speed V_s, and the angle between the upstream magnetic field and the shock normal θ_{Bn} (not shown here) did not show significant differences between solar cycles 23 and 24. Therefore, on average, the distributions of the shock parameters for the shocks reported in ipshocks.fi do not differ substantially between the two solar cycles.

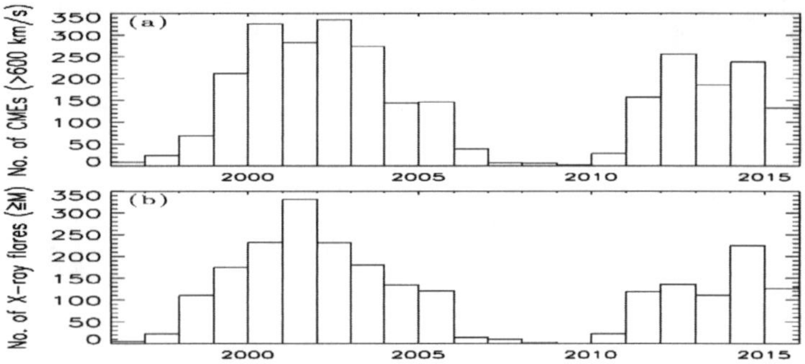

Figure 6. (a) Yearly number of fast (>600 km s^{-1}) CMEs with widths >20° and positive mass as reported in cdaw.gsfc.nasa.gov/CME_list (b) Yearly number of GOES X-ray flares of class M and X.

The distributions of the properties of coronal mass ejections (CMEs) observed by coronagraphs in the two solar cycles shows considerable differences between one solar cycle and the next [e.g., 14, 15]. CMEs in solar cycle 24 were on average significantly slower, less massive and with a lower kinetic energy than in solar cycle 23 [15]. Figure 6a shows the annual number of CMEs from 1996 to 2015 observed by the Large Angle and Spectrometric Coronagraph (LASCO) on board the Solar and Heliospheric Observatory (SOHO) [16] with plane-of-sky speeds above 600 km s^{-1}, angular widths larger than 20°, and positive computed masses as reported in the CME catalog of the Coordinated Data Analysis Workshop (CDAW) posted on cdaw.gsfc.nasa.gov/CME_list/ [17]. Figure 6b shows the annual number of X-ray flares observed by the Geostationary Operational Environmental Satellites (GOES) of class X or M as reported in the Daily Solar Data (DSD) files available at ftp.swpc.noaa.gov/pub/warehouse/. Statistical correlation studies between flares and CMEs over the two solar cycles can be found in [18]. Apart from the dependence with the solar activity (faster CMEs are more frequent during solar maximum), the number of fast CMEs in solar cycle 24 was lower than in solar cycle 23. On the other hand, CMEs in solar cycle 24 showed larger angular widths than in solar cycle 23 [14]. The drop in solar wind density, magnetic field, Alfven speed, and of the sum of magnetic plus thermal pressures have been invoked as the main factors responsible for the larger expansion of the CMEs in solar cycle 24 [14]. The medium where CMEs propagate has also important effects in the strength of the shocks driven by these CMEs. In particular, since the magnetosonic speed of the solar wind showed also a decline in solar cycle 24 (cf. Figure 2 in [13]), the slower CMEs in solar cycle 24 were able to drive shocks with strengths comparable to those in cycle 23. In addition, the faster CMEs of solar cycle 23 propagated in a denser medium, resulting in shocks that weakened faster with helioradius than those in solar cycle 24. The likely result is that the shocks observed at L1 in both solar cycles displayed, on average, similar compression ratios and magnetosonic Mach numbers (cf. Fig. 5 and [13]).

The computation of the shock parameters requires determining the upstream region of the shocks. The catalog ipshocks.fi provides also the plasma parameters of the upstream region used to compute the shock parameters. Figure 7 shows the upstream plasma parameters that showed the largest differences between both solar cycles. In particular, Figures 7a and 7b show the distribution of the upstream magnetic field B_{ups} and Figures 7c and 7d the distribution of the logarithm of the upstream solar wind proton temperature $\log_{10}(T_{ups})$ for the fast forward shocks in solar cycle 23 (top panels) and solar cycle 24 (bottom panels) reported in ipshocks.fi. There is a larger tail of events with large values of B_{ups} in solar cycle 23 yielding a larger average value of B_{ups} in solar cycle 23 than in solar cycle 24. Similarly, Figures 7c and 7d show a slight trend for higher T_{ups} values in solar cycle 23. The decrease of B_{ups} and T_{ups} in solar cycle 24 is a pure consequence of the general drop observed in magnetic field and solar wind parameters that constitute the medium where these shocks propagated [e.g., 1, 9].

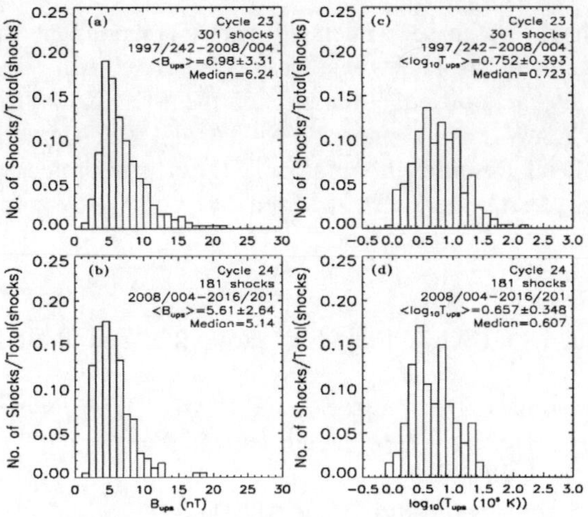

Figure 7. Frequency distribution of (left) upstream magnetic field magnitude B_{ups} and (right) logarithms of the upstream solar wind proton temperature T_{ups} of the fast forward shocks cataloged in ipshocks.fi observed by the Wind spacecraft during solar cycles (a and c) 23 and (b and d) 24.

4. Discussion

Periods when P_{EP} exceeds P_B and P_{PLS} are not infrequent (Fig. 1). When comparing solar cycles 23 and 24, we see that solar cycle 23 showed a larger number of time intervals dominated by P_{EP} than solar cycle 24 (Fig. 4). Although solar cycle 24 displayed a decrease of P_B and P_{PLS} (Fig. 3), the lower energetic particle intensities during this solar cycle (Fig. 2) translated into fewer periods where $P_{EP}>P_B$ and $P_{EP}>P_{PLS}$.

The general decrease in particle intensities observed in solar cycle 24 (Fig. 2) may be caused partly by the fewer fast CMEs occurring in this solar cycle (Fig. 6a), and partly by the fact that the interplanetary magnetic field magnitude was weaker than in solar cycle 23. Slower CMEs are in principle less efficient in driving strong shocks able to accelerate particles. However, the medium where these CMEs propagate is the dominant factor determining the formation and strength of a shock, and the distributions of shock parameters at 1 AU did not show clear differences between solar cycles 23 and 24 (Fig. 5). It has been argued that a weaker magnetic field causes the mechanisms of particle acceleration by shocks to act more slowly; and during the time over which a shock moves from the Sun to 1 AU, the slower acceleration rate results in a lower intensity of high-energy particles [19].

On the other hand, the seed particle population available for acceleration by shocks may have also been reduced in solar cycle 24 [20]. The origin of this seed particle population may include the solar wind itself, particles continuously accelerated in IP space by compression regions in thermally isolated systems [21], by magnetic reconnection in interacting magnetic islands [22, 23] or by solar wind stream interaction regions [24], as well as particles that originated in prior discrete solar events [25]. The decrease in solar wind density in solar cycle 24 would imply a deficit of thermal population available for particle acceleration. The lower level of solar activity in solar cycle 24 (Fig. 6) translates also into a fewer number of discrete SEP events able to fill the heliosphere and hence less suprathermal particles available for the shock acceleration mechanisms. On the other hand, if shocks at 1 AU have similar characteristics in both solar cycles, the lower particle intensities in solar cycle 24 would result from the different properties of the upstream medium where shocks propagate. These properties include (1) lower magnetic field B_{ups} that translates into a slower particle acceleration at shocks, (2) lower proton temperature T_{ups} that implies a less energetic tail of the thermal population available for acceleration, and (3) a meager population of suprathermal particles originated in fewer discrete solar events. That would translate into fewer periods with $P_{EP} > P_B$ and $P_{EP} > P_B$ in solar cycle 24 when compared to solar cycle 23.

Acknowledgments

This paper uses data from the Heliospheric Shock Database, generated and maintained at the University of Helsinki. All data used can be downloaded from www.srl.caltech.edu/ACE/ASC/ and spdf.gsfc.nasa.gov/pub/data/omni/low_res_omni/. We acknowledge the use of the SOHO LASCO CME catalog generated and maintained at the CDAW Data Center by NASA and The Catholic University of America in cooperation with the Naval Research Laboratory. SOHO is a project of international cooperation between ESA and NASA. This work is supported by NASA-HGI grant NNX16AF73G.

References

[1] McComas D J, Angold N, Elliott H A, Livadiotis G, Schwadron N A, Skoug R M and Smith C W 2013 *Astrophys. J.* **779** 2
[2] Kasotakis G, Sarris E, Marhavilas P, Sidiropoulos N, Trochoutsos P, and Daglis I 1999 *Physics and Chemistry of the Earth, Part C: Solar, Terrestrial & Planetary Sci.* **24** 79
[3] Lario D and Decker R B 2002 *Geophys. Res. Lett.* **29** 1393
[4] Marhavilas P and Sarris E 2011 *IEEE Transactions on Plasma Science* **39** 1516
[5] Marhavilas P 2012 *Adv. Space Res.* **49** 1333
[6] Russell C T, Mewaldt R A, Luhmann J G, Mason G M, von Rosenvinge T T, Cohen C M S, Leske R A, Gomez-Herrero R, Klassen A, Galvin A B and Simunac K D C 2013 *Astrophys. J.* **770** 38
[7] Lario D, Decker R B, Roelof E C and Viñas A –F 2015a *Astrophys. J.* **813** 85
[8] Lario D, Decker R B, Roelof E C and Viñas A –F 2015b *J. of Phys. Conf. Ser.* **642** 012014
[9] Zerbo J L and Richardson J D 2015 *J. Geophys. Res.* **120** 10259 10.1002/2015JA021407
[10] Gold R E, Krimigis S M, Hawkins S E, Haggerty D K, Lohr D A, Fiore E, Armstrong T P, Holland G and Lanzerotti L J 1998 *Space. Sci. Rev.* **86** 541-562
[11] Newbury J A, Russell C T, Phillips J T and Gary S P 1998 *J. Geophys. Res.* **103** 9553-9566
[12] Lario D, Aran A and Decker R B 2008 *Space Weather* **6** S1201, 10.1029/2008SW000403
[13] Kilpua E K J, Lumme E, Andreeova K, Isavnin A and Koskinen H E J 2015 *J. Geophys Res.* **120** 4112 10.1002/2015JA021138
[14] Gopalswamy N, Akiyama S, Yashiro S, Xie H, Makela P and Michalek G 2014 *Geophys. Res. Lett.* **41** 2673-2680, 10.1002/2014GL059858
[15] Vourlidas A, Howard R A, Rich N B and Yashiro S 2017 Mass and energy evolution of coronal mass ejections over two solar cycles, *Solar Phys.* (in preparation)
[16] Brueckner G E, Howard R A, Koomen M J, Korendyke C M, Michels D J, Moses J D, Socker D G, Dere K P, Lamy P L, Llebaria A, Bout M V, Schwenn R, Simnett G M, Bedford D K and Eyles C J 1995 *Solar Phys.* **162** 357-402
[17] Yashiro S, Gopalswamy N, Michalek G, St. Cyr O C, Plunkett S P, Rich N B and Howard R A 2004 *J. Geophys Res.* **109** A07105 10.1029/2003JA010282
[18] Compagnino A, Romano P and Zuccarello 2017 *Solar Phys.* **292**:5 10.1007/s11207-016-1029-4
[19] Giacalone J and Mewaldt R A 2016 "Why are there fewer large SEP this solar cycle?" *AGU 2016 Fall Meeting* Abstract SH32A-05
[20] Mewaldt R A, Li G, Cohen C, Mason G M, von Rosenvinge T T, Smith C W and Vourlidas A 2016 "Why are solar energetic particle intensities so much lower in solar cycle 24, espcially at high energies? *AGU 2016 Fall Meeting* Abstract SH41D-01
[21] Fisk L A and Gloeckler G 2012 *Space Sci. Rev.* **173** 433-458
[22] Drake J F, Swisdak M and Fermo R 2013 *Astrophys. J. Lett.* **763** 10.1088/2041-8205/763/1/L5
[23] Zank G P, le Roux J A, Webb G M, Dosch A and Khabarova O 2014 *Astrophys. J.* **797** 28
[24] Mason G M 2000 "Composition and energy spectra of ions accelerated in corotating interaction regions" in Acceleration and Transport of Energetic Particles Observed in the Heliosphere: ACE 2000 Symposium, *AIP Conf. Proc.* **528** 234-241
[25] Dayeh M A, Desai M I, Mason G M, Ebert R W and Farahat A 2017 *Astrophys. J.* **835** 155

IOP Conf. Series: Journal of Physics: Conf. Series **900** (2017) 012013 doi:10.1088/1742-6596/900/1/012013

Acceleration of Solar Energetic Particles at a Fast Traveling Shock in Non-uniform Coronal Conditions

J A le Roux[1,2], and A D Arthur[1]

[1]Department of Space Science, University of Alabama in Huntsville (UAH), 320 Sparkman Drive, Huntsville, Al 35805
[2]Center for Space Plasma and Aeronomic Research (CSPAR), University of Alabama in Huntsville, 320 Sparkman Drive, Huntsville, Al 35805

Abstract. Time-dependent solar energetic particle (SEP) acceleration is investigated at a fast, nearly parallel spherical traveling shock in the strongly non-uniform corona by solving the standard focused transport equation for SEPs and transport equations for parallel propagating Alfvén waves that form a set of coupled equations. This enables the modeling of self-excitation of Alfvén waves in the inertial range by SEPs ahead of the shock and its role in enhancing the efficiency of the diffusive shock acceleration (DSA) of SEPs in a self-regulatory fashion. Preliminary results suggest that, because of the highly non-uniform coronal conditions that the shock encounters, both DSA and wave excitation are highly time-dependent processes. Thus, DSA spectra of SEPs strongly deviate from the simple power-law prediction of standard steady-state DSA theory and initially strong wave excitation weakens rapidly. Consequently, the ability of DSA to produce high energy SEPs in the corona of ~1 GeV, as observed in the strongest gradual SEP events, appears to be strongly curtailed at a fast nearly parallel shock, but further research is needed before final conclusions can be drawn.

1. Introduction

We report preliminary simulation results of time-dependent solar energetic particle (SEP) acceleration at a fast, nearly parallel spherical traveling shock in the corona by taking into account the strong radial dependence in background coronal plasma conditions that the shock encounters during the first hour of shock propagation. For this purpose a simple empirical coronal plasma model [1] is used as a first step. The model involves solving a set of coupled equations for SEPs and parallel propagating Alfvén waves. The standard focused transport equation is solved to simulate SEP acceleration at the traveling shock when SEPs are scattered by Alfvén waves in its vicinity [2,3], and standard transport equations for parallel propagating Alfvén waves are solved to model wave transport near the shock including self-excitation of waves in the inertial range by upstream SEPs undergoing shock acceleration [4,5,6,7,8]. Thus, enhanced scattering of SEPs by Alfvén waves is induced that enhances the efficiency of diffusive shock acceleration (DSA) of SEPs in a self-regulatory fashion at the nearly parallel shock [6,9,10]. For an excellent recent review of the current status of SEP modelling at traveling shocks, see Verkhoglyadova et al. [11]. Our calculations discussed below show that, because of the highly non-uniform coronal conditions that the shock encounters, both the DSA of SEPs and self-excitation of Alfvén waves by SEPs are highly time-dependent coronal processes. Thus, we find that the accelerated spectra of SEPs deviate strongly from the simple power-law prediction of standard steady-state DSA theory. Furthermore, strong initial wave excitation weakens rapidly, thus hampering DSA in producing SEPs with high energies close to ~1 GeV at the nearly parallel shock as observed in the strongest SEP events. The two main factors that contribute to these results are the strong radial decrease, as inferred from the coronal model, in both the SEP source function when modeled as a kappa distribution, and in the SEP parallel diffusion coefficient that the traveling shock encounters in the corona. Thus, a key reason for our results is that both these factors, although in different ways, strongly reduce the efficiency of SEP injection into DSA with time at the traveling shock.

2. The Model

We solve numerically the standard focused transport equation to model time-dependent SEP acceleration at a fast spherical, nearly parallel traveling shock propagating radially outward at a constant speed in the corona. For this purpose it is convenient to transform the equation to the traveling shock frame. Then, the equation can be expressed (shown here for a purely parallel shock for simplicity) as [10,12]

$$\begin{aligned}\frac{\partial f}{\partial t_{sh}} + (U - V_{sh} + v'\mu')\frac{\partial f}{\partial r} - \left[\mu'^2\frac{\partial U}{\partial r} + (1-\mu'^2)\frac{U}{r} + \mu'\left(\frac{1}{v'}\frac{dU}{dt_{sh}} - \frac{qE_{sh}}{p'}\right)\right]p'\frac{\partial f}{\partial p'} \\ + (1-\mu'^2)\left[-\mu'\frac{\partial U}{\partial r} + \frac{v'+U\mu'}{r} - \left(\frac{1}{v'}\frac{dU}{dt_{sh}} - \frac{qE_{sh}}{p'}\right)\right]\frac{\partial f}{\partial \mu'} = \frac{\partial}{\partial \mu'}\left(D_{\mu\mu}\frac{\partial f}{\partial \mu'}\right) + Q\end{aligned} \qquad (1)$$

when neglecting small relativistic particle factors. In equation (1), both the radial solar wind speed U_{sh} and shock normal distance x_{sh} in the shock frame were mapped back to the observer frame using the Galilean transformations $U_{sh} = U - V_{sh}$ and $r = x_{sh} + V_{sh}t_{sh}$, and by applying $\partial/\partial x_{sh} = \partial/\partial r$ valid for a constant shock speed (U is the radial solar wind speed in the observer frame, V_{sh} is the radial shock speed, r is heliocentric distance in the observer frame, and t_{sh}≡t is time measured in the shock frame). Thus the SEP distribution $f(r(x_{sh},t_{sh}), p', \mu', t_{sh})$ is in a mixed coordinate system where particle momentum p' (or particle speed v') and cosine of the particle pitch angle μ' are both measured in the solar wind frame. Furthermore, the acceleration of the solar wind in the shock frame can be expressed as $dU_{sh}/dt_{sh} = \partial U/\partial t_{sh} + (U - V_{sh})\partial U/\partial r = dU/dt_{sh}$ for a constant shock speed, while $D_{\mu\mu}$ represents the standard pitch-angle scattering coefficient for gyro-resonant interaction of SEPs with parallel propagating Alfvén waves in the inertial range [13], modified with an ad-hoc-resonance broadening model [7]. The coefficient describes SEP interaction with forward and backward propagating left-hand, and forward and backward propagating right-hand circularly polarized Alfvén waves, amounting to four wave possibilities. Also, E_{sh} is the cross-shock electric field and Q is the SEP proton source term ahead of the shock determined from specifying a kappa distribution as a source function.

The SEP transport equation is coupled to the standard transport equation for Alfvén waves propagating parallel to the magnetic field in a radially varying radial solar wind outflow based on wave action conservation in a weakly non-uniform plasma medium [7,8]. The wave transport equation, which is solved numerically in the spherical traveling shock frame for all four above-mentioned Alfvén wave possibilities cases, is given by

$$\frac{\partial W_j^m}{\partial t_{sh}} + (U - V_{sh} + jV_A)\frac{\partial W_j^m}{\partial r} - \left[\frac{\partial}{\partial r}(U + jV_A)\right]k'\frac{\partial W_j^m}{\partial k'} = -\frac{1}{2}\frac{1}{r^2}\frac{\partial}{\partial r}(r^2U)W_j^m + 2\gamma_j^m W_j^m \qquad (2)$$

where $W^{jm}(r(x_{sh},t_{sh}), k', t_{sh})$ is total spectral Alfvén wave energy density (sum of kinetic and magnetic energy) as a function of parallel wave number k' in the solar wind frame, $j = \pm 1$ denotes forward (j = +1) and backward (j = -1) wave propagation directions, $m = \pm 1$ indicates left-hand (m = +1) and right-hand (m = -1) circularly polarized Alfvén waves, and γ_j^m is the rate of Alfvén wave excitation/damping by SEPs ahead of the traveling shock following standard quasi-linear theory [7].

Because $\gamma_j^m \propto jV_A \partial f / \partial \mu'$, and SEPs undergoing DSA tend to be beam-like upstream at lower energies so that $\partial f / \partial \mu' > 0$, forward propagating Alfvén waves (j = +1) are generated ahead of the shock by these SEPs. This increases the rate of pitch-angle scattering of SEPs upstream that enhances the efficiency of DSA, and weakens the beam-like SEP distributions and rate of wave generation in a self-regulatory manner [10]. Consistent with shocks driven by coronal mass ejections in the largest SEP events, the shock is modeled to propagate for 1 hour at a high speed of V_{sh} = 2400 km s^{-1} through the corona. The shock, which has a compression ratio of ~4, is initialized at 3.5 solar radii to reach ~16 solar radii during this period, a sufficient distance to sample highly non-uniform coronal plasma conditions in accordance with the coronal model discussed below.

3. Radial Variation of Coronal Parameters

As a first step to model the strong radial variation of background plasma parameters in the corona that the traveling shock encounters, we simply apply the empirical model of Köhnlein [1] which smoothly connects remote sensing observations of the corona with *in situ* measurements in the heliosphere. Future work will involve further refining the coronal model by looking for consistency between the remote sensing observations and coronal models based on first principles, which is beyond the scope of the current paper. In Figure 1(a) and (b) we show the radial decrease in coronal density and temperature, respectively. The vertical lines delineate the radial interval in which we consider the variation of these parameters in our simulations corresponding to the 1 hour shock propagation interval between 3.5 and ~16 solar radii. Using the density and temperature variations as input, we calculate a strong radial decrease of ~4 orders of magnitude in the background SEP proton source function ahead of the traveling shock in the radial domain of shock propagation (see Figure 1(c)). The source function is modeled as a kappa distribution with $\kappa \approx 3$ [14]. We find the background Alfvén speed to be considerably larger than the background solar wind speed for most of the radial interval of shock propagation (Figure 1(d)). However, the assumed high shock speed ensures that, in the shock frame, both forward and backward propagating Alfvén waves upstream are rapidly advected toward the shock, and downstream both waves are advected away from the shock (the Alfvén speed decreases by a factor of ~2 across the strong, nearly parallel traveling shock, thus becoming less than the downstream solar wind speed).

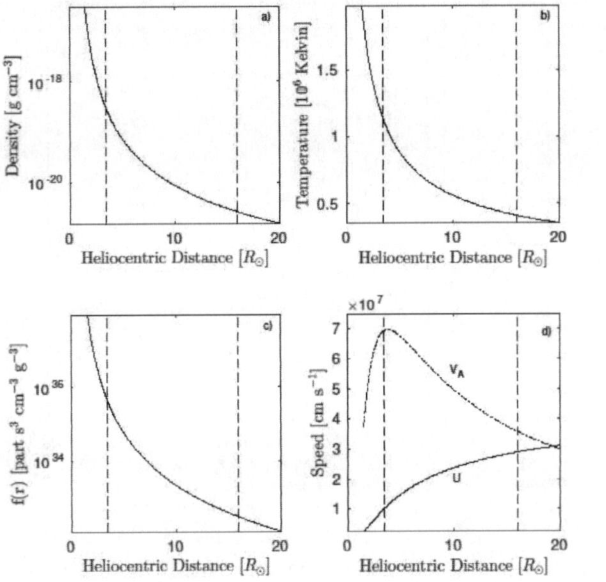

Figure 1. The modeled dependence of (a) density, (b), temperature, (c) SEP proton source function at 10 MeV, and (d) solar wind speed and Alfvén speed on heliocentric distance expressed in solar radii. The dashed vertical lines indicate the radial domain of propagation for a shock with a radial speed of 2400 km s^{-1} during 1 hour as specified in our simulations assuming shock initiation at 3.5 solar radii. The results are based on the empirical Köhnlein coronal model [1].

In Figure 2(a) we illustrate the modeled coronal radial decrease in the background Alfvén wave energy density E_w. In the calculation we employed the radial Alfvén wave energy density transport model of Jacques [15] (based on conservation of wave action in a weakly non-uniform plasma medium) by inserting the radial dependence of the solar wind speed and Alfvén speed shown in Figure 1 into the expression for E_w. Upon combining the radial dependence of E_w with that of the background magnetic field (assumed to be a radial field with strength $B \propto r^{-2}$) in the expression for $D_{\mu\mu}$, we find that in the corona the SEP parallel diffusion coefficient $\kappa_{\parallel}$ decreases by almost 2 orders of magnitude in the considered domain of shock propagation (see Figure 2(b)). As is discussed below in reference to Figure 6, both this decrease and the strong decrease in the SEP source function encountered by the fast traveling shock causes DSA of SEPs to be highly time-dependent in the corona, resulting in accelerated spectra that strongly differ from the predictions of standard steady-state DSA theory.

4. Simulation Results

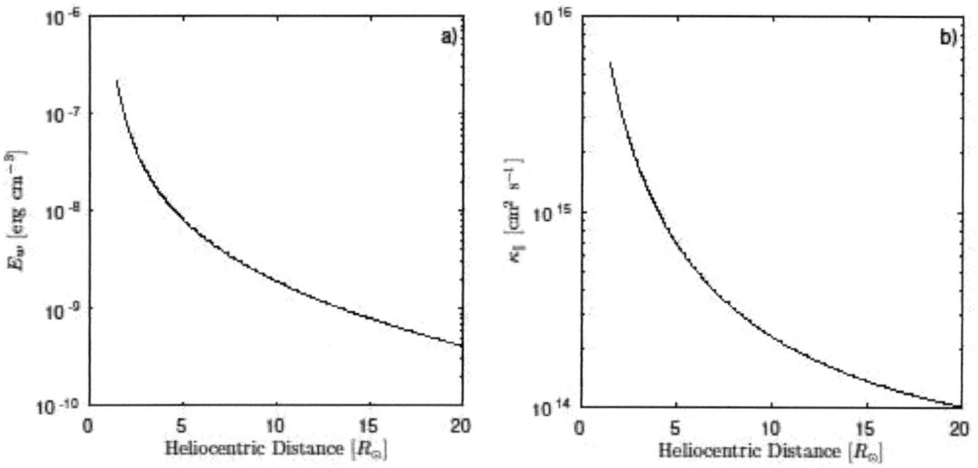

Figure 2. The modeled dependence of (a) the Alfvén wave energy density in erg cm^{-3} and (b) the parallel diffusion coefficient $\kappa_{\parallel}$ of SEPs in cm^2 s^{-1} on heliocentric distance in solar radii in ambient coronal conditions based on the empirical Köhnlein coronal model [1].

In Figure 3 (right panel) simulation results are shown to illustrate the excitation of parallel propagating right-hand circularly polarized Alfvén waves in the inertial range just upstream of a fast spherical shock by SEPs undergoing DSA. The evolution of the wave generation is followed for Alfvén waves propagating in the forward direction (away from the shock) during the first hour of shock propagation in the non-uniform corona as modeled above. The initial background inertial range spectrum (dotted curve) is assumed to be a Kolmogorov $k'^{-5/3}$ spectrum. After 10 minutes of shock propagation the wave spectrum (dashed curve) develops a strong enhancement in spectral energy density with an amplification factor of as much as ~10,000 relative to the background spectrum due to Alfvén wave generation by upstream SEPs undergoing DSA. After 60 minutes, however, the spectral bump (solid curve) is strongly diminished with a maximum amplification factor of only ~65 occurring at smaller wave numbers. The shift is a consequence of DSA producing an increasing number of higher energy SEPs that can resonate with Alfvén waves with longer wavelengths, thus increasing their excitation

IOP Conf. Series:Journal of Physics: Conf. Series **900**(2017) 012013 doi:10.1088/1742-6596/900/

rate. The rapid weakening in Alfvén wave generation can be understood in terms of the rate of Alfvén wave generation $\gamma_j^m \propto \partial f / \partial \mu'$ and how this gradient decreases ahead of the traveling shock. The main contributions to the decrease are: (i) The initially large SEP pitch-angle anisotropy upstream reduces with time due to ongoing pitch-angle scattering of SEPs by Alfvén waves. (ii) The modeled radial decrease of the SEP source function in the corona implies that the propagating shock encounters rapidly increasingly less SEPs that can generate Alfven waves upstream. (iii) When higher energy SEPs generate Alfvén waves at later times, there are simply less of them available for this purpose compared to when lower energies SEPs generate Alfvén waves during earlier times (negative SEP spectral slope).

In Figure 4 we display simulation results confirming a rapid decrease in the SEP pitch-angle anisotropy ahead of the traveling shock. The results in Figure 4(a) refer to upstream SEP pitch-angle distributions for selected particle energies after 10 minutes of shock propagation in the corona.

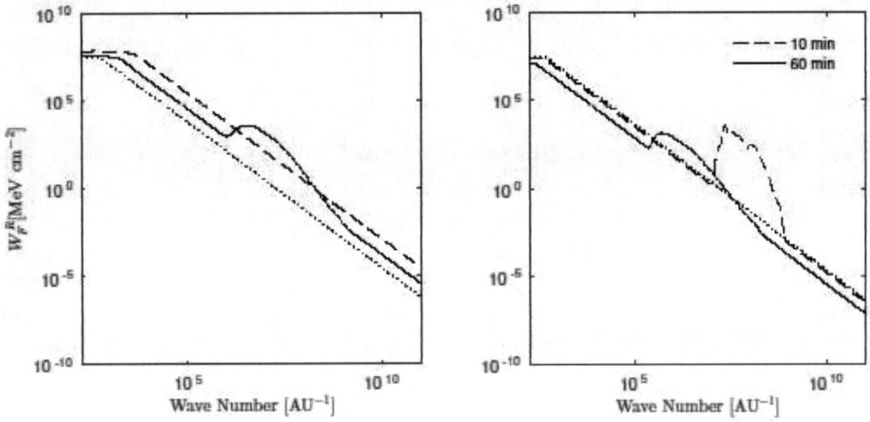

Figure 3. (right panel) Upstream inertial range spectral wave energy density in MeVcm^{-2} as a function of wave number in AU^{-1} for forward parallel propagating, right-hand circularly polarized Alfvén waves in a non-uniform corona. The dotted line is the initial background spectral wave energy density without a shock, the dashed curve is the spectral wave energy density after 10 minutes of shock propagation, and the solid curve is the spectral energy density after 60 minutes of shock propagation. (left panel) The same forward propagating Alfvén waves as right panel, but downstream of shock.

Especially at lower SEP energies (see solid curve for 0.3 MeV protons) these initial distributions are beam-like with most SEPs propagating upstream with 0° degree pitch angle away from the shock as SEPs with larger pitch angles find it difficult to stay ahead of the fast traveling shock. This explains the strong initial forward propagating Alfvén wave growth ahead of the shock at larger wave numbers as shown in Figure 3 (dashed curve). Compared to Figure 4(a), Figure 4(b) indicates rapid weakening of the initial beam-like SEP pitch-angle distributions upstream after 1 hour of shock propagation in the corona. The beam-like distributions in the forward hemisphere ($0 < \mu' \leq 1$) largely dissipated in response to ongoing pitch-angle scattering of SEPs by Alfvén waves. However, still some wave generation takes place (solid curve in Figure 3) due to the significant positive $\partial f / \partial \mu'$-gradient in the μ'-interval around $\mu'=0$ (transition between $\mu'>0$ (forward propagating), and $\mu'<0$ (backward propagating) SEPs ahead of the shock). The rapid dissipation of the beam-like SEP distributions by pitch-angle scattering benefitted from the strong initial wave growth and the radial increase in the background pitch-angle scattering coefficient (decrease in $\kappa_{\parallel}$) encountered by the traveling shock.

Displayed in Figure 5 is the simulated evolution of the radial variation in the total Alfvén energy density across the traveling shock during the first hour of shock propagation in the corona. The energy density is calculated by integrating over all wave numbers in the inertial range and by summing up energy densities for all four Alfvén wave possibilities. After 10 minutes of shock propagation, there is a clear radial bump in the wave energy density ahead of the shock as a consequence of strong initial wave excitation by upstream SEPs undergoing DSA due to their beam-like distributions. After 20 minutes or more, the radial bump disappears as wave excitation contributes negligibly to the total energy density in the inertial range due to the rapid weakening in upstream wave generation as discussed above. The upstream wave energy density keeps decreasing with time as the traveling shock

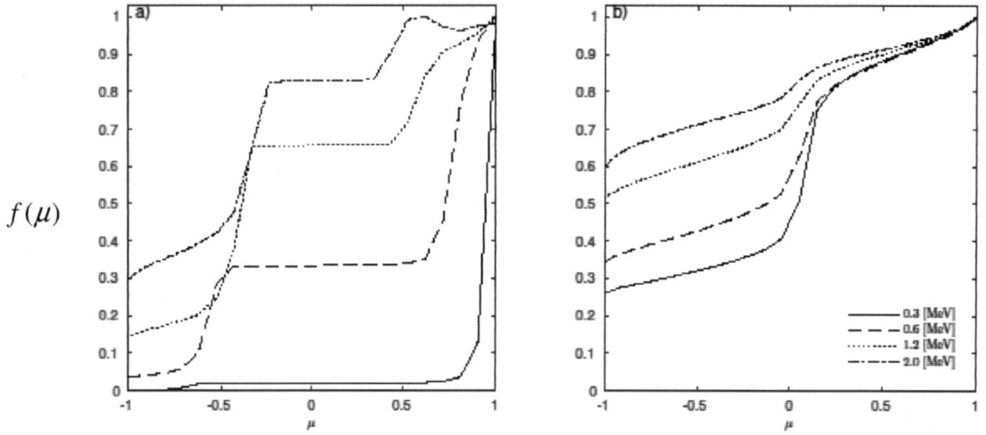

Figure 4. (a) Simulated SEP distributions $f(\mu')$ just ahead of the traveling shock as a function of μ' ($\mu'=\cos\theta'$) where θ' is the particle pitch angle. The distributions are normalized to a value of 1 at $\mu'=1$. SEPs with $\mu'=1$ indicate particles propagating with 0° pitch angle away from the shock whereas $\mu'=-1$ refer to SEPs propagating with 180° pitch angle toward the shock. The distributions are shown after 10 minutes of shock propagation in the corona for different SEP kinetic energies ranging from 0.3-2 MeV as indicated in the legend in (b). (b) As (a), but after 60 minutes of shock propagation.

encounters less energy density due to the modeled negative radial gradient in the background wave energy density in the corona (Figure 2(a)). The shock propagates much faster than Alfvén waves so that both forward and backward propagating waves get compressed across the shock. We find that the total Alfvén wave energy density jumps by a factor of ~20 across the shock initially. At later times the cross-shock jump levels off to a factor of ~10. It should be noted that the Alfvén wave transport equation that we solve across the shock (equation (2)) is simplified in the sense that wave reflection and conversion into different wave modes, such as magnetosonic waves, are not included. Nonetheless, the simulated jump in wave energy density across the shock is reasonable because it is of the same order of magnitude as observed across fast traveling shocks at 1 AU in ground level enhancement SEP events [16]. The jump can be roughly approximated as s^2, where s is the shock compression ratio. Behind the nearly parallel traveling shock we see growth in the radial extent of the compressed wave energy density as downstream Alfvén waves, with an Alfvén speed reduced by a factor of 2 compared to upstream, cannot keep pace with the shock. Notice formation of a peak in

IOP Conf. Series:Journal of Physics: Conf. Series **900**(2017) 012013 doi:10.1088/1742-6596/900

wave energy density at the trailing edge of the growing wave enhancement region downstream. The peak forms as the strong initial wave growth feature driven upstream by SEP anisotropy crosses the shock to downstream and fails to keep up with the shock. It would be of interest to investigate in future work the repercussions of the downstream sheath of enhanced Alfvén wave energy density for SEP acceleration to higher energies. Interesting related questions that can be addressed are: (i) To what extent can SEPs undergoing DSA experience additional 2nd order Fermi acceleration by the enhanced Alfvén wave intensity downstream? [17] (ii) Considering a "twin" shock scenario where our traveling shock is followed by closely spaced second traveling shock [18], how much can the leading shock contribute to provide a pre-accelerated SEP population that improves injection of SEPs into DSA at the trailing shock, and (iii) to what extent can the enhanced wave turbulence behind the leading shock speed up DSA at the trailing shock by reducing $\kappa_{\parallel}$ upstream of the trailing shock [19]?

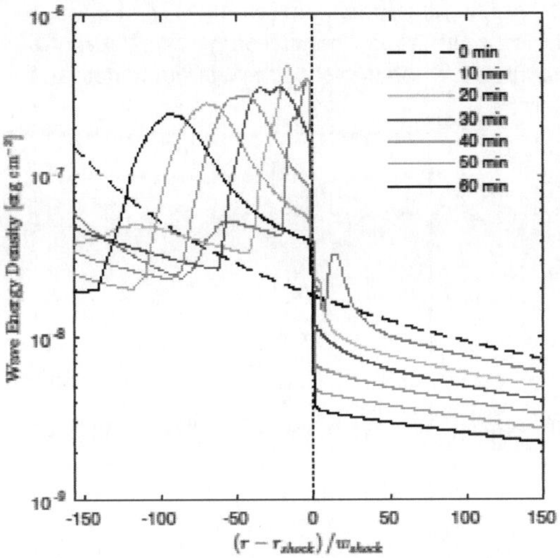

Figure 5. The total Alfvén wave energy density in erg cm^{-3}, integrated over all wave numbers in the inertial range and summed up for all 4 parallel propagating wave mode cases (forward and backward left and right hand circularly polarized modes), as a function of radial distance relative to the spherical traveling shock normalized to the shock width in the simulation. a fast spherical traveling shock in a non-uniform corona. Positive relative radial distances indicate positions ahead of the shock whereas negative values represent positions behind the shock. The dashed curve shows the radial decrease in the wave energy density in the absence of a shock, whereas the solid curves indicate the evolution of the wave energy density across the shock in 10 minute time intervals.

Simulated SEP proton spectra just behind the traveling shock during the first hour of shock propagation in the non-uniform corona are presented in Figure 6(a). For reference, the initial upstream proton source function (kappa distribution with $\kappa = 3$) is included as the dotted curve. The dashed curve is the downstream spectrum after 10 minutes of shock acceleration and the solid curve illustrates its evolution after 1 hour of shock acceleration. At lower energies the SEP spectra downstream have positive slopes with a peak at ~40 keV. Comparison with the upstream source function (dotted curve), reveals strong pre-acceleration of lower energy SEPs by the shock. This is predominantly a consequence of the shock-frame acceleration of the plasma flow across the shock dU_{sh} / dt_{sh} (see expression below equation (1)) that acts as a non-inertial force on SEPs in the downstream direction. Thus SEPs crossing the shock to downstream will be heated whereas those SEPs returning back upstream will be cooled. Since SEPs experience pitch-angle scattering back and forth across the shock, energy gain from this mechanism is stochastic in nature and can be thought of as a second order Fermi pre-heating mechanism [20]. This pre-acceleration mechanism is especially strong at low energies for a fast traveling shock [12], thus dominating the heating effects of the cross-shock electric field and adiabatic compression across the shock which are also included in the simulation. Beyond ~ 40 keV, the dashed SEP spectrum cuts off abruptly because the dU_{sh} / dt_{sh}-pre-heating mechanism is inefficient at higher SEPs energies ($(1/p')dp'/dt_{sh} \propto (1/v')dU_{sh}/dt_{sh}$) and after 10 minutes there is little sign of DSA of SEPs at higher energies. This is consistent with Figure 2(a) where after 10 minutes upstream SEP distributions are

still strongly beam-like, thus showing little sign of the pitch-angle scattering needed for DSA to be efficient. After 60 minutes of shock acceleration (solid curve in Figure 6(a)), the DSA spectrum is in the process of unfolding at higher energies above ~40 keV. However, different from the prediction of standard steady-state DSA theory, there is no sign of a power law. In fact, the spectrum tends to be concave, and is harder than expected compared to the prediction of standard steady-state DSA theory at a fast traveling shock with a compression ratio of ~4. Our analysis shows that the spectral shape is a direct result of the strongly non-uniform coronal conditions that the shock encounters in two ways: (i) The DSA spectrum evolves to become concave because the traveling shock encounters a decreasing $\kappa_{\parallel}$ upstream with time in response to the radial decrease of $\kappa_{\parallel}$ as modeled in the corona (see Figure 2(b)). Thus, the injection speed for DSA increases with time as increasingly higher SEP energies are required to enable SEPs to diffuse ahead of the shock. Since less SEPs are available for injection at higher energies, the rate of SEP injection decreases during the injection threshold shift to higher SEP energies with time. (ii) The DSA spectrum also becomes harder than predicted by standard theory. This occurs because the shock encounters a strongly decreasing upstream proton source function in response to the estimated radial decrease of the source function in the corona (see Figure 1(c)). Accordingly, the rate of SEP injection into DSA decreases with time at a given injection threshold.
We conclude that DSA at a fast traveling, nearly parallel acceleration shock in a strongly non-uniform

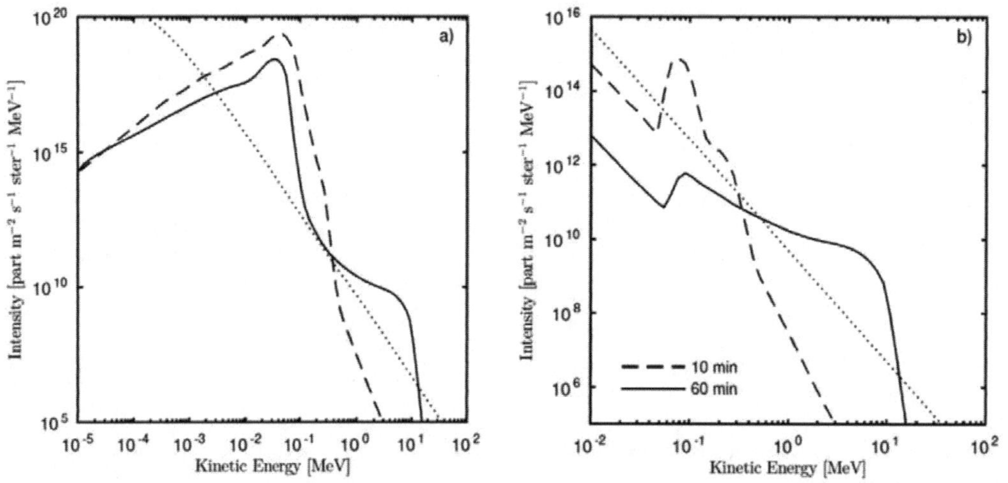

Figure 6. (a) The simulated evolution of the SEP proton spectrum downstream of a fast spherical shock in the corona during the first 60 minutes of shock propagation. The spectrum is plotted in terms of differential intensity in particles m^{-2} s^{-1} sr^{-1} as a function of kinetic energy in MeV. The dotted curve is the initial upstream proton source spectrum, the dashed curve is the accelerated spectrum downstream after 10 minutes, and the solid curve show the same spectrum after 60 minutes of shock propagation. (b) The same as (a), except that the dashed and solid curves are upstream spectra.

corona might be highly time-dependent so that the resultant SEP spectrum potentially strongly deviates from the predictions of steady-state DSA theory early on in the SEP event. We also note that the maximum energy that SEP protons reach after 1 hour of DSA is ~10-20 MeV in our simulation. This is considerably lower than in typical strong SEP events that can generate significant amounts of SEPs above 1 GeV within a few hours [21]. These results form a contrast to a previous focused transport simulation that produced 1 GeV SEPs within 10 minutes [10]. We note that recently Afansiev et al. [22], with the aid of a self-consistent Monte Carlo simulation model, reached maximum SEP energies of ~4 MeV after 10 minutes. However, linear extrapolation of their results assuming higher SEP injection efficiencies into DSA suggest that maximum energies of ~300 MeV in 10 minutes should be possible. A main factor in failing to reach 1 GeV energies in our case is the rapid decline in the rate of Alfvén wave excitation by upstream SEPs as discussed above. Nonetheless, self-excitation of Alfvén waves by SEPs ahead of the shock considerably improves the efficiency of DSA at a fast, nearly parallel traveling shock in our simulations. Without wave excitation, the maximum SEP energy reached in our simulations is only ~100 keV.

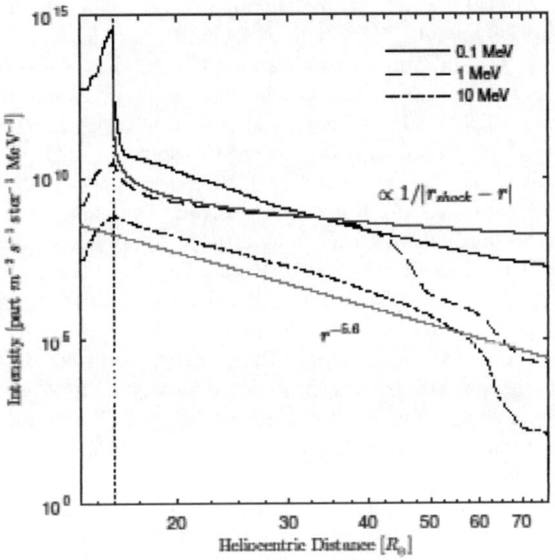

Figure 7. Simulated Differential intensity of SEP protons in particles m^{-2} s^{-1} sr^{-1} MeV^{-1} at selected particle energies as a function of heliocentric distance r in solar radii. The blue curve, which varies according to the relationship $\propto | r_{sh} - r |^{-1}$ (r_{sh} is the shock position), is compared to the dashed curve for 1 MeV SEPs. The vertical line indicates the shock position.

Consider the large step-decrease transition in the downstream SEP spectrum between lower-energy SEPs experiencing the effect of the cross-shock acceleration of the solar wind flow, and higher energy SEPs undergoing DSA in the solid curve in Figure 6(a). We think the size of the step is a reflection of the dominance of the acceleration of the flow as an energization mechanism at lower energies that enforces a strong cutoff in the spectrum close to the minimum injection threshold (minimum energy SEPs need to stay ahead of the shock which is ~20-30 keV). We expect this step decrease to become considerably weaker when the shock travels beyond the corona where the cross-shock acceleration of the solar wind will weaken as the shock continues to slow down. Furthermore, the pressure in SEPs accelerated by the cross-shock solar wind flow is considerable, thus raising the question of modification of the shock structure by their pressure gradient. It could well be that a self-consistent calculation will produce weaker cross-shock acceleration of the solar wind, thus leading to a smaller step transition. Also note that in our simulations particle momentum is measured in the solar wind flow frame. Thus, the SEP spectrum will look quite different in the vicinity of step the transition when transformed to the observer frame.

Finally, we show simulation results in Figure 7 to illustrate the radial variation of the SEP differential intensity at selected SEP energies ahead of the fast, nearly parallel traveling shock after 60 minutes of shock propagation. It is interesting that especially the curves for 1 and 10 MeV SEPs indicate a power-law decrease in intensity with increasing radial distance ahead of the shock before their respective rollovers to the background proton distribution. The rollover occurs further ahead of the shock for 10 MeV SEPs than at 1 MeV, thus illustrating the more efficient diffusion of higher energy SEPs upstream. The power-law decay is in contrast to the familiar prediction of standard steady-state DSA theory of an exponential intensity decrease ahead of the shock. The 1 MeV curve upstream agrees well with the relationship for SEP intensity $\propto 1/(r - r_{sh})$ when $r > r_{sh}$ sufficiently (r_{sh} is the shock position). This relationship was predicted by Bell [23] in a steady-state theory that includes self-generation of Alfvén waves assuming strong resonance broadening. Thus, it appears that a power-law intensity decrease upstream is a direct consequence of Alfvén wave excitation by SEPs undergoing DSA at quasi-parallel shocks (see also Afanasiev [19] for a similar discussion). Therefore, at quasi-parallel shocks upstream wave excitation provides an alternative explanation for such power laws thought by some to be produced by anomalous diffusion of energetic particles across traveling shocks [24].

5. Summary and Conclusions

Results were presented to illustrate simulations of SEP acceleration at a fast spherical, nearly parallel shock propagating with a constant speed V_{sh} = 2400 km s^{-1} in the corona taking into account (i) Alfvén wave excitation upstream by SEPs undergoing DSA, and (ii) the highly non-uniform radial plasma conditions that the shock may encounter. The non-uniform corona was modelled using a simple empirical coronal model [1] as a first step, so that our results should be considered as preliminary. The main idea was to investigate the interplay between DSA and upstream wave excitation during the first hour of fast shock propagation in a strongly radially dependent corona. For this purpose a coupled system of spherically-symmetric time-dependent equations involving the focused transport equation and wave transport equations for parallel propagating Alfvén waves were solved. The main results were: (i) Low-energy SEPs below the injection threshold for DSA was efficiently pre-accelerated stochastically across injection threshold. Above the injection threshold this acceleration mechanism became inefficient rapidly with increasing SEP energy. (ii) The simulated DSA spectrum for SEPs downstream became concave and harder than expected in comparison to the power-law prediction of standard steady-state DSA theory after 1 hour of acceleration. These differences occurred predominantly because the traveling shock encountered both a SEP source function upstream, modeled as a kappa distribution, and a SEP parallel diffusion coefficient upstream that decreased strongly with increasing radial distance in the corona. Both these factors contributed in different ways to a rapid decrease in the rate at which SEPs were injected into DSA, thus resulting in a distorted accelerated spectrum. We concluded that DSA at a fast traveling acceleration shock in the strongly non-uniform corona might be highly time-dependent and thus quite different from the predictions of standard steady-state DSA theory. (iii) The upstream wave excitation by SEPs, initially strong, became weaker rapidly during the first hour of shock propagation in response to the fast decreasing SEP injection rate as explained above, combined with ongoing pitch-angle scattering that reduced the magnitude of the SEP pitch-angle anisotropy upstream responsible for wave excitation. Consequently, during the first hour of shock acceleration, maximum energies of 10-20 MeV, considerably below maximum SEP energies > 1 GeV possible in the strongest SEP events, were reached. Nonetheless, upstream wave excitation strongly enhanced the efficiency of the DSA of SEPs at the nearly parallel traveling shock because without it the maximum energy of SEPs was found to be ~100 keV. (iv) The boost in the total Alfvén energy density upstream resulting from strong initial wave excitation upstream, ended up later during the first hour of shock propagation as a prominent peak in total wave energy density downstream after advection and compression of the waves across the shock. This raised interesting questions for further research such as (a) to what extent can shock accelerated SEPs experience additional 2nd order Fermi acceleration by the enhanced Alfvén wave intensity downstream? [17] (b) Assuming a "twin" shock scenario [18], how much can the leading shock contribute to provide a pre-accelerated SEP population that improves injection of SEPs into DSA at the trailing shock, and (c) to what extent can the enhanced wave turbulence behind the leading shock speed up DSA at the trailing shock by reducing $\kappa_{\parallel}$ upstream of the trailing shock [19]? (v) A consequence of Alfvén wave excitation upstream was that the SEP radial distribution ahead of a quasi-parallel shock became a power law [22,23] instead of an exponential as predicted by standard DSA theory. Therefore, at quasi-parallel shocks upstream wave excitation provided an alternative explanation for such power laws thought to by some to be produced by anomalous diffusion of energetic particles across traveling shocks [24].

6. References

[1] Köhnlein W 1996 *Solar Phys.* **169** 209

[2] Skilling J 1975 *Month. Not. Royal Astron. Soc.* **172** 155
[3] Isenberg P A 1997 *J. Geophys. Res.* **102** 4719
[4] Lee M A 1983 *J. Geophys. Res.* **88** 6109
[5] Zank G P, Rice W K M, and Wu C C 2000 *J. Geophys. Res.* **105** 25079
[6] Rice W K M, Zank G P, and Li G 2003 *J. Geophys. Res.* **108** 1369
[7] Ng C K, Reames D V and Tylka A J 2003 *Astrophys. J.* **591** 461
[8] Sokolov I V, Roussev I I, Skender M, Gombosi T I, Usmanov A V 2009 *Astrophys. J.***696** 261
[9] Gordon B E, Lee M A, Möbius E and Trattner K H 1999 *J. Geophys. Res.* **104** 28263
[10] Ng C K and Reames D V 2008 *Astrophys. J. Lett.* **686** L123
[11] Verkhoglyadova O V, Zank G P and Li G 2015 *Phys. Reports* **557** 1
[12] le Roux J A and Webb G M 2012 *Astrophys. J.* **746** 104
[13] Schlickeiser R 1989 *Astrophys. J.* **336** 243
[14] Scudder J D 1992 *Astrophys. J.* **398** 299
[15] Jacques S A 1977 *Astrophys. J.* **215** 942
[16] Bamert K, Kallenbach R, Ness N F, Smith C W, Terasawa T, Hilchenbach M, Wimmer-Schweingruber R F and Klecker B 2004 *Astrophys. J. Lett.* **601** L99
[17] Afanasief A, Vainio R and Kocharov L 2014 *Astrophys. J.* **790** 36
[18] Li G, Moore R, Mewaldt R A, Zhao L and Labarador A W 2012 *Space Sci. Rev.* **171** 141
[19] Li G and Zank G P 2005 *Proc. 29th Int. Cosmic Ray Conf.* (Pune, India) vol 1 (Mumbai: Tata Institute of Fundamental Research) p 173
[20] Webb G M 1989 *Astrophys. J.* **340** 1112
[21] Mewaldt R A, Looper M D, Cohen C M S, Mason G M, Haggerty D K, Desai M I, Labrador A W and Leske R A 2005 *Proc. 29th Int. Cosmic Ray Conf.* (Pune, India) vol 1 (Mumbai: Tata Institute of Fundamental Research) p 111
[22] Afanasiev A, Batterbee M and Vainio R. 2015 *Astron. Astrophys.* **584** A81
[23] Bell R A 1978 *Mon. Not. Royal Astron. Soc.* **182** 147
[24] Perri S, Zimbardo G, Effenberger F and Fichtner H 2015 *Astron Astrophys.* **578** A2

Acknowledgments
The work was supported by NASA grants NNX14AF43G, NNX15AI65G, and NNX14AJ53G.

IOP Conf. Series: Journal of Physics: Conf. Series **900** (2017) 012014 doi:10.1088/1742-6596/900/1/012014

Statistical origin and properties of kappa distributions

George Livadiotis

Southwest Research Institute, USA
glivadiotis@swri.edu

Abstract. Classical particle systems reside at thermal equilibrium with their velocity distribution function stabilized into a Maxwell distribution. On the contrary, collisionless and correlated particle systems, such as space and astrophysical plasmas, are characterized by a non-Maxwellian behavior, typically described by kappa distributions, or combinations thereof. Empirical kappa distributions have become increasingly widespread across space and plasma physics. A breakthrough in the field came with the connection of kappa distributions to non-extensive statistical mechanics. Understanding the statistical origin of kappa distributions was the cornerstone of further theoretical developments and applications, namely, (i) the concept of temperature; (ii) the physical meaning of the kappa index; (iii) the *N*-particle description of kappa distributions; and the (iv) the generalization to phase-space kappa distribution of a Hamiltonian with non-zero potential.

1. Introduction

Classical particle systems are de facto characterized by the absence of correlations among the individual particle energies. The statistical behavior of these systems has been successfully described by the Boltzmann-Gibbs (BG) statistical mechanics and the Maxwellian distribution of velocities. Space plasmas are collisionless and correlated particle systems, characterized by stationary states out of thermal equilibrium that exhibit a non-Maxwellian behavior; this is typically described by kappa distributions and combinations thereof.

The kappa distributions have been used to describe numerous space plasma populations. Single types of kappa distributions are usually sufficient to satisfactorily model these plasma populations (e.g., [1-3]). Occasionally, more complicated models of kappa distributions are employed (e.g., [4-6]).

Among many others, some examples of space plasma applications are the following: (i) the *inner heliosphere*, including solar wind (e.g., [4, 7-18), solar spectra (e.g., [19,20]), solar corona (e.g., [21-24]), solar energetic particles (e.g., [25,26]), corotating interaction regions (e.g., [27]), and solar flares related (e.g., [28-31]); (ii) the *planetary magnetospheres*, including magnetosheath (e.g., [32,33]), near magnetopause (e.g., [34]), magnetotail (e.g., [35]), ring current (e.g., [36]), plasma sheet (e.g., [37-39]), ionosphere (e.g., [40]), magnetospheric substorms (e.g., [41]), magnetospheres of giant planets, such as Jovian (e.g., [42,43]), Saturnian (e.g., [2,44,45]), Uranian (e.g., [46]), Neptunian (e.g., [47]), magnetospheres of planetary moons, such as Io (e.g., [48]) and Enceladus (e.g., [49]), or cometary magnetospheres (e.g., [50,51]); (iii) the *outer heliosphere and the inner heliosheath* (e.g., [12-15,52-66); (iv) *beyond the heliosphere*, including HII regions (e.g., [67]), planetary nebula (e.g., [68,69]), and supernova magnetospheres (e.g., [70]); or *other space plasma-related analyses* (e.g., [57,58,71-93). (See also: [6,94,95], and references therein.)

Empirical kappa distributions have been introduced in mid-60's by Binsack (1966) [96], Olbert (1968) [97], and Vasyliūnas (1968) [1], while the statistical origin of kappa distributions is determined within the framework of non-extensive statistical mechanics [94]. Their connection with statistical mechanics was shown and studied in detail in about half century later (see [98], and references therein). Non-extensive statistical mechanics is a consistent generalization of the classical statistical mechanics, which is based on a mono-parametric (q-index) entropic formula [99]. The theoretical q-exponential distribution, which results from the maximization of entropy in the canonical ensemble, and the empirical kappa distribution, have identical formulation under the transformation of their indices, $q = 1+1/\kappa$.

Having understood the physical underpinnings of the statistical mechanics connected with kappa distributions, it is straightforward to employ all the statistical well-grounded equations and capabilities for analyses that seek to apply kappa distributions in datasets, simulations, modeling, and theory of space plasmas. In Section 2 we describe the statistical origin of kappa distributions in detail. In Section 3 we show: (i) the concept of temperature; (ii) the physical meaning of the kappa index; (iii) the N-particle description of kappa distributions; and the (iv) the generalization to phase-space kappa distribution of a Hamiltonian with non-zero potential. Finally, Section 4 briefly summarizes the conclusions.

2. Statistical origin

2.1. Motivation

Classical particle systems reside in a special stationary state, the thermal equilibrium - the concept that any flow of heat (thermal conduction, thermal radiation) is in balance. At thermal equilibrium particles were described by the Boltzmann distribution of energies or the Maxwell distribution of velocities. This was the only known stationary state by the time BG statistical mechanics was built and applied. However, space plasmas have been observed in other stationary states. Since these are described by non-Maxwellian behavior and the Maxwell distribution was known as the only distribution interwoven with thermal equilibrium, the space plasma stationary states were called "out of thermal equilibrium". They were also called "non-thermal", but this is fairly unjustified, because these new types of stationary states are quite thermalized and surely described by a temperature.

Once we realized that non-equilibrium stationary states exist, the challenge is to understand their statistics and the involved statistical mechanics. They are characterized by a non-Maxwellian behavior, but the Maxwell distribution is still showing the right path to describe and express the new distributions and their statistics. Indeed, one way to understand this is via the variation of the temperature: If the particles are characterized by a Maxwellian distribution, where its temperature is not fixed but varies under certain statistics, it can lead to other stationary states (for details, see: [95], Chapter 6). Moreover, here we show some other way to understand how the Maxwell distribution and BG statistical mechanics can lead to the non-equilibrium stationary states observed in space plasmas.

Two basic functions are involved in the formulation of BG statistical mechanics: The exponential and logarithmic functions; the former describes the canonical distribution and the latter the entropic function. Both functions can be written using the Euler's limit,

$$\exp(x) = \lim_{\kappa\to\infty}\left(1-\frac{x}{\kappa}\right)^{-\kappa} \Leftrightarrow \ln(x) = \lim_{\kappa\to\infty}\left[\kappa\cdot\left(1-x^{-\frac{1}{\kappa}}\right)\right]. \tag{1}$$

Setting the definitions

$$\exp_\kappa(x) \equiv \left(1-\frac{x}{\kappa}\right)^{-\kappa}, \; \ln_\kappa(x) \equiv \kappa\cdot\left(1-x^{-\frac{1}{\kappa}}\right), \tag{2}$$

known as deformed exponential and logarithmic functions, respectively, we have

$$\exp(x) = \lim_{\kappa\to\infty} \exp_\kappa(x) \Leftrightarrow \ln(x) = \lim_{\kappa\to\infty} \ln_\kappa(x) \ . \tag{3}$$

Note that these functions are inverse to each other for any value of the kappa index.

The exponential function is used in the Boltzmann distribution, that is, the (canonical) distribution maximizes the BG entropy under the constraints of the canonical ensemble. This can be written as

$$p_i \propto \exp\left(-\frac{\varepsilon_i}{k_B T}\right) = \lim_{\kappa\to\infty} \exp_\kappa\left(-\frac{\varepsilon_i}{k_B T}\right) = \lim_{\kappa\to\infty}\left(1+\frac{1}{\kappa}\cdot\frac{\varepsilon_i}{k_B T}\right)^{-\kappa} . \tag{4}$$

On the other hand, the logarithmic function is used in the entropy formulation. This is expressed in terms of a discrete probability distribution $\{p_i\}$, which can be written as

$$S = \sum_i p_i \cdot \ln(1/p_i) = \lim_{\kappa\to\infty} \sum_i p_i \cdot \ln_\kappa(1/p_i) = \lim_{\kappa\to\infty} \sum_i \left\{\kappa \cdot \left[p_i - p_i^{1+\frac{1}{\kappa}}\right]\right\} . \tag{5}$$

At this point, we note that while space physics community is used in the notation of the kappa index, the statistical mechanics community is using the q-index; both the indices are simply related:

$$\kappa = \frac{1}{q-1} \Leftrightarrow q = 1+\frac{1}{\kappa} . \tag{6}$$

Then, the canonical distribution and entropy are written in a limit form as follows:

$$p_i \propto \exp\left(-\frac{\varepsilon_i}{k_B T}\right) = \lim_{\kappa\to\infty}\left(1+\frac{1}{\kappa}\cdot\frac{\varepsilon_i}{k_B T}\right)^{-\kappa} = \lim_{q\to 1}\left[1+(q-1)\cdot\frac{\varepsilon_i}{k_B T}\right]^{-\frac{1}{q-1}} , \tag{7a}$$

$$S = \sum_i p_i \cdot \ln(1/p_i) = \lim_{\kappa\to\infty} \sum_i \left\{\kappa \cdot \left[p_i - p_i^{1+\frac{1}{\kappa}}\right]\right\} = \lim_{q\to 1} \sum_i \left(\frac{p_i - p_i^q}{q-1}\right) . \tag{7b}$$

We define the kappa dependent distribution and entropy,

$$p_{\kappa i} \equiv \frac{1}{Z}\cdot\left(1+\frac{1}{\kappa}\cdot\frac{\varepsilon_i}{k_B T}\right)^{-\kappa} , \ S_\kappa = \sum_i \left\{\kappa \cdot \left[p_i - p_i^{1+\frac{1}{\kappa}}\right]\right\} . \tag{8}$$

(Z is an auxiliary partition function, e.g., see: [100]). Hence, we have that

$$p_i = \lim_{\kappa\to\infty} p_{\kappa i} , \ S = \lim_{\kappa\to\infty} S_\kappa . \tag{9}$$

Therefore, the special stationary state at thermal equilibrium can be described by a specific value of the kappa index of the distribution $p_{\kappa i}$ and entropy S_κ, that is, $\kappa \to \infty$. However, what is the physical meaning of the states that correspond to finite values of the kappa index? As we will see, these finite kappa indices describe the non-equilibrium stationary states observed in space plasmas. The formed $p_{\kappa i}$ and S_κ constitute the kappa distribution and its associated Tsallis entropy.

2.2. *Entropy maximization*

Consider a system described by a discrete energy distribution, that is, energy ε_1 with probability p_1, energy ε_2 with probability p_2, ..., energy ε_W with Probability p_W. The classical BG entropy is given by

$$S = S(p_1, p_2, ..., p_W) = -k_B \cdot \sum_{k=1}^{W} p_k \ln(p_k) , \tag{10}$$

while the Tsallis entropy is

$$S = S(p_1, p_2, ..., p_W) = k_B \cdot \frac{1}{q-1} \cdot \left(1 - \sum_{k=1}^{W} p_k{}^q\right) = k_B \cdot \frac{1}{q-1} \cdot \sum_{k=1}^{W} \left(p_k - p_k{}^q\right). \quad (11)$$

We expand the entropic form in terms of (q-1):

$$p_k{}^q = p_k \cdot e^{(q-1)\cdot \ln p_k} = p_k \cdot \left\{1 + (q-1) \cdot \ln p_k + \tfrac{1}{2}(q-1)^2 \cdot \ln^2 p_k + O[(q-1)^3]\right\}$$

$$\Rightarrow \sum_{k=1}^{W} p_k{}^q \cong 1 + (q-1) \cdot \sum_{k=1}^{W} p_k \ln p_k + \tfrac{1}{2}(q-1)^2 \cdot \sum_{k=1}^{W} p_k \ln^2 p_k + O[(q-1)^3]$$

$$\Rightarrow \frac{1}{q-1} \cdot \left(1 - \sum_{k=1}^{W} p_k{}^q\right) \cong -\sum_{k=1}^{W} p_k \ln p_k - \tfrac{1}{2}(q-1) \cdot \sum_{k=1}^{W} p_k \ln^2 p_k + O[(q-1)^2] .$$

Hence, the Tsallis entropy can be approximated by the BG entropy and expanding (q-1) terms:

$$S = k_B \cdot \frac{1}{q-1} \cdot \left(1 - \sum_{k=1}^{W} p_k{}^q\right) \cong -k_B \cdot \sum_{k=1}^{W} p_k \ln p_k - k_B \cdot \tfrac{1}{2}(q-1) \cdot \sum_{k=1}^{W} p_k \ln^2 p_k + O[(q-1)^2] . \quad (12)$$

The expansion to infinite (q-1) terms is

$$S = k_B \cdot \frac{1}{q-1} \cdot \left(1 - \sum_{k=1}^{W} p_k{}^q\right) = -k_B \cdot \sum_{k=1}^{W} p_k \ln p_k - k_B \cdot \sum_{n=2}^{\infty} \left[\tfrac{1}{n!}(q-1)^n \cdot \sum_{k=1}^{W} p_k \ln^n p_k\right] . \quad (13)$$

Therefore, Tsallis entropy recovers to BG entropy for $q \to 1$.

The Gibbs' path (1902) [101] for the maximization of the entropy $S(p_1, p_2, ..., p_W)$ under the constraints of canonical ensemble, i.e., (i) normalization $1 = \sum_{k=1}^{W} p_k$, and (ii) fixed internal energy $U = \sum_{k=1}^{W} p_k \varepsilon_k$, involves maximizing the functional

$$G(p_1, p_2, ..., p_W) = S(p_1, p_2, ..., p_W) + \lambda_1 \sum_{k=1}^{W} p_k - \lambda_2 \sum_{k=1}^{W} p_k \varepsilon_k . \quad (14)$$

that is, setting

$$\frac{\partial}{\partial p_j} G(p_1, p_2, ..., p_W) = 0 \ , \ \forall \ j: 1, ..., W . \quad (15)$$

Next, we examine the BG and Tsallis entropic formulations. For the BG entropy we find

$$p_j = \exp(\lambda_1 - 1) \cdot \exp(-\lambda_2 \varepsilon_j) \ , \quad (16)$$

or using the mean-less energies, $\varepsilon_j - U$,

$$p_j = \exp(\lambda_1 - 1 - \lambda_2 U) \cdot \exp[-\lambda_2 (\varepsilon_j - U)] \ . \quad (17)$$

The case of Tsallis entropy is rather more complicated. The non-extensive statistical mechanics is interwoven with the concept of escort probabilities [102]. The escort probability distribution $\{P_k\}_{k=1}^{W}$ is constructed from the ordinary probability distribution, $\{p_k\}_{k=1}^{W}$, as $P_k \propto p_k{}^q$, $\forall \ k = 1, ..., W$. The interpretation for the internal energy U is given by the escort expectation value of energy, $U = \sum_{k=1}^{W} P_k \ \varepsilon_k$, which complicates the functional G that needs to be maximized.

Another, fairly simpler way of approaching the entropy maximization within the framework of non-extensive statistical mechanics is given by avoiding the notion of escort probabilities [100]. Then, a different Q-index characterizes the canonical distribution, e.g., $Q = q^{-1}$,

$$p_j = \left[1+(1-Q)\cdot(\lambda_1-1)\right]^{-\frac{Q}{Q-1}} \cdot \left[1+(Q-1)\cdot\frac{\lambda_2\varepsilon_j}{1+(1-Q)\cdot(\lambda_1-1)}\right]^{-\frac{Q}{Q-1}} . \quad (18)$$

We enter the mean-less energies, $\varepsilon_j - U$,

$$p_j = \left[1+(1-Q)\cdot(\lambda_1-1+\lambda_2 U)\right]^{-\frac{Q}{Q-1}} \cdot \left[1+(Q-1)\cdot\frac{\lambda_2(\varepsilon_j-U)}{1+(1-Q)\cdot(\lambda_1-1+\lambda_2 U)}\right]^{-\frac{Q}{Q-1}} . \quad (19)$$

Then, using the typical notation of the kappa index, $\kappa \equiv (Q-1)^{-1}$, we construct the kappa distribution,

$$p_j \propto \left[1+\frac{1}{\kappa}\cdot\beta(\varepsilon_j-U)\right]^{-\kappa-1} , \quad (20)$$

where we set the (inverse) temperature as $\beta \equiv \lambda_2/[1+(1-Q)\cdot(\lambda_1-1+\lambda_2 U)]$.

2.3. General entropy

Here we assume that the entropic function is given in terms of an unknown function $f(x)$:

$$S = S(p_1, p_2, ..., p_W) = k_B \cdot \sum_{k=1}^{W} f(p_k) . \quad (21)$$

We claim that the exact form of $f(x)$ can be determined by considering basic physical properties like the macroscopic additivity among energies and entropies.

First, the maximization of entropy gives

$$f'(p_j)+\lambda_1-\lambda_2\varepsilon_j = 0 , \quad (22)$$

and given the energy additivity among two parts of the system, A and B, i.e., $\varepsilon_{ij}^{A+B} = \varepsilon_i^A + \varepsilon_j^B$, we obtain (for constant λ_1 and λ_2):

$$f'\left(p_{ij}^{A+B}\right)-\lambda_1 = f'\left(p_i^A\right)+f'\left(p_j^B\right) . \quad (23)$$

Then, applying $\sum_{i=1}^{W}\sum_{j=1}^{W} p_{ij}^{A+B}\times$ on both sides of Eq.(23), we derive

$$\sum_{i,j}^{W}\left[\tfrac{1}{-\lambda_1}f'\left(p_{ij}^{A+B}\right)-1\right]p_{ij}^{A+B} = \sum_{i}^{W}\left[\tfrac{1}{-\lambda_1}f'\left(p_i^A\right)-1\right]p_i^A + \sum_{j}^{W}\left[\tfrac{1}{-\lambda_1}f'\left(p_j^B\right)-1\right]p_j^B . \quad (24)$$

Second, the entropy additivity among the same two parts of the system, $S^{A+B} = S^A + S^B$, gives

$$\sum_{i,j}^{W} f\left(p_{ij}^{A+B}\right) = \sum_{i}^{W} f\left(p_i^A\right) + \sum_{j}^{W} f\left(p_j^B\right) . \quad (25)$$

The only solution for having both relations (24) and (25) is to have the proportionality

$$f(x) \propto \left[\tfrac{1}{-\lambda_1}f'(x)-1\right]\cdot x . \quad (26)$$

Solving this differential equation we find

$$f(x)=\lambda_1\cdot\frac{x-x^{\frac{\lambda_1}{c}}}{\frac{\lambda_1}{c}-1}\ , \tag{27}$$

where c is the proportionality constant. Setting $\lambda_1\equiv k_{\rm B}=1$ and $q\equiv\frac{\lambda_1}{c}$, we end up with the Tsallis entropy, as given by Eq.(11),

$$S(p_1,p_2,...,p_W)=k_{\rm B}\cdot\sum_{k=1}^{W}f(p_k)\ \text{with}\ f(x)=\frac{1}{q-1}\cdot\left(x-x^q\right). \tag{28}$$

(For more details, see: [103].)

3. Statistical properties

3.1. The concept of temperature

The standard formulation of the kinetic energy is

$$P(\varepsilon_{\rm K})\propto\left(1+\frac{1}{\kappa-\frac{d_{\rm K}}{2}}\cdot\frac{\varepsilon_{\rm K}}{k_{\rm B}T}\right)^{-\kappa-1}\varepsilon_{\rm K}{}^{\frac{1}{2}d_{\rm K}-1}\ , \tag{29}$$

while sometimes a different formulation of a kappa distribution is also used,

$$P(\varepsilon_{\rm K})\propto\left(1+\frac{1}{\kappa}\cdot\frac{\varepsilon_{\rm K}}{k_{\rm B}T_\kappa}\right)^{-\kappa-1}\varepsilon_{\rm K}{}^{\frac{1}{2}d_{\rm K}-1}\ , \tag{30}$$

where the parameter T_κ is not the actual temperature of the particle system, but a quantity dependent on both the actual temperature T and the kappa index.

The temperature can kinetically defined by the mean kinetic energy,

$$\langle\varepsilon_{\rm K}\rangle\equiv\tfrac{1}{2}d_{\rm K}\,k_{\rm B}T\ , \tag{31}$$

where $d_{\rm K}$ is the kinetic degrees of freedom. If a thermal parameter other than the temperature is used for characterizing the kappa distribution, it must be expressed in terms of the temperature (and possibly, the kappa index). For example, in Eq.(30), the parameter T_κ can be derived from the mean kinetic energy,

$$\langle\varepsilon_{\rm K}\rangle=\tfrac{1}{2}d_{\rm K}\cdot k_{\rm B}T_\kappa\cdot\kappa/(\kappa-\tfrac{1}{2}d_{\rm K})\ . \tag{32}$$

Comparing Eqs.(31,32) we derive the relation $(\kappa-\frac{d_{\rm K}}{2})\cdot k_{\rm B}T=\kappa\cdot k_{\rm B}T_\kappa$; hence, we find that T_κ (and not T) depends on the kappa index, i.e., $T_\kappa=T_\kappa(T,\kappa)$ (see: [6]; [95], Chapter 1).

The correct definition and use of temperature in the formulation of the kappa distribution can be helpful for understanding the properties of this distribution. For example, we may ask, are there types of kappa distributions with different exponents? No, the generated types would be equivalent with the standard kappa distribution: Let's assume that the kappa distribution can be generalized to have an arbitrary exponent,

$$P(\varepsilon)\propto\left(1+\frac{1}{\kappa-\frac{3}{2}}\cdot\frac{\varepsilon}{k_{\rm B}T}\right)^{-\kappa-a}\ ,\ \tfrac{5}{2}-a<\kappa\leq\infty\ . \tag{33}$$

How can this be equivalent to the standard kappa distribution, i.e.,

$$P(\varepsilon) \propto \left(1+\frac{1}{\kappa-\frac{3}{2}}\cdot\frac{\varepsilon}{k_{\mathrm{B}}T}\right)^{-\kappa-1} \;,\; \tfrac{3}{2}<\kappa\leq\infty \;. \tag{34}$$

The parameter T in Eq.(33) is not the actual temperature but a quantity depended on the actual temperature (and the kappa index). By substituting the parameter T in Eq.(33) with the actual temperature T_{real}, derived from the mean kinetic energy,

$$<\varepsilon>\equiv \tfrac{3}{2}k_{\mathrm{B}}T_{real} = \ldots = \frac{\kappa-\frac{3}{2}}{\kappa+a-\frac{5}{2}}\cdot\tfrac{3}{2}k_{\mathrm{B}}T \;, \tag{35}$$

we find

$$P(\varepsilon) \propto \left[1+\frac{1}{(\kappa+a-1)-\frac{3}{2}}\cdot\frac{\varepsilon}{k_{\mathrm{B}}T_{real}}\right]^{-(\kappa+a-1)-1} \;,\; \tfrac{3}{2}<\kappa+a-1\leq\infty \;. \tag{36}$$

Then, we note that the kappa index κ appears always in the quantity $\kappa+a-1$. Hence, we redefine the kappa index, so that $\kappa+a-1\rightarrow\kappa$, thus, the distribution (33) becomes

$$P(\varepsilon) \propto \left(1+\frac{1}{\kappa-\frac{3}{2}}\cdot\frac{\varepsilon}{k_{\mathrm{B}}T_{real}}\right)^{-\kappa-1} \;,\; \tfrac{3}{2}<\kappa\leq\infty \;, \tag{37}$$

which is the standard kappa distribution formulation.

3.2. The physical meaning of the kappa index

The kappa index depends on the kinetic degrees of freedom, d_{K}:

$$\kappa=\kappa(d_{\mathrm{K}})=\text{constant}+\tfrac{1}{2}d_{\mathrm{K}}\text{, with } \kappa_0\equiv\kappa(d_{\mathrm{K}})-\tfrac{1}{2}d_{\mathrm{K}} = \text{invariant kappa index} \;. \tag{38}$$

Therefore, the d_{K}-dimensional kappa distribution of the velocities of 1-particle can be written using the dependent kappa index, κ, or the invariant kappa index, κ_0,

$$P(\vec{u};T;\kappa) \propto \left(1+\frac{1}{\kappa-\frac{d_{\mathrm{K}}}{2}}\cdot\frac{u^2}{\theta^2}\right)^{-\kappa-1} \rightarrow P(\vec{u};T;\kappa_0) \propto \left(1+\frac{1}{\kappa_0}\cdot\frac{u^2}{\theta^2}\right)^{-\kappa_0-1-\frac{1}{2}d_{\mathrm{K}}} \;. \tag{39}$$

The physical meaning of the kappa index is interwoven with the statistical correlation between the energy of the particles. It has been shown that a simple relation exists between the correlation and the kappa index, $R=\frac{1}{2}d_{\mathrm{K}}/\kappa=\frac{1}{2}d_{\mathrm{K}}/(\kappa_0+\frac{1}{2}d_{\mathrm{K}})$ [59,78] (see also: [95], Chapter 5]). The largest value of the kappa index is infinity, corresponding to the system residing at thermal equilibrium. This is a special stationary state for which the particles are characterized by zero correlation. The whole structure of classical statistical mechanics is based on this ideal property that the energies of any two particles are uncorrelated (i.e., zero covariance). On the other hand, the smallest possible value of the kappa index is $\kappa_0\rightarrow 0$ (or $\kappa\rightarrow\frac{1}{2}d_{\mathrm{K}}$) and corresponds to the furthest state from thermal equilibrium, a state called anti-equilibrium [14].

The classical, single stationary state at equilibrium is generalized into a whole set of different non-equilibrium stationary states, each labeled by a separate kappa index, an arrangement called the κ-spectrum (Figure 1). Hence, the parameter kappa provides a suitable measure of the "thermodynamic distance" of stationary states from equilibrium.

IOP Conf. Series:Journal of Physics: Conf. Series **900**(2017) 012014 doi:10.1088/1742-6596/900

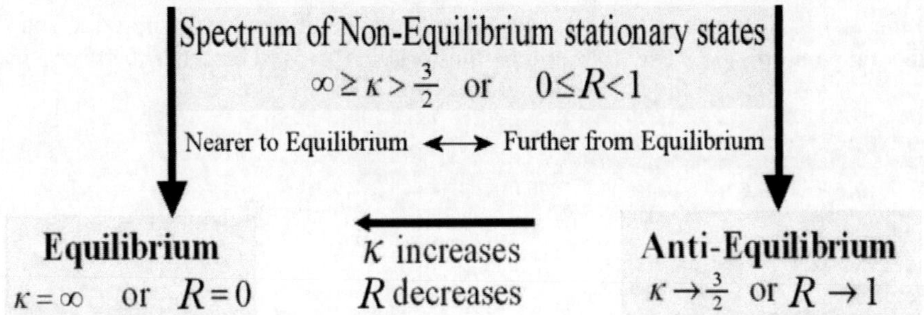

Figure 1. The κ-spectrum. For $\kappa \to \infty$, the system resides at thermal equilibrium; for $\kappa \to \frac{3}{2}$, the furthest possible stationary state from thermal equilibrium is attained, the anti-equilibrium. This is shown comparing with the correlation coefficient *R*. The classical case of systems residing in thermal equilibrium ($\kappa \to \infty$) corresponds to zero correlation ($R \to 0$), while the other extreme state of "anti-equilibrium" ($\kappa \to \frac{3}{2}$) indicates a maximum correlation ($R \to 1$). (Adopted from [95], Chapter 5)

Furthermore, observations showed an empirical separation of the κ-spectrum that characterizes space plasmas [12,15,59,95]: While the inner heliosphere plasmas reside in near equilibrium states with large kappa indices, generally κ>2.5, analysis of the IBEX ENA-spectra showed that inner heliosheath resides in states far from equilibrium with small kappa indices κ<2.5 (Figure 2). The distinction in near-/far- equilibrium regions can be explained by the entropy associated with kappa distributions, which exhibits different functional behavior within the two regions (for details, see: [57] and [95], Chapters 2 and 10).

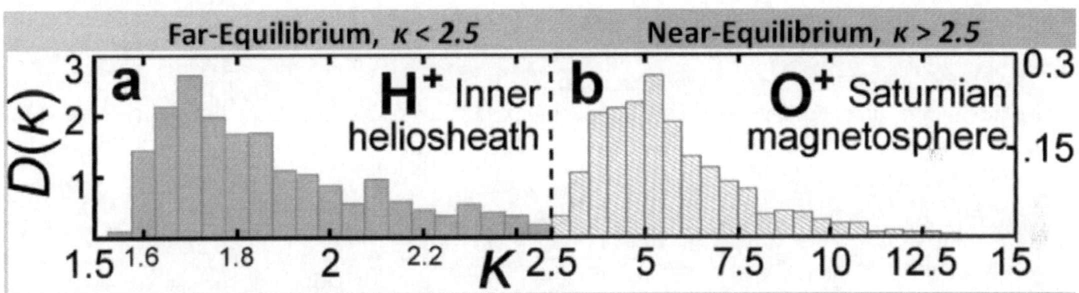

Figure 2. The distribution of kappa indices, $P(\kappa)$, constructed for (a) inner heliosheath protons (green) [60-63], and (b) Saturnian magnetosphere oxygen O^+ (red) (kappa indices derived by [2]). While the kappa indices of the inner heliosheath are distributed only in the far-equilibrium region, the kappa indices of the Saturnian magnetosphere are distributed only in the near-equilibrium region. (Adopted from [15])

Density, temperature and kappa index are independent parameters. Various correlations and trends appear in several space plasmas do not constitute universal relations, as these are subject to the location, time, and plasma parameters. This behavior is expected, as there are many factors causing correlations between those three thermodynamic parameters (e.g., potential energy, e.g., [77] and [95], Chapters 3 and 5). This is similar to the classical case at thermal equilibrium, where density and temperature are independent parameters, but polytropes apply correlations and trends among their values.

Figure 3 illustrates the parameter values of space plasmas. In particular, it plots the representative values of density *n*, temperature *T*, and kappa index κ of ~40 different space plasmas. The kappa indices were collected from the results of one or more published analyses. We also calculate the measure M=1/(κ-0.5), an alternative of the kappa index κ. This is a measure of how far the system resides from thermal equilibrium [12,59,80]. We observe a general trend characterizing space plasmas: The higher the temperature, the closer to thermal equilibrium, that is, the higher the kappa index.

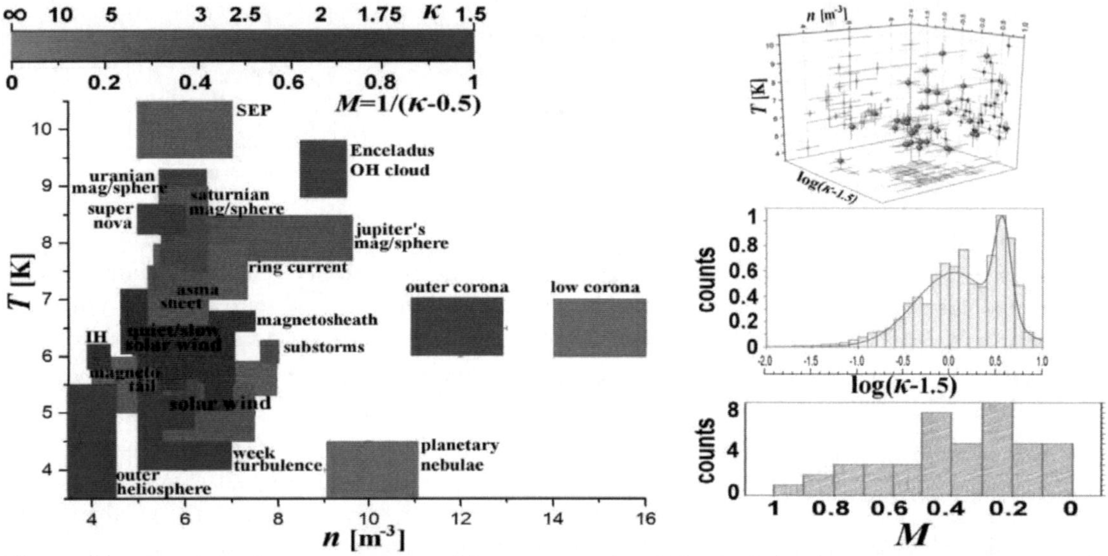

Figure 3. Examples of space plasmas with representative values of density n, temperature T, and kappa index κ. (a) A color-map based on the measure $M=1/(\kappa-0.5)$ in the n-T plane. This measure M takes values from $M=0$ (thermal equilibrium) to $M=1$ (anti-equilibrium, the furthest state from thermal equilibrium). (b) 3-D scatter map of the values of (n,T,M). (c) Histogram of the M- values. (Adopted from [6])

3.3. The N-particle description of kappa distributions

The advantage of using the invariant kappa index, κ_0, instead of the dimensionality dependent kappa index, κ, is that we can express the N-particle kappa distribution, that is, the $(d_K \cdot N)$-dimensional kappa distribution of the velocities of N-particles

$$P(\vec{u}_1,\ldots,\vec{u}_N)=(\pi\kappa_0\theta^2)^{-\frac{1}{2}d_K N}\cdot\frac{\Gamma(\kappa_0+1+\frac{d_K}{2}N)}{\Gamma(\kappa_0+1)}\cdot\left[1+\frac{1}{\kappa_0}\cdot\frac{1}{\theta^2}\sum_{n=1}^{N}(\vec{u}_{(n)}-\vec{u}_b)^2\right]^{-\kappa_0-1-\frac{1}{2}d_K N} . \quad (40)$$

The N-particle kappa distribution in terms of the kinetic energies, $\varepsilon_{K(n)}=\frac{1}{2}m(\vec{u}_{(n)}-\vec{u}_b)^2$, is

$$P(\varepsilon_{K(1)},\ldots,\varepsilon_{K(N)})=(\kappa_0 k_B T)^{-\frac{1}{2}d_K N}\cdot\frac{\Gamma(\kappa_0+1+\frac{d_K}{2}N)}{\Gamma(\kappa_0+1)\cdot\Gamma(\frac{d_K}{2})^N}\cdot\left(1+\frac{1}{\kappa_0}\cdot\frac{1}{k_B T}\sum_{n=1}^{N}\varepsilon_{K(n)}\right)^{-\kappa_0-1-\frac{1}{2}d_K N}\cdot\prod_{n=1}^{N}\varepsilon_{K(n)}^{\frac{1}{2}d_K-1} . \quad (41)$$

For example, the two-particle velocity distribution, with d_K=3 degrees of freedom per particle, is

$$P(\vec{u}_1,\vec{u}_2)=\pi^{-3}\theta^{-6}\cdot\frac{(\kappa_0+1)(\kappa_0+2)(\kappa_0+3)}{\kappa_0^{3}}\cdot\left[1+\frac{1}{\kappa_0}\cdot\frac{(\vec{u}_1-\vec{u}_b)^2+(\vec{u}_2-\vec{u}_b)^2}{\theta^2}\right]^{-\kappa_0-4} , \quad (42a)$$

or, in terms of the kinetic energy,

$$P(\varepsilon_{K(1)},\varepsilon_{K(2)})=\frac{4}{\pi}(k_B T)^{-3}\cdot\frac{(\kappa_0+1)(\kappa_0+2)(\kappa_0+3)}{\kappa_0^{3}}\cdot\left(1+\frac{1}{\kappa_0}\cdot\frac{\varepsilon_{K(1)}+\varepsilon_{K(2)}}{k_B T}\right)^{-\kappa_0-4}\cdot\sqrt{\varepsilon_{K(1)}\varepsilon_{K(2)}} . \quad (42b)$$

Therefore, the variance, co-variance, and (Pearson's) correlation R, are given by

$$\sigma^2_{(1)(1)} = \left\langle \varepsilon_{\mathrm{K}(1)}{}^2 \right\rangle - \left\langle \varepsilon_{\mathrm{K}(1)} \right\rangle^2 = \tfrac{3}{2} \frac{\kappa_0 + \frac{3}{2}}{\kappa_0 - 1} , \tag{43a}$$

$$\sigma^2_{(1)(2)} = \left\langle \varepsilon_{\mathrm{K}(1)} \varepsilon_{\mathrm{K}(2)} \right\rangle - \left\langle \varepsilon_{\mathrm{K}(1)} \right\rangle \left\langle \varepsilon_{\mathrm{K}(2)} \right\rangle = \tfrac{9}{4} \frac{1}{\kappa_0 - 1} , \tag{43b}$$

$$R = \frac{\sigma^2_{(1)(2)}}{\sigma_{(1)(1)} \sigma_{(2)(2)}} = \frac{\frac{3}{2}}{\kappa_0 + \frac{3}{2}} = \frac{\frac{3}{2}}{\kappa} . \tag{43c}$$

3.4. Non-zero potential energy

So far we have presented and discussed kappa distributions of the velocities or kinetic energies, considering the absence of any potential energy. In the presence of a potential energy Φ, however, the kappa distribution depends on both the velocities (kinetic energies) and the positions, thus it describes the particle phase-space. Indeed, the phase-space kappa distribution in given by

$$P(\vec{r},\vec{u}\,) \propto \left[1 + \frac{1}{\kappa_0} \cdot \frac{\varepsilon_{\mathrm{K}}(\vec{u}) + \Phi(\vec{r})}{k_{\mathrm{B}} T} \right]^{-\kappa_0 - 1 - \frac{1}{2} d_{\mathrm{K}} - \frac{1}{2} d_{\Phi}} , \quad \Phi(\vec{r}) > 0 , \tag{44a}$$

$$P(\vec{r},\vec{u}) \propto \left[\frac{|\Phi(\vec{r})| - \varepsilon_{\mathrm{K}}(\vec{u})}{\kappa_0 k_{\mathrm{B}} T} - 1 \right]_+^{\kappa_0 - 1 - \frac{1}{2} d_{\mathrm{K}} + \frac{1}{2} d_{\Phi}} , \quad \Phi(\vec{r}) < 0 , \tag{44b}$$

(the operator "+" means $[x]_+=0$ when $x \leq 0$), where the kinetic and potential degrees of freedom are

$$\tfrac{1}{2} d_{\mathrm{K}} = \frac{\langle \varepsilon_{\mathrm{K}} \rangle}{k_{\mathrm{B}} T} , \; \tfrac{1}{2} d_{\Phi} \equiv \frac{\langle \Phi \rangle}{k_{\mathrm{B}} T} . \tag{45}$$

If the phase-space kappa distribution is integrated over the positions, we arrive at the standard kappa distribution of velocities

$$\int d\vec{r} \;\rightarrow\; P(\vec{u}\,) \propto \left[1 + \frac{1}{\kappa_0} \cdot \frac{\varepsilon_{\mathrm{K}}(\vec{u})}{k_{\mathrm{B}} T} \right]^{-\kappa_0 - 1 - \frac{1}{2} d_{\mathrm{K}}} . \tag{46}$$

If the phase-space kappa distribution is integrated over the velocities, we derive the new type of positional kappa distribution

$$\int d\vec{u} \;\rightarrow\; P(\vec{r}) \propto \left[1 + \frac{\Phi(\vec{r})}{\kappa_0 k_{\mathrm{B}} T} \right]^{-\kappa_0 - 1 - \frac{1}{2} d_{\Phi}} , \quad \Phi(\vec{r}) > 0 , \tag{47a}$$

$$\int d\vec{u} \;\rightarrow\; P(\vec{r}) \propto \left[\frac{|\Phi(\vec{r})|}{\kappa_0 k_{\mathrm{B}} T} - 1 \right]_+^{-\kappa_0 - 1 + \frac{d_{\Phi}}{2}} , \quad \Phi(\vec{r}) < 0 . \tag{47b}$$

We proceed with central attractive potentials, $\Phi(\vec{r}) = \Phi(r)$. Two characteristic examples are the attractive power-laws with positive or negative exponents, corresponding to oscillation and gravitation types, respectively:

(i) Oscillation type, $\Phi(r) = +\frac{1}{b} k \cdot r^{+b}$:

- Phase-Space distribution:

$$P(\vec{r},\vec{u}\,)\propto\left\{1+\frac{1}{\kappa_0}\cdot\left[\frac{\varepsilon_{\rm K}(\vec{u})}{k_{\rm B}T}+\tfrac{d_r}{b}\left(\frac{|\vec{r}|}{r_0}\right)^b\right]\right\}^{-\kappa_0-1-\frac{1}{2}d_{\rm K}-\frac{1}{b}d_r} . \tag{48a}$$

- Positional distribution:

$$P(\vec{r})\propto\left[1+\frac{\frac{d_r}{b}}{\kappa_0}\cdot\left(\frac{|\vec{r}|}{r_0}\right)^b\right]^{-\kappa_0-1-\frac{1}{b}d_r} . \tag{48b}$$

(ii) Gravitation type, $\Phi(r)=-\frac{1}{b}k\cdot r^{-b}$:

- Phase-Space distribution:

$$P(\vec{r},\vec{u}\,)\propto\left\{\frac{1}{\kappa_0}\cdot\left[\tfrac{d_r}{b}\left(\frac{r_0}{|\vec{r}|}\right)^b-\frac{\varepsilon_{\rm K}}{k_{\rm B}T}\right]-1\right\}_+^{-\kappa_0-1-\frac{1}{2}d_{\rm K}+\frac{1}{b}d_r} . \tag{49a}$$

- Positional distribution:

$$P(\vec{r})\propto\left[\frac{\frac{d_r}{b}}{\kappa_0}\cdot\left(\frac{r_0}{|\vec{r}|}\right)^b-1\right]^{-\kappa_0-1+\frac{1}{b}d_r} . \tag{49b}$$

- Involved parameters for $\Phi(r)=\pm\frac{1}{b}k\,r^{\pm b}$:

$$r_0\equiv\left\langle r^{\pm b}\right\rangle^{\pm\frac{1}{b}}=\left(\frac{\frac{d_r}{b}k_{\rm B}T}{\frac{1}{b}k}\right)^{\pm\frac{1}{b}} ,\ \tfrac{1}{2}d_\Phi\equiv\pm\frac{\langle\Phi\rangle}{k_{\rm B}T}=\frac{\frac{1}{b}k}{k_{\rm B}T}r_0^{\pm b}=\tfrac{1}{b}d_{\rm r} . \tag{49c}$$

Below, we present two examples.

3.4.1. *Generalized barometric formula*

The potential energy is $\Phi(z)\propto z$, e.g., $\Phi(z)=mg\cdot z$ or $\Phi(z)=qE\cdot z$. The Hamiltonian is $H(z,\vec{u})=\frac{1}{2}m(\vec{u}-\vec{u}_{\rm b})^2+mg\cdot z$. We find:

- Phase-Space distribution:

$$P(z,\varepsilon_{\rm K};\kappa_0,T)\propto\left[1+\frac{1}{\kappa_0}\cdot\left(\frac{\varepsilon_{\rm K}}{k_{\rm B}T}+\frac{z}{z_0}\right)\right]^{-\kappa_0-2-\frac{1}{2}d_{\rm K}}\varepsilon_{\rm K}^{\frac{1}{2}d_{\rm K}-1} . \tag{50a}$$

- Positional distribution:

$$P(z;\kappa_0,T)\propto\left(1+\frac{1}{\kappa_0}\cdot\frac{z}{z_0}\right)^{-\kappa_0-2}\quad\text{with } z_0\equiv\frac{k_{\rm B}T}{mg} . \tag{50b}$$

- Density, n:

$$n(z)\propto P(z)\ ,\ n(z)=n(0)\cdot\left(1+\frac{1}{\kappa_0}\cdot\frac{z}{z_0}\right)^{-\kappa_0-2} . \tag{50c}$$

- Local Temperature $T(z)$:

$$\tfrac{d}{2}k_B T(z) = \frac{\int_0^\infty P(z,\varepsilon_K)\varepsilon_K d\varepsilon_K}{\int_0^\infty P(z,\varepsilon_K)d\varepsilon_K} = \tfrac{d}{2}k_B T\cdot\left(1+\frac{1}{\kappa_0}\cdot\frac{z}{z_0}\right) \quad T(z)=T(0)\cdot\left(1+\frac{1}{\kappa_0}\cdot\frac{z}{z_0}\right). \tag{50d}$$

Note that the spatial average of the local temperature, $T(z)$, is the global temperature, T.

- Thermal Pressure, $p(z) = n(z)\cdot k_B T(z)$:

The generalized *barometric formula* is given by

$$p(z;\kappa_0) = p(0)\cdot\left(1+\frac{1}{\kappa_0}\cdot\frac{mg\,z}{k_B T}\right)^{-\kappa_0-1} \quad \text{with } p(z;\kappa_0\to\infty) = p(0)\exp\left(-\frac{mg\,z}{k_B T}\right). \tag{50e}$$

3.4.2. Oscillation

Let the 1-dimensional oscillation described by the Hamiltonian: $H(z,u_z) = \tfrac{1}{2}m(u_z - u_{b_z})^2 + \tfrac{1}{2}k\cdot z^2$.

- Phase-Space distribution:

$$P(z,u_z)\propto\left\{1+\frac{1}{\kappa_0}\cdot\left[\frac{\varepsilon_K(u_z)}{k_B T}+\tfrac{1}{2}\left(\frac{z}{z_0}\right)^2\right]\right\}^{-\kappa_0-2} \quad \text{or } P(z,u_z)\propto\left[1+\frac{1}{\kappa_0}\cdot\frac{u_z^{\,2}+(k/m)\cdot z^2}{\theta^2}\right]^{-\kappa-2}. \tag{51a}$$

- Positional distribution:

$$P(z)\propto\left[1+\frac{1}{2\kappa_0}\cdot\left(\frac{z}{z_0}\right)^2\right]^{-\kappa_0-\frac{3}{2}} \quad \text{or } P(z)\propto\left[1+\frac{1}{\kappa_0}\cdot\frac{(k/m)\cdot z^2}{\theta^2}\right]^{-\kappa_0-\frac{3}{2}}, \text{ with } z_0=\sqrt{\frac{k_B T}{k}}. \tag{51b}$$

These kappa distributions were used by [49] to describe the 1-dimensional speed and altitude of OH in the plasma near Enceladus. Interestingly, the kappa distribution used for describing the OH velocities and the kappa distribution used for describing the OH positions had the same kappa index and temperature, as it is predicted by the theory (Figure 4).

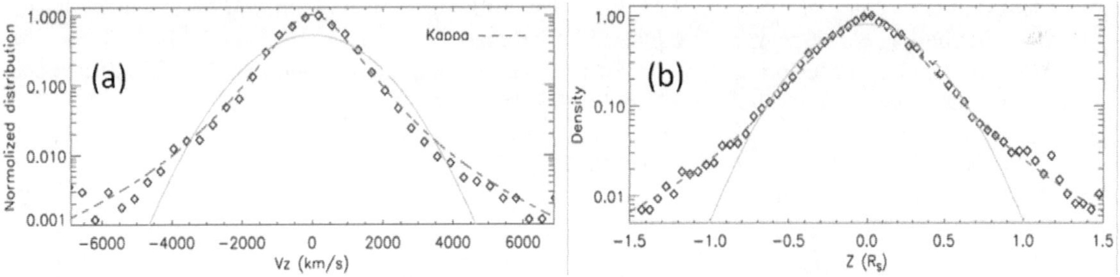

Figure 4. Density and velocity distribution of the OH cloud at Saturn. The vertical density profile is well fit to a positional kappa distribution (see: Eq.(51b); [95], Chapter 3). The velocity distribution of individual particles is non-Maxwellian, and also well fit to a kappa distribution. (Taken from [49])

4. Conclusions

Empirical kappa distributions have become increasingly widespread across space and plasma physics. Space plasmas from the solar wind to planetary magnetospheres and the outer heliosphere are systems out of thermal equilibrium, described by kappa distributions. A breakthrough in the field came with the connection of kappa distributions to non-extensive statistical mechanics. This paper presented important aspects of the statistical origin of kappa distributions, as well as theoretical developments and applications, such as (i) the concept of temperature; (ii) the physical meaning of the kappa index;

(iii) the *N*-particle description of kappa distributions; and the (iv) the generalization to phase-space kappa distribution of a Hamiltonian with non-zero potential.

References

[1] Vasyliūnas, V. M. 1968 A survey of low-energy electrons in the evening sector of the magnetosphere with OGO 1 and OGO 3. *J. Geophys. Res.* **73**, 2839–2884.
[2] Dialynas, K., S. M. Krimigis, D. G. Mitchell, D. C. Hamilton, N. Krupp, and P. C. Brandt 2009 Energetic ion spectral characteristics in the Saturnian magnetosphere using Cassini/MIMI measurements. *J. Geophys. Res.* **114**, A01212.
[3] Yoon, P. H.; L.F. Ziebell, R. Gaelzer, R.P. Lin, L. Wang, 2012 Langmuir turbulence and suprathermal electrons, *Space Sci. Rev.* **173**, 459–489.
[4] Štverák, S., Maksimovic M., Travnicek P. M., Marsch E., Fazakerley A. N., and Scime E. E. 2009 Radial evolution of nonthermal electron populations in the low-latitude solar wind: Helios, Cluster, and Ulysses Observations. *J. Geophys. Res.* **114**, A05104.
[5] Lazar, M., V. Pierrard, R. Schlickeiser and S. Poedts 2012 Modeling space plasma dynamics with anisotropic Kappa distributions, *Astrophys. Space Sci.* **33**, 97-107.
[6] Livadiotis, G. 2015 Statistical background and properties of kappa distributions in space plasmas. *J. Geophys. Res.* **120**, 1607–1619.
[7] Maksimovic, M., V. Pierrard and J. Lemaire 1997 A kinetic model of the solar wind with Kappa distributions in the corona. *Astron. Astrophys.* **324**, 725-734.
[8] Pierrard, V., M. Maksimovic, and J. Lemaire 1999 Electron velocity distribution function from the solar wind to the corona. *J. Geophys. Res.* **104**, 17021-17032.
[9] Mann, G., H. T. Classen, E. Keppler, and E. C. Roelof 2002 On electron acceleration at CIR related shock waves. *Astron. Astrophys.* **391**, 749–756.
[10] Marsch, E. 2006 Kinetic physics of the solar corona and solar wind, *Living Rev. Sol. Phys.* **3**, 1.
[11] Zouganelis, I. 2008 Measuring suprathermal electron parameters in space plasmas: Implementation of the quasi-thermal noise spectroscopy with kappa distributions using in situ Ulysses/URAP radio measurements in the solar wind. *J. Geophys. Res.* **113**, A08111.
[12] Livadiotis, G. and McComas, D.J. 2010 Exploring transitions of space plasmas out of equilibrium. *Astrophys. J.* **714**, 971.
[13] Livadiotis, G. and McComas, D.J. 2011 The influence of pick-up ions on space plasma distributions. *Astrophys. J.* **738**, 64.
[14] Livadiotis, G. and McComas, D.J. 2013 Understanding kappa distributions: A toolbox for space science and astrophysics. *Space Sci. Rev.* **75**, 183–214.
[15] Livadiotis, G. and McComas, D.J. 2013 Near-equilibrium heliosphere – Far-equilibrium heliosheath. *AIP Conf. Proc.* **1539**, 344-350.
[16] Yoon, P. H. 2014 Electron kappa distribution and quasi-thermal noise. *J. Geophys. Res.* **119**, 7074.
[17] Pierrard, V. and Pieters, M. 2015 Coronal heating and solar wind acceleration for electrons, protons, and minor ions, obtained from kinetic models based on kappa distributions. *J. Geophys. Res.* **119**, 9441.
[18] Pavlos, G.P.; O.E. Malandraki, E.G. Pavlos, A.C. Iliopoulos, L.P. Karakatsanis 2016 Non-extensive statistical analysis of magnetic field during the March 2012 ICME event using a multi-spacecraft approach. *Physica A* **464**, 149–181.
[19] Dzifčáková, E., and J. Dudík 2013 H to Zn ionization equilibrium for the non-Maxwellian electron κ-distributions: Updated calculations, *Astrophys. J. Suppl. Ser.* **206**, 6.
[20] Dzifčáková, E., J. Dudík, P. Kotrč, F. Fárník and A. Zemanová 2015 KAPPA: A package for synthesis of optically thin spectra for the non-Maxwellian κ-distributions based on the Chianti database. *Astrophys. J. Suppl. Ser.* **217**, 14.
[21] Owocki, S.P., and J.D. Scudder 1983 The effect of a non-Maxwellian electron distribution on oxygen and iron ionization balances in the solar corona. *Astrophys. J.* **270**, 758-768.

[22] Vocks, C., G. Mann, and G. Rausche 2008 Formation of suprathermal electron distributions in the quiet solar corona. *Astron. Astrophys.* **480**, 527–536.

[23] Lee, E., D. R. Williams, and G. Lapenta 2013 Spectroscopic indication of suprathermal ions in the solar corona, arXiv:1305.2939v.

[24] Cranmer, S. R. 2014 Suprathermal electrons in the solar corona: Can nonlocal transport explain heliospheric charge states? *Astrophys. J. Lett.* **791**, L31.

[25] Xiao, F., C. Shen, Y. Wang, H. Zheng, and S. Whang 2008 Energetic electron distributions fitted with a kappa-type function at geosynchronous orbit. *J. Geophys. Res.* **113**, A05203.

[26] Laming, J. M., J. D. Moses, Y.-K. Ko, C. K. Ng, C. E. Rakowski, and A. J. Tylka (2013), On the remote detection of suprathermal ions in the solar corona and their role as seeds for solar energetic particle production. *Astrophys. J.* **770**, 73.

[27] Chotoo, K., et al. 2000 The suprathermal seed population for corotaing interaction region ions at 1AU deduced from composition and spectra of H+, He++, and He+ observed by Wind. *J. Geophys. Res.* **105**, 23107–23122.

[28] Mann, G., A. Warmuth, and H. Aurass 2009 Generation of highly energetic electrons at reconnection outflow shocks during solar flares. *Astron. Astrophys.* **494**, 669-675.

[29] Livadiotis, G. and McComas, D.J. 2013 Evidence of large scale phase space quantization in plasmas. *Entropy* **15**, 1118-1132.

[30] Bian, N., G.A. Emslie, D.J. Stackhouse, and E.P. Kontar 2014 The formation of a kappa-distribution accelerated electron populations in solar flares. *Astrophys. J.* **796**, 142.

[31] Jeffrey, N. L. S., Fletcher, L., and Labrosse, N. 2016 First evidence of non-Gaussian solar flare EUV spectral line profiles and accelerated non-thermal ion motion. *Astron. Astrophys.* **590**, A99.

[32] Formisano, V., G. Moreno, F. Palmiotto, and P. C. Hedgecock 1973 Solar Wind Interaction with the Earth's Magnetic Field 1. Magnetosheath, *J. Geophys. Res.* **78**, 3714-3730.

[33] Ogasawara, K., Angelopoulos, V., Dayeh, M.A., Fuselier, S.A., Livadiotis, G., McComas, D.J., McFadden, J.P. 2013 Characterizing the dayside magnetosheath using ENAs: IBEX and THEMIS observations. *J. Geophys. Res.* **118**, 3126-3137.

[34] Ogasawara, K., Dayeh, M.A., Funsten, H.O., Fuselier, S.A., Livadiotis, G., and McComas, D.J. 2015 Interplanetary magnetic field dependence of the suprathermal energetic neutral atoms originated in subsolar magnetopause. *J. Geophys. Res.* **120**, 964-972.

[35] Grabbe, C. 2000 Generation of broadband electrostatic waves in Earth's magnetotail. *Phys. Rev. Lett.* **84**, 3614.

[36] Pisarenko, N. F., E.Yu. Budnik, Yu.I. Ermolaev, I.P. Kirpichev, V.N. Lutsenko, E.I. Morozova, and E.E. Antonova 2002 The ion differential spectra in outer boundary of the ring current: November 17, 1995 case study. *J. Atm. Solar-Terr. Phys.* **64**, 573 – 583.

[37] Christon, S. P. 1987 A comparison of the Mercury and earth magnetospheres: electron measurements and substorm time scales. *Icarus* **71**, 448-471.

[38] Wang, C.-P., L. R. Lyons, M. W. Chen, R. A. Wolf, and F. R. Toffoletto 2003 Modeling the inner plasma sheet protons and magnetic field under enhanced convection. *J. Geophys. Res.* **108**, 1074.

[39] Kletzing, C. A., J. D. Scudder, E. E. Dors, and C. Curto 2003 Auroral source region: Plasma properties of the high latitude plasma sheet. *J. Geophys. Res.* **108**, 1360.

[40] Ogasawara, K., Livadiotis, G., Grubbs, G.A., Jahn, J.-M., Michell, R., Samara, M., Sharber, J. R., and Winningham, J. D. 2017 Properties of suprathermal electrons associated with discrete auroral arcs, *Geophys. Res. Lett.* In Press.

[41] Hapgood, M., C. Perry, J. Davies and M. Denton 2011 The role of suprathermal particle measurements in CrossScale studies of collisionless plasma processes, *Planet. Space Sci.* **59**, 618–629.

[42] Collier, M.R., and D.C. Hamilton 1995 The relationship between kappa and temperature in the energetic ion spectra at Jupiter. *Geophys. Res. Lett.* **22**, 303-306.

[43] Mauk, B. H., D. G. Mitchell, R. W. McEntire, C. P. Paranicas, E. C. Roelof, D. J. Williams, S. M. Krimigis, and A. Lagg 2004 Energetic ion characteristics and neutral gas interactions in Jupiter's magnetosphere. *J. Geophys. Res.* **109**, A09S12.

[44] Livi, R., Goldstein, J., Burch, J. L., Crary, F., Rymer, A. M., Mitchell, D. G., and Persoon, A. M. 2014 Multi-instrument analysis of plasma parameters in Saturn's equatorial, inner magnetosphere using corrections for spacecraft potential and penetrating background radiation. *J. Geophys. Res.* **119**, 3683.

[45] Carbary, J. F., M. Kane, B. H. Mauk and S. M. Krimigis 2014 Using the kappa function to investigate hot plasma in the magnetospheres of the giant planets. *J. Geophys. Res.* **119**, 8426–8447.

[46] Mauk, B. H., S. M. Krimigis, E. P. Keath, A. F. Cheng, T. P. Armstrong, L. J. Lanzerotti, G. Gloeckler, and D. C. Hamilton 1987 The hot plasma and radiation environment of the Uranian magnetosphere. *J. Geophys. Res.* **92**, 15283.

[47] Krimigis, S. M., et al. 1989 Hot plasma and energetic particles in Neptune's magnetosphere. *Science* **246**, 1483.

[48] Moncuquet, M., F. Bagenal, and N. Meyer-Vernet 2002 Latitudinal structure of the outer Io plasma torus. *J. Geophys. Res.* **108**, 1260.

[49] Jurac, S., M. A. McGrath, R. E. Johnson, J. D. Richardson, V. M. Vasyliunas, and A. Eviatar 2002 Saturn: Search for a missing water source. *Geophys. Res. Lett.* **29**, 2172.

[50] Broiles, T.W., Livadiotis, G., Burch, J.L., Chae, K., Clark, G., Cravens, T.E., Davidson, R., Eriksson, A., Frahm, R.A., Fuselier, S.A., Goldstein, J, Goldstein, R, Henri, P, Madanian, H, Mandt, KE, Mokashi, P., Pollock, C., Rahmati, A., Samara, M., Schwartz, S.J. 2016 Characterizing cometary electrons with kappa distributions. *J. Geophys. Res.* **121**, 7407-7422.

[51] Broiles, T.W., Burch, J.L., Chae, K., Clark, G., Cravens, T.E., Eriksson, A., Fuselier, S.A., Frahm, R.A., Gasc, S., Goldstein, R, Henri, P, Koenders, C., Livadiotis, G., Mandt, K.E., Mokashi, P., Nemeth, Z., Rubin, M. and Samara, M. 2016 Statistical analysis of suprathermal electron drivers at 67P/Churyumov-Gerasimenko. *MNRAS* **462**, S312-S322.

[52] Decker, R. B. and Krimigis, S. M. 2003 Voyager observations of low-energy ions during solar cycle 23. *Adv. Space Res.* **32**, 597–602.

[53] Decker, R. B., et al. 2005 Voyager 1 in the foreshock, termination shock, and heliosheath. *Science* **309**, 2020–2024.

[54] Heerikhuisen, J., Pogorelov, N. V., Florinski, V., Zank, G. P. and le Roux, J. A. 2008 The effects of a k-distribution in the heliosheath on the global heliosphere and ENA flux at 1 AU. *Astrophys. J.* **682**, 679-689.

[55] Heerikhuisen, J., Zirnstein, E., Pogorelov, N. 2015 κ-distributed protons in the solar wind and their charge-exchange coupling to energetic hydrogen. *J. Geophys. Res.* **120**, 1516–1525.

[56] Zank, G. P.; J. Heerikhuisen, N. V. Pogorelov, R. Burrows, D. J. McComas 2010 Microstructure of the heliospheric termination shock: implications for energetic neutral atom observations. *Astrophys. J.* **708**, 1092.

[57] Livadiotis, G. 2014 Lagrangian temperature: Derivation and physical meaning for systems described by kappa distributions. *Entropy* **16**, 4290-4308.

[58] Livadiotis, G. 2016, Curie law for systems described by kappa distributions. *EPL* **113**, 10003.

[59] Livadiotis, G. and McComas, D.J. 2011 Invariant kappa distribution in space plasmas out of equilibrium. *Astrophys. J.* **741**, 88.

[60] Livadiotis, G. and McComas, D.J. 2012 Non-equilibrium thermodynamic processes: Space plasmas and the inner heliosheath. *Astrophys. J.* **749**, 11.

[61] Livadiotis, G., McComas, D.J, Dayeh, M.A., Funsten, H.O. and Schwadron, N.A. 2011 First sky map of the inner heliosheath temperature using IBEX spectra. *Astrophys. J.* **734**, 1.

[62] Livadiotis, G., McComas, D. J., Randol, B., Möbius, E., Dayeh, M. A., Frisch, P. C., Funsten, H. O., Schwadron, N. A. and Zank, G. P. 2012 Pick-up ion distributions and their influence

on ENA spectral curvature. *Astrophys. J.* **751**, 64.

[63] Livadiotis, G., McComas, D. J., Schwadron, N. A., Funsten, H. O. and Fuselier, S. A. 2013 Pressure of the proton plasma in the inner heliosheath. *Astrophys. J.* **762**, 134.

[64] Fuselier, S. A., Allegrini, F., Bzowski, M., Dayeh, M. A., Desai, M., Funsten, H. O., Galli, A., Heirtzler, D., Janzen, P., Kubiak, M. A., Kucharek, H., Lewis, W., Livadiotis, G., McComas, D. J., Möbius, E., Petrinec, S. M., Quinn, M., Schwadron, N., Sokół, J. M., Trattner, K. J., Wood, B. E., Wurz, P. 2014 Low energy neutral atoms from the heliosheath. *Astrophys. J.*, **784**, 89.

[65] Zirnstein, E. J. and McComas, D. J. 2015 Using kappa functions to characterize outer heliosphere proton distributions in the presence of charge-exchange. *Astrophys. J.* **815**, 31.

[66] Zank, G. P. 2015 Faltering steps into the galaxy: The boundary regions of the heliosphere. *Ann. Rev. Astron. Astrophys.* **53**, 449.

[67] Nicholls, D. C., Dopita, M. A., Sutherland, R. S. 2012 Resolving the Electron Temperature Discrepancies in H II Regions and Planetary Nebulae: κ-distributed Electrons. *Astrophys. J.* **752**, 148.

[68] Nicholls, D. C., Dopita, M. A., Sutherland, R. S., Kewley, L. J., Palay, E. 2013 Measuring nebular temperatures: the effect of new collision strengths with equilibrium and κ-distributed electron energies. *Astrophys. J. Supp.* **207**, 21.

[69] Zhang, Y., Liu, X.-W. and Zhang B. 2014 H-I free-bound emission of planetary nebulae with large abundance discrepancies: Two-component models vs. κ-distributed electrons. *Astrophys. J.* **780**, 93.

[70] Raymond, J. C., P. F. Winkler, W. P. Blair, J.-J. Lee, and S. Park 2010 Non-Maxwellian Hα profiles in Tycho's supernova remnant, *Astrophys. J.* **712**, 901.

[71] Milovanov A. V., and L. M. Zelenyi 2000 Functional background of the Tsallis entropy: "coarse-grained" systems and "kappa" distribution functions, *Nonlin. Processes Geophys.* **7**, 211–221.

[72] Saito, S., F. R. E. Forme, S. C. Buchert, S. Nozawa, R. Fujii 2000 Effects of a kappa distribution function of electrons on incoherent scatter spectra, *Ann. Geophys.* **18**, 1216-1223.

[73] Du, J. 2004 The nonextensive parameter and Tsallis distribution for self-gravitating systems. *EPL* **67**, 893–899.

[74] Yoon, P. H.; T. Rhee, C. M. Ryu, 2006 Self-consistent formation of electron κ distribution: 1. Theory, *J. Geophys. Res.* **111**, A09106.

[75] Raadu, M. A., and Shafiq, M. 2007 Test charge response for a dusty plasma with both grain size distribution and dynamical charging. *Phys. Plasmas* **14**, 012105.

[76] Livadiotis, G. 2009 Approach on Tsallis statistical interpretation of hydrogen-atom by adopting the generalized radial distribution function. *J. Math. Chem.* **45**, 930-939.

[77] Livadiotis, G. 2015 Kappa distribution in the presence of a potential energy. *J. Geophys. Res.* **120**, 880.

[78] Livadiotis, G. 2015 Kappa and *q* indices: Dependence on the degrees of freedom. *Entropy* **17**, 2062.

[79] Tribeche, M., Mayout, S. and Amour, R. 2009 Effect of ion suprathermality on arbitrary amplitude dust acoustic waves in a charge varying dusty plasma. *Phys. Plasmas* **16**, 043706.

[80] Livadiotis, G. and McComas, D.J. 2010 Measure of the departure of the q-metastable stationary states from equilibrium. *Phys. Scr.* **82**, 035003.

[81] Livadiotis, G. and McComas, D.J. 2014 Electrostatic shielding in plasmas and the physical meaning of the Debye length. *J. Plasma Phys.* **80**, 341-378.

[82] Baluku, T. K.; M. A. Hellberg, I. Kourakis, N. S. Saini, 2010 Dust ion acoustic solitons in a plasma with kappa-distributed electrons. *Phys. Plasmas* **17**, 053702.

[83] Le Roux, J. A.; G. M. Webb, A. Shalchi, G. P. Zank 2010 A generalized nonlinear guiding center theory for the collisionless anomalous perpendicular diffusion of cosmic rays.

Astrophys. J. **716**, 671-692.

[84] Eslami, P., Mottaghizadeh, M., and Pakzad, H. R. 2011 Nonplanar dust acoustic solitary waves in dusty plasmas with ions and electrons following a q-nonextensive distribution. *Phys. Plasmas* **18**, 102303.

[85] Kourakis, I., Sultana, S. Hellberg, M. A. 2012 Dynamical characteristics of solitary waves, shocks and envelope modes in kappa-distributed non-thermal plasmas: an overview. *Plasma Phys. Control. Fusion* **54**, 124001.

[86] Randol, B. M. and Christian, E.R. 2016 Coupling of charged particles via Coulombic interactions: Numerical simulations and resultant kappa-like velocity space distribution functions. *J. Geophys. Res.* **121**, 1907–1919.

[87] Varotsos, P.A., Sarlis, N.V., Skordas, E.S., 2014 Study of the temporal correlations in the magnitude time series before major earthquakes in Japan. *J. Geophys. Res.* **119**, 9192–9206

[88] Nicolaou, G., Livadiotis, G., and Moussas, X. 2014 Long term variability of the polytropic Index of solar wind protons at ~1AU. *Sol. Phys.* **289**, 1371-1378.

[89] Fisk, L. A., and G. Gloeckler 2014 The case for a common spectrum of particles accelerated in the heliosphere: Observations and theory. *J. Geophys. Res.* **119**, 8733.

[90] Viñas, A. F., P. S. Moya, R. E. Navarro, J. A. Valdivia, J. A. Araneda, and V. Muñoz 2015 Electromagnetic fluctuations of the whistler-cyclotron and firehose instabilities in a Maxwellian and Tsallis-kappa-like plasma, *J. Geophys. Res.* **120**, 3307-3317.

[91] Ourabah, K., Ait Gougam, L., and Tribeche, M. 2015 Nonthermal and suprathermal distributions as a consequence of Superstatistics. *Phys. Rev. E.* **91**, 012133.

[92] Dos Santos, M. S., L. F. Ziebell, and R. Gaelzer 2016 Ion firehose instability in a dusty plasma considering product-bi-kappa distributions for the plasma particles. *Phys. Plasmas* **23**, 013705.

[93] Nicolaou, G. and Livadiotis, G. 2016 Misestimation of temperature when applying Maxwellian distributions to space plasmas described by kappa distributions. *Astrophys. Space Sci* **361**, 359.

[94] Tsallis, C. 2009 Introduction to Nonextensive Statistical Mechanics, New York: Springer.

[95] Livadiotis, G. 2017 *Kappa distribution: Theory & Applications in plasmas* Netherlands, UK, US: Elsevier.

[96] Binsack, J. H. 1966 Plasma studies with the IMP-2 satellite, Ph.D. Thesis, MIT.

[97] Olbert, S. 1968 Summary of experimental results from M.I.T. detector on IMP-1, in Physics of the Magnetosphere, R. L. Carovillano, J. F. McClay, and H. R. Radoski (Eds), New York: Springer, 641.

[98] Livadiotis, G. and McComas, D.J. 2009 Beyond kappa distributions: Exploiting Tsallis statistical mechanics in space plasmas. *J. Geophys. Res.* **114**, A11105.

[99] Tsallis, C. 1988 Possible generalization of Boltzmann-Gibbs statistics. *J. Stat. Phys.* **52**, 479.

[100] Livadiotis, G. 2017 On the simplification of statistical mechanics for space plasmas. Submitted.

[101] Gibbs, J.W., 1902, Elementary Principles in Statistical Mechanics, Scribner's sons, New York.

[102] Beck C. and Schlögl F. 1993 *Thermodynamics of Chaotic Systems*, Cambridge University Press.

[103] Livadiotis, G. 2017 Derivation of the entropic formula for the statistical mechanics of space plasmas. Submitted.

Acknowledgments
The work was supported in part by the project NNX17AB74G of NASA's HGI Program.

IOP Conf. Series: Journal of Physics: Conf. Series **900** (2017) 012015 doi:10.1088/1742-6596/900/1/012015

The Interaction of Coronal Mass Ejections with Alfvénic Turbulence

Ward Manchester IV and Bart van der Holst
University of Michigan, Space Research Lab, 2455 Hayward Drive, Ann Arbor, Michigan
E-mail: chipm@umich.edu

Abstract. We provide a first attempt to understand the interaction between Alfvén wave turbulence, kinetic instabilities and temperature anisotropies in the environment of a fast coronal mass ejection (CME) near the Sun. The impact of a fast CME on the solar corona causes turbulent energy, thermal energy and dissipative heating to increase by orders of magnitude, and produces conditions suitable for a host of kinetic instabilities. We study these CME-induced effects with the recently developed Alfvén Wave Solar Model, with which we are able to self-consistently simulate the turbulent energy transport and dissipation as well as isotropic electron heating and anisotropic proton heating. Furthermore, the model also offers the capability to address the effects of fire hose, mirror mode, and cyclotron kinetic instabilities on proton energy partitioning all in a global-scale numerical simulation. We find amplified turbulent energy in the CME sheath, along with strong wave reflection at the shock combine to cause wave dissipation rates to increase by more than a factor of 100. In contrast, wave energy is greatly diminished by adiabatic expansion in the flux rope. Finally, we find proton temperature anisotropies are limited by kinetic instabilities to a level consistent with solar wind observations.

1. Introduction

Alfvén waves are regularly measured *in situ* in the solar wind, [2] and have more recently been remotely observed in the solar corona [6], where their energy is sufficient to heat and accelerate the solar wind. Based on these and similar observations, theories have been developed, which describe the evolution and transport of Alfvénic turbulence, e.g., [28, 20, 30, 31]. To self-consistently describe the heating and acceleration of the solar wind in response to turbulence, several extended magnetohydrodynamical (MHD) have been developed [25, 24, 4, 26, 12]. The recently developed Alfvén Wave Solar Model (AWSoM), [23, 27, 21] extends prior simulation capability with a three-dimensional solar corona/solar wind model that incorporates low-frequency Alfvén wave turbulence with three-temperature thermodynamics that captures the electron temperature in addition to resolving the anisotropic proton temperature into components parallel and perpendicular to the magnetic field.

AWSoM has already been used to study thermal structure of the steady state corona, and has been shown capable of explaining coronal non-thermal line-broadening with the turbulent motions of Alfvén waves [22]. The same work has shown it possible to explain the charge state composition of the slow wind with temperature evolution of plasma on open field lines at coronal hole boundaries. Manchester et al. [17] and Jin et al. [9] have employed AWSoM (in two-temperature mode [26, 27]) to study the coupled electron-proton temperature structure of coronal mass ejections (CMEs). These works found protons heated by the shock and energy

IOP Conf. Series: Journal of Physics: Conf. Series **900** (2017) 012015 doi:10.1088/1742-6596/900/1/012015

transferred to electrons by Coulomb collisions low in the corona where the collision frequency is high. Brief thermal coupling allowed electrons to attain no more than 10% of the maximum temperature of the protons, while electron thermal conduction allowed this heat to rapidly propagate ahead of the CME on open magnetic field lines forming a precursor of hot electrons ahead of the shock.

Here, we simulate the effects of CMEs and CME-driven shocks on the transport and dissipation of Alfvén wave turbulence as well as shock heating and Joule heating with the three-temperature version of AWSoM [27, 21]. We consider low-frequency Alfvén waves, which are assumed to dissipate below the ion cyclotron frequency. Wave amplitudes are prescribed at the lower boundaries to match observed wave motions in the low corona [6]. Our work follows that of Adhikari et al. [1] who first applied a detailed one-dimensional steady state turbulent model [29] to simulate quasi-parallel and quasi-perpendicular interplanetary shocks. In contrast to Adhikari et al., our turbulent transport is simpler with a phenomenological treatment of the wave dissipation, with a prescribed correlation length inversely proportional to the magnetic field strength as opposed to a time evolving formulation. The wave spectrum is not captured, so we only evolve the total forward and backward propagating wave energy densities. The partitioning of dissipated wave energy between electrons and protons is fixed so information about the wave spectrum is not required. The only turbulence parameters are the wave energy densities, the correlation length and the reflection rate. Our wave reflection model is essential the same formulation of Matthaeus et al. [20]. In this case, the energy of the dominant wave is transferred to the counter-propagating minor wave, with the reflection coefficient controlled by the gradient of the Alfvén speed.

We these simplifications, we extend the Adhikari et al. study in several significant ways. First we determine how turbulence depends on enhanced wave reflection at the shock surface, which increases turbulent dissipation. Second, our model addresses temperature anisotropies caused by preferential perpendicular proton heating at shocks and current sheets. Third, we include the effects of three kinetic instabilities: fire hose, mirror mode, and cyclotron instabilities. These modes are included to limit temperature anisotropies with thresholds that are dependent on the proton temperature ratio and plasma β. Finally our three-dimensional model includes the entire structure of the CME-driven disturbance, including sheath region and magnetic flux rope. This is the first time such kinetic physics has been incorporated into a global numerical model of a CME propagating through the solar corona, which allows us to address both particle heating, Alfvén wave damping, and their nonlinear coupled interaction.

2. Model Description

2.1. Coronal Model: AWSoM

The simulations are performed with the Alfvén Wave Solar Model (AWSoM), a global magnetohydrodynamic model from the upper chromosphere to the corona and the heliosphere in which the coronal heating and solar wind acceleration are driven with low-frequency Alfvénic turbulence [27]. AWSoM is able to address the coupled evolution of Alfvén turbulence with the magnetized plasma of the corona and solar wind in a global context. The model extends to a temperature minimum of 50,000K so that energy in the lower corona can be transported back to the upper chromosphere via electron heat conduction where it is lost via radiative cooling. The electron and proton plasma is described with a single fluid with three temperatures, an isotropic electron temperature and an anisotropic proton population with separate temperatures parallel and perpendicular to the magnetic field. The model employs heat conduction only for electrons with constant adiabatic index ($\gamma = 5/3$) and adiabatic indices of two and three for parallel and perpendicular pressures.

Alfvénic turbulence is described with two wave populations that propagate parallel and anti-parallel to the magnetic field with enhanced turbulent dissipation where populations coexist at balanced levels. At the inner boundary, the Poynting flux of the outbound Alfvén waves is assumed to be proportional to the magnetic field strength. These outward propagating waves experience partial reflection on field-aligned Alfvén speed gradients and vorticity of the background. The nonlinear interaction between these oppositely propagating Alfvén waves results in an energy cascade through the inertial range to the smaller perpendicular gyro-radius scales, where the dissipation takes place. The apportioning of the dissipated wave energy to the electron and proton temperatures depends on kinetic physics that is not included in our model. The AWSoM model does possess a partitioning strategy based on the dissipation of kinetic Alfvén waves (KAWs). In particular the model employs the stochastic heating mechanism for the heating of the protons [3]. However, for this study, we use a simpler uniform partitioning strategy in which 40% of the wave dissipation is transferred to electron heating while the remaining 60% is used to heat the perpendicular proton temperature. Future work will examine the consequences of the self-consistent energy partitioning.

In the AWSoM model, we solve the MHD equations with proton temperature anisotropy augmented with equations for the low-frequency Alfvén wave turbulence. We employ a simplified, phenomenological form of the full turbulence transport equations that were first introduced by Zank et al. [28] and subsequently developed over the years [20, 29, 30, 31]. The following are a subset of the turbulence transport equations. For a full description of the numerical model, the reader is referred to van der Holst et al. [27]:

- The time evolution of the proton pressure is solved for the parallel proton pressure $P_{i\parallel}$ and the averaged ion pressure $P_i = (2P_{i\perp} + P_{i\parallel})/3$

$$\frac{\partial}{\partial t}\left(\frac{P_i}{\gamma-1} + \frac{\rho u^2}{2} + \frac{B^2}{2\mu_0}\right) + \nabla\cdot\left[\left(\frac{\rho u^2}{2} + \frac{P_i}{\gamma-1} + \frac{B^2}{\mu_0}\right)\mathbf{u} + \mathbf{P}_i\cdot\mathbf{u} - \frac{\mathbf{B}(\mathbf{u}\cdot\mathbf{B})}{\mu_0}\right]$$
$$= -\mathbf{u}\cdot\nabla(P_e + P_A) + \frac{N_i k_B}{\tau_{ei}}(T_e - T_i) + Q_i - \rho\frac{GM_\odot}{r^3}\mathbf{r}\cdot\mathbf{u}, \qquad (1)$$
$$\frac{\partial P_{i\parallel}}{\partial t} + \nabla\cdot(P_{i\parallel}\mathbf{u}) + 2P_{i\parallel}\mathbf{b}\cdot(\nabla\mathbf{u})\cdot\mathbf{b} = \frac{\delta P_{i\parallel}}{\delta t} + 2\frac{N_i k_B}{\tau_{ei}}(T_e - T_{i\parallel}) + 2Q_{i\parallel}, \qquad (2)$$

  where $T_{i\parallel}$ is the parallel proton temperature obtained from the equation of state $P_{i\parallel} = N_i k_B T_{i\parallel}$ and $\mathbf{P}_i = P_{i\perp}I + (P_{i\parallel} - P_{i\perp})\mathbf{bb}$ is the ion pressure tensor. The second term on the right hand sides of Equations (1) and (2) are the collisional energy exchanges with the electrons. The third term on the right hand sides are the heating functions Q_i and $Q_{i\parallel}$ for the averaged ion and parallel ion pressure, respectively. The first term on the right hand side of Equation (2), $\delta P_{i\parallel}/\delta t$, is for the relaxation of the pressure anisotropy by the parallel firehose, mirror, and ion-cyclotron instabilities [21]

- The electron pressure

$$\frac{\partial}{\partial t}\left(\frac{P_e}{\gamma-1}\right) + \nabla\cdot\left(\frac{P_e}{\gamma-1}\mathbf{u}\right) + P_e\nabla\cdot\mathbf{u} = -\nabla\cdot\mathbf{q}_e + \frac{N_i k_B}{\tau_{ei}}(T_i - T_e) - Q_{\rm rad} + Q_e, \qquad (3)$$

  in which Q_e is the electron coronal heating.

- The wave energy density $w_\pm$ (+ parallel to **B**, - antiparallel)

$$\frac{\partial w_\pm}{\partial t} + \nabla\cdot[(\mathbf{u}\pm\mathbf{V}_A)w_\pm] + \frac{w_\pm}{2}(\nabla\cdot\mathbf{u}) = \mp\mathcal{R}\sqrt{w_- w_+} - \Gamma_\pm w_\pm, \qquad (4)$$

  where

$$\Gamma_\pm = \frac{2}{L_\perp}\sqrt{\frac{w_\mp}{\rho}}, \qquad (5)$$

IOP Conf. Series: Journal of Physics: Conf. Series **900** (2017) 012015 doi:10.1088/1742-6596/900/1/012015

is the phenomenological dissipation rate. The signed reflection rate $\mathcal{R}$ is derived as

$$\mathcal{R} = \min\left[\mathcal{R}_{\text{imb}}, \max(\Gamma_{\pm})\right] \begin{cases} \left(1 - 2\sqrt{\frac{w_-}{w_+}}\right) & \text{if} \quad 4w_- \leq w_+ \\ 0 & \text{if} \quad 1/4w_- < w_+ < 4w_- \\ \left(2\sqrt{\frac{w_+}{w_-}} - 1\right) & \text{if} \quad 4w_+ \leq w_- \end{cases}, \tag{6}$$

$$\mathcal{R}_{\text{imb}} = \sqrt{\left[(\mathbf{V}_A \cdot \nabla) \log V_A\right]^2 + (\mathbf{b} \cdot [\nabla \times \mathbf{u}])^2}. \tag{7}$$

For this particular simulation, we model the solar corona for Carrington Rotation number 2052, corresponding to the time period 2006 December 12 to 2007 January 08. We the use GONG synoptic magnetograms to specify the magnetic field in the inner boundary of the model. To describe the time evolution of the turbulence, the following parameters are set. The Poynting flux at the inner boundary is proportional to the magnetic field, with proportionality constant 3×10^5 J/m^{2}sT. The outward propagating wave energy density is specified to match this energy flux in accordance with the local flow and Alfvén speeds. The single prescribed correlation length is set equal to $1.5 \times 10^5 \sqrt{(B)}$ m (where B is in Tesla), such that it scales with the diameter of a given flux tube.

2.2. CME Model: EEGGL

In order to simulate the CME, we employ the Gibson-Low flux rope model [7] with parameters specified by the Eruptive Event Generator Gibson-Low (EEGGL). This tool was developed by Jin et al. [11] and recently installed at the Community Coordinated Modeling Center (CCMC). Parameters are chosen to produce a fast CME with a speed of 1600 km/s certain to produce a fast-mode shock in the surrounding corona and solar wind. Without regard to any observed event, we select an active region from which to launch the CME. For these circumstances, EEGGL provides the following parameters: longitude =45.5°, latitude =-6.5°, orientation =330.7°, rope radius =0.22 R_a, and magnetic strength of 146 Gauss. The grid is uniformly refined in a spherical wedge in the path of the CME extending form 1.15 to 22 R_s and -30°to 60°latitude and 20°to 90°longitude. This background grid provides 1.0 degree resolution in latitude and longitude and radial resolution that varies from 0.01 to 0.25 Rs. Two additional levels of grid refinement (increasing resolution by a factor of four) are applied to the CME's path of propagation, which is sufficient to capture the shock and sheath structures.

3. Results

The flux rope and its entrained plasma are linearly superimposed in the closed field of the active region in a state of force imbalance. The flux rope expands rapidly and is ejected from the corona where it produces a fast CME that drives a fast-mode shock through the corona. This simulation is similar to earlier work [16, 14, 17, 9, 18, 10], however here we consider for the first time the interaction of the CME with the Alfvénic turbulence and proton temperature anisotropy.

We begin here with a detailed description of the interaction of the CME-disturbance with the low-frequency Alfvén turbulence. We give a representative description of the simulation results at time t = 2.0 hours after CME initiation, when the CME-driven shock has reached approximately 15 R_s. The model at this time is shown in Figure 1 where the CME disturbance is shown on the meridional cut plane. The plasma velocity shown in Panel (a) reveals the shock front extending into the fast wind over the pole and protruding into the slow-speed wind at low latitude where the CME propagates. Panels (b) through (d) show the reflection rate, total (sum of forward and counter-propagating) wave energy density, and wave dissipation rate respectively. Inspection reveals the remarkable impact the CME has on the Alfvénic turbulence. We find complex structured enhancements in wave energy, as well as reflection and dissipation

IOP Conf. Series: Journal of Physics: Conf. Series **900** (2017) 012015 doi:10.1088/1742-6596/900/1/012015

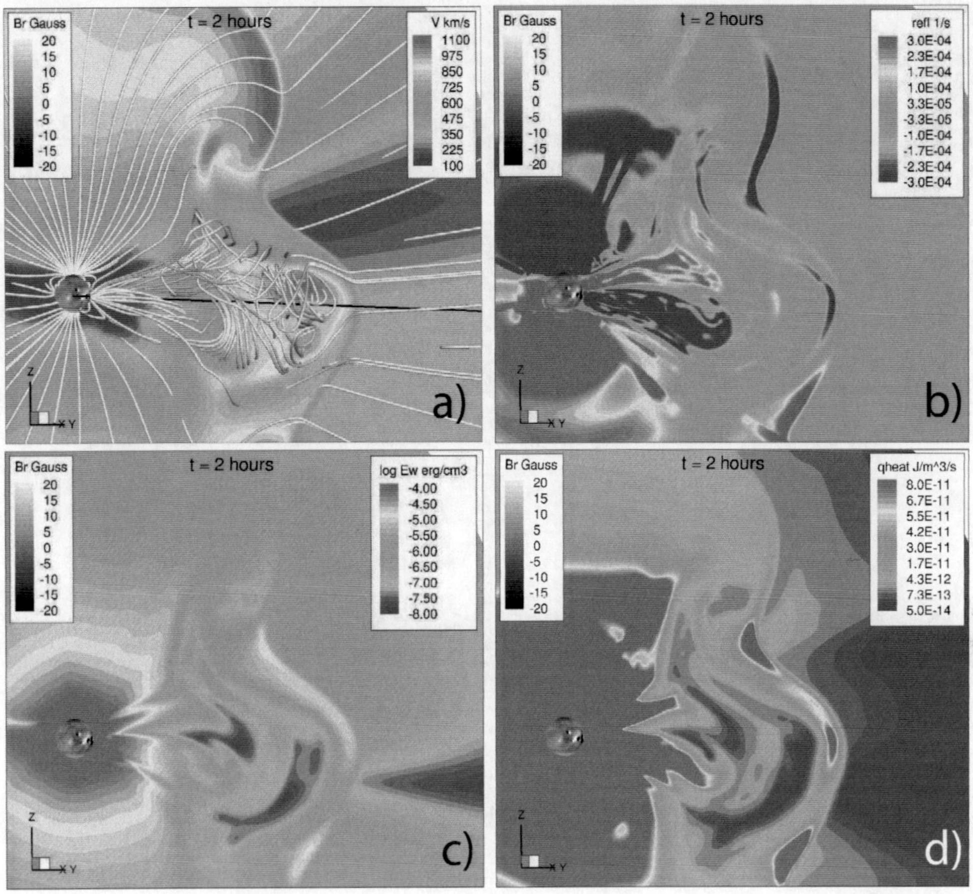

Figure 1. Wave related quantities are shown in color contours on a meridional cut plane passing though the center of the CME. two hours after initiation. Panel (a) shows the plasma velocity, which clearly illustrates the shock front propagating ahead of the CME and extending in to fast wind. Three-dimensional magnetic field lines are shown with the twisted and distorted field of the flux rope clearly evident. The black line is where data is extracted for line plots. Panel (b) shows the magnitude of the wave reflection coefficient, which is greatly enhanced at the shock front and in the CME. The total (forward and counter-propagating) wave energy density and wave dissipation rate are shown in panels (c) and (d) respectively. Levels of wave energy and dissipation are greatly enhanced in the sheath and suppressed in the flux rope.

rates, while also producing regions of depleted wave energy in the ejected flux rope. Panel (b) shows the enhancements in wave reflection are greatest at the shock front at the flanks of the CME and in the flux rope. The logarithmic representation in Panel (c) shows variation in wave energy density that covers 4 orders of magnitude. A broad region of the sheath wrapping around the flux rope contains energy at level of 10^{-5} erg/cm^3 with the exception of the region connected to the heliospheric current sheet. Panel (d) shows that the dissipation rate with a pattern of enhancement that roughly matches the elevated levels of wave reflection.

To better understand the relationship of the Afvénic turbulence with the CME disturbance, we provide Figure 2, which shows color contour images of the wave related quantities (Panels (b)-(d) respectively) divided by the pre-eruption ambient values. Panel (a) shows the mass density ratio, which provides a measure of the compression going through the shock. As expected for

IOP Conf. Series: Journal of Physics: Conf. Series **900** (2017) 012015 doi:10.1088/1742-6596/900/1/012015

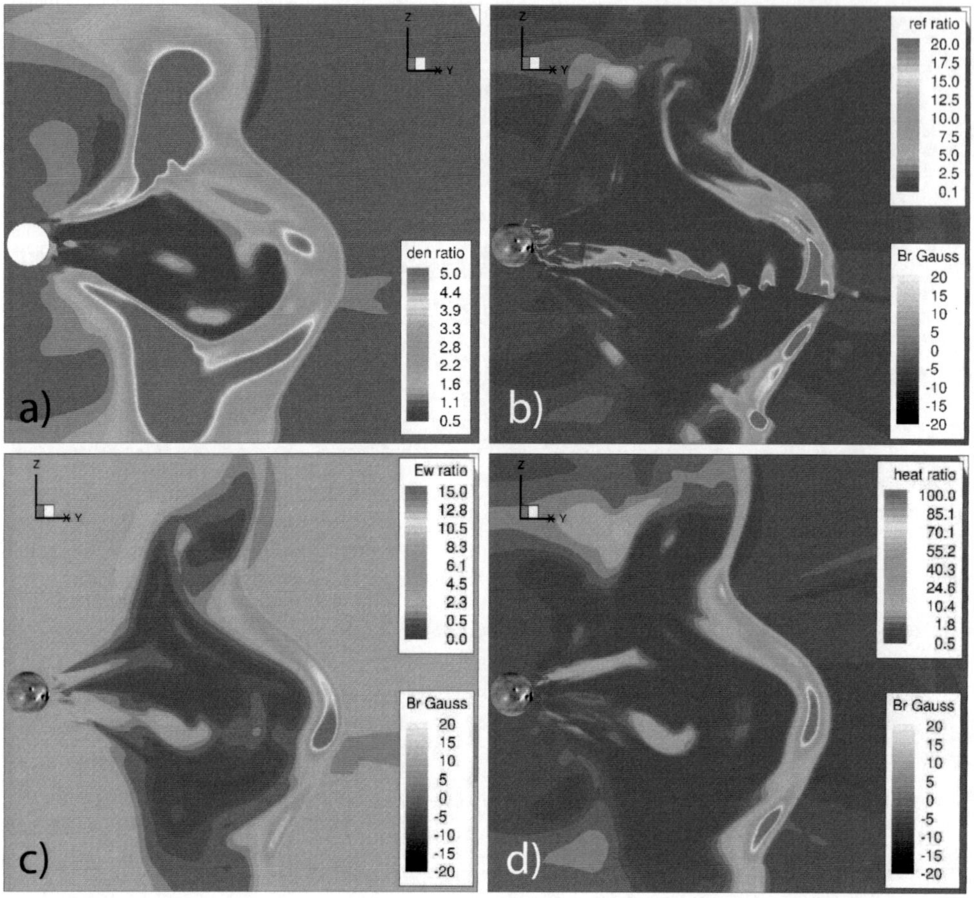

Figure 2. The same wave related quantities from Figure 1 are shown divided by the ambient values. Panel (a) shows the plasma mass density, which shows the shock compression that varies between three and four for this strong shock. Panel (b) shows the wave reflection ration, which shows the large increases at the shock and at the heliospheric current sheet. The wave energy and dissipation ratios are shown in panels (c) and (d) respectively. The wave energy in the sheath typically increases by a factor of ten, while peaking at a value of 25 at the nose of the CME. In conjunction wave dissipation rates rates increase in the sheath by roughly a factor of 50, reaching a peak of 140 at the nose of the CME.

a strong shock, the compression ratio varies between three and four, while a low density cavity is found in the ejected flux rope. The wave reflection ratio shows a large increase at the shock front that varies by factor of 10 to 25, while a region of enhancement cuts through the center of the flux rope, which marks the location of the pre-eruption current sheet. This magnitude of the reflection is mathematically expressed in equations 6 and 7, which is dependent on the gradient in the Alfvén speed. The reflection in this case is occurring where the density increase at the shock resulting in a dramatic decrease in the Alfvén speed over a short distance. The opposite is true at the current sheet, where magnetic field was once nonexistent and is then replaced by the flux rope. The wave energy is typically enhanced in the sheath region by a factor of 8-10, which is consistent with adiabatic compression at the shock. At the nose of the shock, the Alfvén energy increases by a factor of 25, which corresponds a high rate of reflection

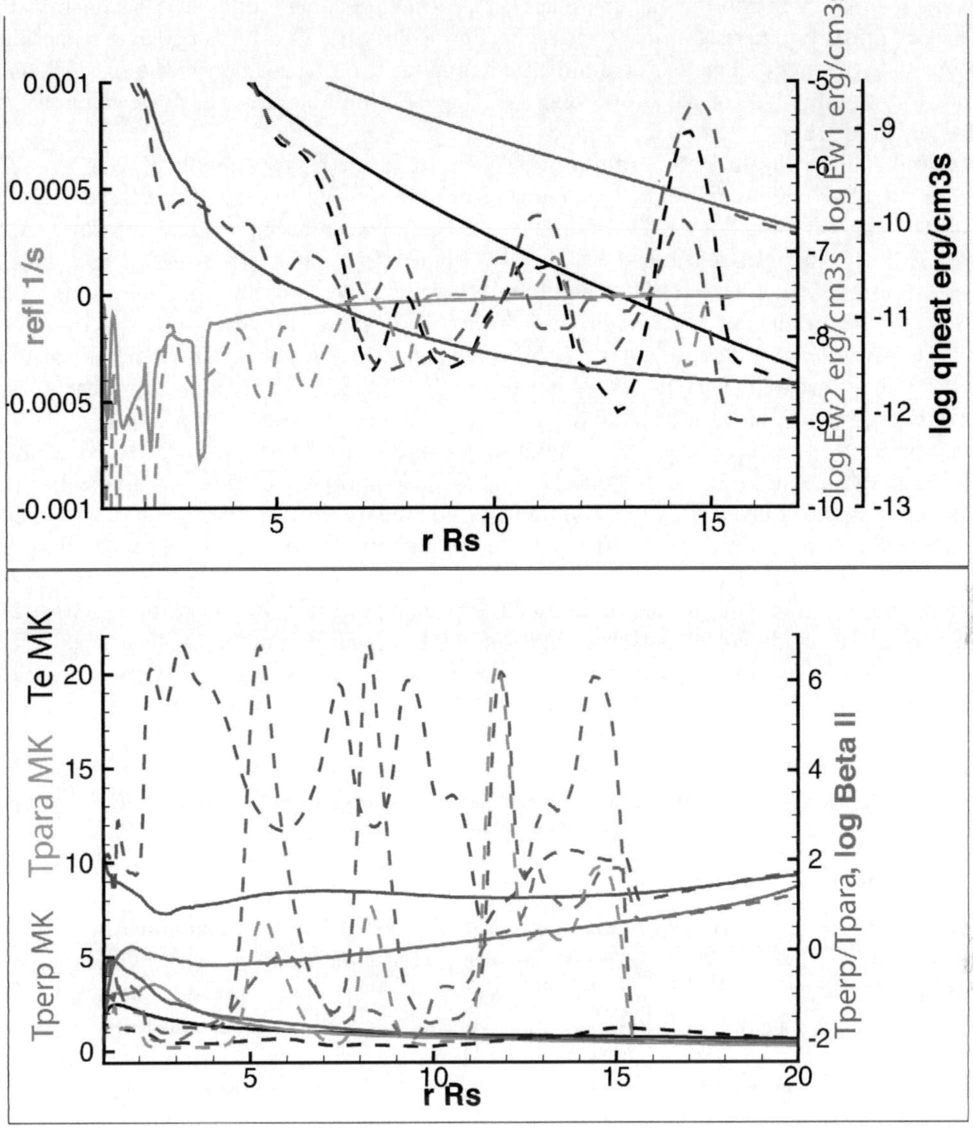

Figure 3. Plots of wave related and plasma thermal quantities from data extracted on the black line shown in Panel (a) of Figure 1. For reference, solid lines show ambient values at time$t = 0$ while dashed lines show CME-perturbed quantities at $t = 2$ hours. The top panel shows forward Ew_1 and counter-propagating Ew_2 wave energy densities in blue and red respectively. On this field line pointing toward the sun, the Ew_2 wave propagates away from the Sun while Ew_1 is directed toward the Sun. The reflection and dissipation rates are plotted in green and black respectively. Note the large increases in wave energy associate with reflection from the moving shock wave. Plotted in the bottom panel are the electron and proton temperatures (perpendicular and parallel) the temperature ratio, T_{perp}/T_{para}, and the parallel plasma beta in lines colored black, red, green, blue and purple. Note the inverse relationship between the temperature ratio and plasma beta.

transporting wave energy to the current sheet region that previously had very low energy. Panel (d) shows the greatest increases in wave dissipation occur where both energy and reflection are simultaneously increased. Due to the quadratic dependence on wave energy, the dissipation rate increases by roughly a factor of 50 over most of the sheath and reaches a peak value of 140 at the nose of the CME.

Next we evaluate the line plots of the wave quantities in the top panel of Figure 3. These values are extracted along a straight line, shown in Figure 1, that extends from the center of the Sun and passing through the CME. Here, the forward and counter-propagating wave energies are plotted with blue and red lines respectively, where direction is with regard to a field line pointing toward the Sun. Green and black lines show the reflection rate and energy dissipation rate respectively. In all four cases, solid and dashed lines show values at time $t = 0$ and $t = 2$ hours respectively. At time $t = 2$ hours, the wave reflection peaks at the shock with a value that is 42.7 times greater than the ambient value. This strong reflection is most clearly seen in the wave energy reflected back toward the Sun, Ew_1, which increased by a factor of 68.5 over the ambient value. In comparison, Ew_2 increases by only a factor of 13.6 here. Reflection does not not create wave energy but only converts one form to another. In this example gain in Ew_2 comes from a large compression, while the much large increase in Ew_1 is attributed to reflection converting Ew_2 to Ew_1. The waves interact to dissipate energy at a rate 111 times above the ambient value.

We next discuss the heating and cooling of the plasma and the temperature anisotropy. For this purpose, we provide color contour images of thermodynamic quantities in Figure 4, which are shown on the same time and meridional slice as the previous Figures. Here, the electron temperature, shown in Panel (a), is elevated at the shock from roughly 0.7 MK to 1.2 MK, with a significant precursor of hot electrons extending ahead of the shock. The proton parallel, perpendicular temperatures, and their ratio, T_{per}/T_{par}, are shown in Panels (b) through (d) respectively. To complement this Figure, the bottom panel of Figure 3 provides line plots of the electron temperature, and proton perpendicular and parallel temperatures, their ratio, T_{per}/T_{par} and the parallel plasma β. Examining the line plots we find that prior to the shock, the parallel temperature is 0.4 MK, which then rises to 9.7 MK as the density increases by a factor of 3.6 at the shock. In contrast, the perpendicular temperature rises nearly twice as much from from 0.5 MK to 19.8 MK. These elevated temperatures are greatest at the nose of the CME where the shock passes through the slow wind with the highest Mach number. In the flux rope, we find a complex mix of hot plasma caused by Joule heating at magnetic current sheets surrounded by cold plasma cooled by adiabatic expansion. The electrons receive no heating from either shock or magnetic dissipation, and thus form a large cool cavity at a temperature near 0.27 MK.

In reality, the dissipative heating at shock and current sheets is distributed to both parallel and perpendicular temperatures by a host of complex plasma instabilities. For our model, the MHD simulation is designed to account for the effects of the fire hose, mirror mode, and cyclotron plasma instabilities. These instabilities are invoked to limit the temperature anisotropy in accordance with their respective thresholds for onset, which are prescribed analytically. The thresholds for these instabilities are determined by the parallel and perpendicular plasma βs, which vary strongly over the range of the CME-disturbed solar wind. Examining the bottom panel of Figure 3, we find in the sheath, high temperatures and mass density produce parallel β that ranges from roughly 10 to 60. In contrast, the plasma beta in the flux rope is much lower, ranging from roughly 0.5 to 0.01. The affects of plasma instabilities can be seen in the inverse relationship between T_{per}/T_{par} and the parallel β, which is indicative of the thresholds for the cyclotron and mirror mode instabilities, which limits T_{per}/T_{par} to a value approaching 7 in the lowest β portions of the flux rope.

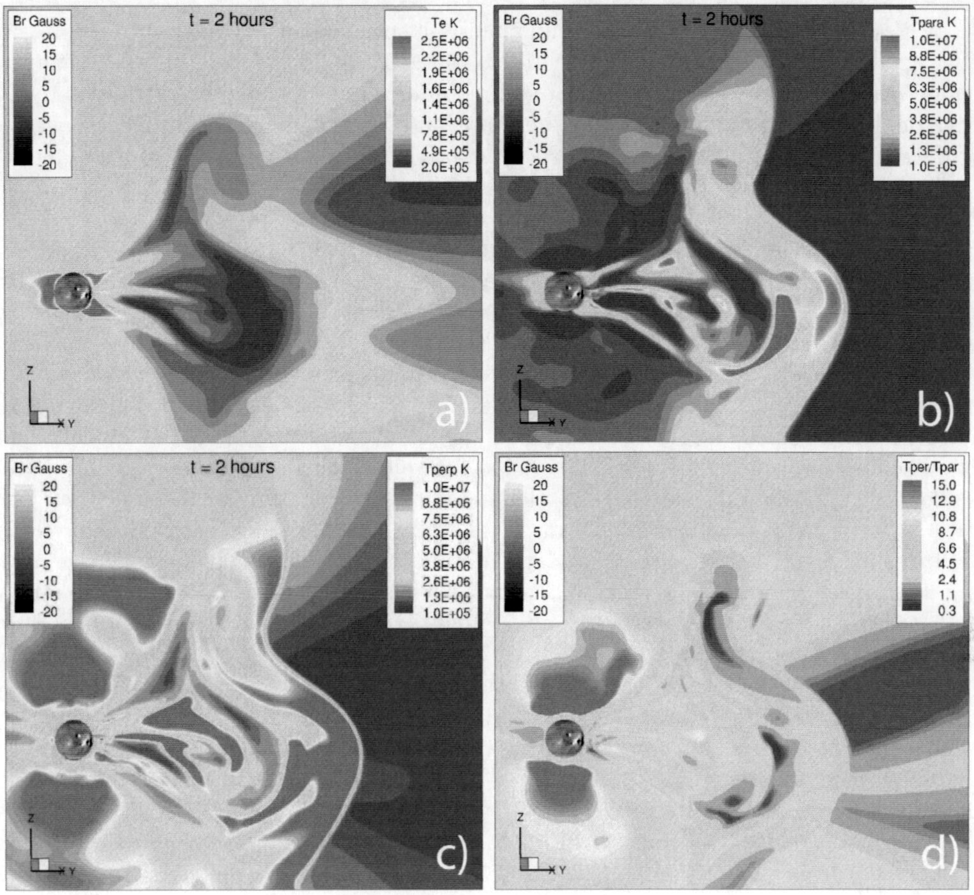

Figure 4. Plasma temperatures are shown in color contours at two hours after CME initiation on the same cut plane shown in Figures 1 and 2. Panel (a) shows the electron temperature, which exhibits only mild elevation behind the shock with a hot precursor formed by heat conduction. Parallel and perpendicular temperatures are shown in panels (b) and (c) which show high degrees of shock heating, while Panel (d) shows the temperature ratio T_{perp}/T_{para}.

4. Conclusions

We simulate the propagation of a CME with the AWSoM coronal model to provide an understanding of the driven interaction between Alfvén wave turbulence, kinetic instabilities and temperature anisotropies in the environment close to the Sun. In the impacted solar corona, turbulence, temperatures, and dissipative heating increases by orders of magnitude providing conditions suitable for a host of kinetic instabilities. Significant enhancements of wave turbulence are found primarily in the CME sheath region while a significant depression in wave energy forms in the magnetic flux rope that at its lowest is at 1% of the level of the background level. In passing through the strong fast-mode shock, the total turbulent energy increases by factors ranging from 10 to 25 while wave energy reflected by the shock toward the Sun reaches a factor of nearly 100. Over most of the sheath, wave dissipation rates increase by a factor of 50 while peaking at 140 times the ambient value at the nose of the CME.

In spite of these large increases, wave heating is relatively insignificant on the time-scale of encountered by a fast CME close to the Sun. The conspicuous signs of heating we find here are driven by dissipative heating at the shock and at current sheets as well as adiabatic compressional heating, and adiabatic expansion being the only source of cooling. At the shock, we find very strong anisotropic proton heating with proton perpendicular and parallel temperatures reaching 20 MK and 10 MK respectively. The temperature ratio (perpendicular/parallel) remains at a value of roughly two over a broad range of the sheath region. Localized temperature peaks occur in the flux rope rope where the temperature ratio is much higher reaching values approaching 7.

The AWSoM is designed such that proton heating from dissipative processes, either turbulence, shocks or current sheets preferentially goes to the proton perpendicular thermal energy. This ansatz alone would produce enormous temperature anisotropies that would incite kinetic instabilities whose primary effect is to redistribute energy to the parallel thermal reservoir of the protons. It is known that elevated perpendicular temperatures will drives mirror-mode and cyclotron instabilities, which has been made manifest at CME-driven shocks by the presence of increased temperature anisotropy and mirror mode waves [13]. Our MHD model can not describe the development of such instabilities, but where temperature ratios exceed the stability threshold, temperatures can relax back to the marginal stable state with the time scale of the relaxation set by the corresponding growth rates of the fire hose mirror mode and cyclotron instabilities. The end result is that the temperature ratio scales inversely with the parallel plasma beta, which is consistent with solar wind observations [8, 19] and ACE observation of CME sheaths with elevated perpendicular proton temperature enhancements capable of driving the mirror mode instability [13].

Understanding this complex interaction between the CME-affected plasma and turbulence is essential to the acceleration of solar energetic particles that are scattered by turbulent structures, which strongly affects their transport and energization from multiple shock crossings. Such work is also essential to understanding the impact of the CMEs on the distribution and dissipation of Alfvén waves, which is particularly timely as we anticipate predicting observations of Alfvén wave turbulence to be made by Solar Orbiter and Solar Probe Plus. Our future will also involve more physics of turbulent dissipation including the effects of the inertial range and stochastic heating.

4.1. Acknowledgments

W. Manchester was supported by NASA grant NNX16AL12G and NSF grants AGS1322543 and AGS1408789. B. van der Holst was supported by NASA grant NNX16AL12G, NSF grant AGS1459862 and the European Union's Horizon 2020 research and innovation program under grant agreement No. 637302 PROGRESS. The authors wish to acknowledge high-performance computing support provided with the Yellowstone Supercomputer operated by NCAR's Computational and Information Systems Laboratory, and the Pleiades supercomputer operated by NASA's Advanced Supercomputing Division. Our work utilizes synoptic magnetogram data obtained by the Global Oscillation Network Group (GONG) Program, managed by the National Solar Observatory.

5. References

[1] Adhikari l, Zank G P, Hunana P & Hu Q 2016 *Astrophys. J.* **833** 218
[2] Belcher J W, & Davis L Jr 1971 *J. Geophys. Res.* **76** 3534
[3] Chandran B D G, Dennis T J, Quataert E, & Bale, S D 2011 *Astrophys. J.* **743** 197
[4] Cranmer S R 2009 *Living Rev. Sol. Phys.* **6** 1
[5] Cranmer S R 2010 *Astrophys. J.* **710** 676
[6] De Pontieu B, McIntosh S W, Carlsson M, et al. 2007 *Science* **318** 1574
[7] Gibson S. E. & Low 2008 *Astrophys. J.* **493** 460-473
[8] Hellinger P, Travnicek P, Kasper J C, Lazarus A 2006 *Geophys. Res. Lett.* **33**, 9101
[9] Jin M, Manchester W B IV, van der Holst B, et al. 2013 *Astrophys. J.* **773** 50
[10] Jin M, Manchester W B IV, van der Holst B, et al. 2017a *Astrophys. J.* **834** 172
[11] Jin, M, Manchester W B IV, van der Holst B, et al. 2017b *Astrophys. J.* **834** 173
[12] Lionello R, Velli M, Downs C, et al. 2014 *Astrophys. J.* **784** 120
[13] Liu Y, Richardson J D, Belcher J W, Kasper J C, Skoug R M 2006 *J. Geophys. Res.* **111** A09108
[14] Lugaz N, Manchester W B IV, Roussev I, et al. 2007 *Astrophys. J.* **659** 788
[15] Manchester W B IV, Gombosi, T I, Roussev I, et al. 2004 *J. Geophys. Res.* **109** A01102
[16] Manchester W B IV, Gombosi T I, Ridley A., et al. *J. of Geophys. Res.*, **109**, A02107,
[17] Manchester W B IV, van der Holst B, Tóth G 2012 *Astrophys. J.* **756** 81
[18] Manchester W B IV, van der Holst B, & Lavraud B 2014 *Plasma Phys. Control. Fusion 56* 64006
[19] Maruca B A, Kasper J C, & Bale S D 2011 *Phys. Rev. Lett.* **107** 201101
[20] Matthaeus W H, Zank G P, Oughton S, Mullan D J, & Dmitruk P 1999 *Astrophys. J. Lett.* **523** L93
[21] Meng X, Tóth G, Sokolov I V, et al. 2015 *Mon. Not. Roy. Astro. Soc.* **454** 3697-3709
[22] Oran R, Landi E, van der Holst B, et al. 2015 *Astrophys. J. 806* 55
[23] Sokolov I, van der Holst B, Oran R, et al. 2013 *Astrophys. J.* **764** 23
[24] Suzuki T K, & Inutsuka S I 2005 *Astrophys. J. Lett.* **632** L49
[25] Usmanov A V, Goldstein M L, Besser B P, & Fritzer J M 2000 *J. Geophys. Res.* **105** 12675
[26] van der Holst B, Manchester W B IV, Frazin R A, et al. 2010 *Astrophys. J.* **725** 1373
[27] van der Holst B, Sokolov I V, Meng X, et al. 2014 *Astrophys. J. Supp.* **782** 81
[28] Zank G P, Matthaeus W H, & Smith C W 1996 *J. Geophys. Res.* **101** 17093
[29] Zank G P, Dosch A, Hunana P, et al. 2012 *Astrophys. J.* **745** 35
[30] Zank, G P 2014 *Transport Processes in Space Physics and Astrophysics*, Lecture Notes in Physics, Berlin Springer Verlag, **877**
[31] Zank G P, Adhikari L, Hunana P, et al. 2017 *Astrophys. J.* **835** 147

IOP Conf. Series: Journal of Physics: Conf. Series **900** (2017) 012016 doi:10.1088/1742-6596/900/1/012016

Shock Wave Structure in the Presence of Energetic Particles

Parisa Mostafavi[1,2], Gary P. Zank[1,2], and Gary M. Webb[2]

[1] Department of Space Science, University of Alabama in Huntsville, Huntsville, AL 35899, USA

[2] Center for Space Plasma and Aeronomic Research (CSPAR), University of Alabama in Huntsville, Huntsville, AL 35899, USA

E-mail: psm0011@uah.edu

Abstract. Energetic particles that are not equilibrated with the thermal plasma (such as pickup ions (PUIs), anomalous cosmic rays (ACRs) and solar energetic particles (SEPs)) can modify the structure of collisionless shock waves. This is relevant to the inner and outer heliosphere and the Very Local Interstellar Medium (VLISM) where observations of shock waves in the e.g., the inner heliosphere show that the energetic particle component pressure is greater than the both the magnetic field and thermal gas pressure (Lario et al., 2015). Voyager 2 observations revealed that the heliospheric termination shock (HTS) is very broad and mediated by energetic particles. PUIs and SEPs contribute both a collisionless heat flux and a higher-order viscosity. We show that the incorporation of both effects can completely determine the structure of collisionless shocks mediated by energetic ions. Since the reduced form of the PUI mediated plasma model is structurally identical to the classical cosmic ray two-fluid model (Axford et al., (1982)), we note that the presence of viscosity at least formally eliminates the need of a gas sub-shock in the classical two-fluid model, including in that regime where three are possible. By considering parameters upstream of the heliospheric termination shock (HTS), we show that the thermal gas remains relatively cold and the shock is mediated by PUIs. We determine the structure of the weak interstellar shock observed by Voyager 1. We consider the inclusion of the thermal heat flux and viscosity to address the most general form of an energetic particle-thermal plasma two-fluid model.

1. Introduction

Energetic particles such as pickup ions (PUIs), solar energetic particles (SEPs), and anomalous cosmic rays (ACRs, i.e., very energetic PUIs) can modify the structure of shock waves. Recent observations show that some shocks in the heliosphere can be mediated by energetic particles. Lario et al. (2015) [1] observed that the energetic particle pressure is much greater than thermal gas and magnetic field pressure at several interplanetary shocks within 1 AU. The large energetic particle pressure suggests that these interplanetary shocks are mediated by energetic particles and their characteristics are different than expected from a simple MHD description. Voyager 2 measured the properties of the heliospheric termination shock (HTS), finding it was very broad and mediated by energetic particles [2, 3]. A numerical simulation of the HTS was presented by Florinski et al. (2009) [4] which suggested that the HTS is mediated by ACRs since it has a precursor extending about 0.3 AU upstream of the shock. Voyager 1 is now measuring the properties of the very local interstellar medium (VLISM, for a precise definition of the VLISM,

IOP Conf. Series: Journal of Physics: Conf. Series **900** (2017) 012016 doi:10.1088/1742-6596/900/1/012016

see [5]) region. Burlaga et al. (2013) [6] made the first observations of a shock outside of the heliosphere. The shock was very broad and weak and had a structure completely different than typical shocks in the heliosphere.

Zank et al (2014) [5] estimated the equilibration times for PUIs and thermal gas particles in three different regions (the supersonic solar wind beyond 10 AU, the inner heliosheath, and the VLISM). In these regions, PUIs are not equilibrated with the background plasma and should be treated as a separate component in the system. The approach presented here is applicable to all non-equilibrated plasma systems in which energetic particles such as SEPs, cosmic rays, and ACRs are included [7, 8]. Energetic particles introduce dissipation terms such as a collisionless heat flux and a collisionless stress tensor or viscosity into the plasma system. Dissipation typically determines the structure of shocks. The model developed by Zank et al. (2014) and (2016) [5, 8] is identical structurally to the cosmic ray two-fluid model [9, 10]. Therefore, we compare briefly our PUI modified shock model with the cosmic ray two-fluid model. Drury and Völk (1982) [10] showed that there are no physical solutions in some Mach number regimes connecting an upstream state to a downstream state. In the papers discussed above, the energetic particles are assumed to be nearly isotropic and a heat-conduction like term $\nabla.(K.\nabla P_c)$ is present as a dissipation term. The inclusion of the heat conduction term only is sometimes not enough to determine the shock structure and higher order terms should be included. In the case that a non-physical solution exists, a gas sub-shock has to be added [9] which means that the thermal gas is responsible for the dissipation of the system and is heated more than energetic particles at the sub-shock. However, this is not always physical, as appears to be the case at the HTS. Zank et al. (1996) [11] predicted that the thermal gas remains relatively cold with PUIs being heated through the shock. Voyager 2 confirmed the [11] predictions by observing that the thermal gas remains relatively cold [2]. A further limitation of the cosmic ray two-fluid model is that in some Mach number regimes, three different downstream states exist for one specific upstream state, and it is unclear how to choose a unique transition.

The effect of cosmic ray viscosity on the structure of a cold thermal gas shock was first studied by Jokipii & Williams (1992) [12]. However, they considered only a cold plasma in the absence of a magnetic field and all the shock transitions for a cold plasma are smooth, meaning there is no need for a gas sub-shock.

In this paper, we use the Zank et al. (2014, 2016) model [5, 7] and consider the dissipative role of both the PUI collisionless heat flux and collisionless viscosity in determining the structure of shock waves. We show that the incorporation of both PUI heat flux and viscosity is sufficient to smooth and so determine the shock structure. The comparison of the results from a PUI mediated plasma model with the Voyager 2 observations shows a good agreement. During the HTS transition the PUIs are heated much more than the thermal gas. Both the HTS and a shock in the VLISM are numerically simulated and compared with the Voyager 2 and 1 observations. For the HTS, we show that almost all the internal energy is converted to PUIs and so the thermal gas remains relatively cold during the transition. The flow remains supersonic with respect to the thermal gas downstream of the HTS. However, the flow changes from supersonic to subsonic with respect to the combined thermal gas and PUI Mach number. The HTS results are consistent with observations of the HTS ([2, 13]). The shock in the VLISM is found to be very weak and broad. We also consider a general case where both thermal gas and PUIs are weakly non-isotropic and have included thermal gas dissipation terms (i.e., heat flux and viscosity). The numerical simulation of the general case is discussed in section 5. Finally, we summarize our results about shock structure when mediated by energetic particles such as PUIs, ACRs, and SEPs.

2. Pickup Ion Mediated Plasma Model

PUIs in the supersonic solar wind beyond 10 AU, subsonic solar wind in the inner heliosheath (IHS), and the VLISM are not equilibrated with the background thermal plasma and should be treated as a separate component. Therefore, a multi-component plasma model is required to consider thermal solar wind electrons and ions, and a superthermal ion component (e.g., PUIs, SEPs, or ACRs). Complete details of the model were presented in [5, 7, 8]. The continuity, momentum and energy equations governing the electrons and protons are assumed to have Maxwellian distribution functions and are given by

$$\frac{\partial n_{e,s}}{\partial t} + \nabla \cdot (n_{e,s}\mathbf{u}_{e,s}) = 0; \tag{1}$$

$$m_{e,s}n_{e,s}(\frac{\partial \mathbf{u}_{e,s}}{\partial t} + \mathbf{u}_{e,s} \cdot \nabla \mathbf{u}_{e,s}) = -\nabla(P_{e,s}) - en_{e,s}(\mathbf{E} + \mathbf{u}_{e,s} \times \mathbf{B}); \tag{2}$$

$$\frac{\partial P_{e,s}}{\partial t} + \mathbf{u}_{e,s} \cdot \nabla P_{e,s} + \gamma_{e,s}P_{e,s}(\nabla \cdot \mathbf{u}_{e,s}) = 0, \tag{3}$$

where $n_{e,s}$, $P_{e,s}$, $\mathbf{u}_{e,s}$, and $\gamma_{e,s}$ are the macroscopic fluid variables for the electron (e) and proton (s) number density, pressure, velocity, and the adiabatic index, respectively. $\mathbf{B}$ and $\mathbf{E}$ are the magnetic field and electric field.

On the other hand, PUIs or energetic particles drive streaming instabilities and experience pitch angle scattering, which introduces a collisionless heat flux and a collisionless stress tensor or viscosity. The energetic particle distribution function can be written up to a second order correction as

$$f \simeq f_0 + \mu f_1 + \frac{1}{2}(3\mu^2 - 1)f_2, \tag{4}$$

where μ is the cosine of the particle pitch angle and the f_1 and f_2 are functions of the isotropic distribution, f_0. After some calculations, the pressure tensor and heat flux terms can be expressed as

$$(P_{ij}) = P_p\,(\delta_{ij}) + \begin{pmatrix} -2 & 0 & 0 \\ 0 & 1 & 0 \\ 0 & 0 & 1 \end{pmatrix} \frac{\eta_{k\ell}}{2}\left(\frac{\partial U_k}{\partial x_\ell} + \frac{\partial U_\ell}{\partial x_k} - \frac{2}{3}\delta_{k\ell}\frac{\partial U_m}{\partial x_m}\right) \equiv P_p\mathbf{I} + \mathbf{\Pi}_p; \tag{5}$$

$$q_i(x,t) = -\frac{1}{2}K_{p,ij}\frac{\partial P_p}{\partial x_j}, \tag{6}$$

where the pressure tensor is the sum of an isotropic PUI scalar pressure P_p and a stress tensor $\mathbf{\Pi_p}$ (i.e., collissonless viscosity) and K_p represents the diffusion coefficient (ie. the collisionless heat flux). The inclusion of $\mathbf{\Pi_p}$ and K_p is due to the non-isotropic PUIs or energetic particles present in the system.

The continuity, momentum, energy, and pressure equations governing the energetic particles or PUIs are

$$\frac{\partial \rho_p}{\partial t} + \nabla \cdot (\rho_p\mathbf{U}_p) = 0; \tag{7}$$

$$\frac{\partial}{\partial t}(\rho_p\mathbf{U}_p) + \nabla \cdot (\rho_p\mathbf{U}_p\mathbf{U}_p + IP_p + \mathbf{\Pi_p}) = en_p(\mathbf{E} + \mathbf{U}_p \times \mathbf{B}); \tag{8}$$

$$\frac{\partial}{\partial t}\left(\frac{1}{2}\rho_p\mathbf{U}_p^2 + \frac{P_p}{\gamma_p - 1}\right) + \nabla \cdot \left[\frac{1}{2}\rho_p\mathbf{U}_pU_p^2 + \frac{\gamma_p}{\gamma_p - 1}P_p\mathbf{U}_p + \mathbf{\Pi_p}\cdot\mathbf{U}_p - \frac{\mathbf{K_p}}{3(\gamma_p - 1)}\cdot\nabla P_p\right] = en_p\mathbf{U}_p\cdot\mathbf{E}; \tag{9}$$

$$\frac{\partial P_p}{\partial t} + \mathbf{U}_p \cdot \nabla P_p + \gamma_pP_p(\nabla \cdot \mathbf{U}_p) = \frac{1}{3}\nabla \cdot (\mathbf{K}_p \cdot \nabla P_p) - (\gamma_p - 1)\mathbf{\Pi_p} : (\nabla\mathbf{U}_p), \tag{10}$$

where the γ_p is the adiabatic index for PUIs.

Equations (1)-(3) and (7)-(9) or (10), together with Maxwell's equations describe the multi-fluid system of equations. However, in most cases solving the multi-fluid model is difficult and by making some assumptions it can be reduced to a single-fluid-like model (see [5, 7] for more details). The reduced single-fluid-like model equations together with Maxwell's equations may be expressed as

$$\frac{\partial \rho}{\partial t} + \nabla \cdot (\rho \mathbf{U}) = 0; \tag{11}$$

$$\rho(\frac{\partial \mathbf{U}}{\partial t} + \mathbf{U} \cdot \nabla \mathbf{U}) = -\nabla(P_g + P_p) - \nabla \cdot \mathbf{\Pi_p} + \mathbf{J} \times \mathbf{B}; \tag{12}$$

$$\frac{\partial P_g}{\partial t} + \mathbf{U} \cdot \nabla P_g + \gamma_g P_g \nabla \cdot \mathbf{U} = 0; \tag{13}$$

$$\frac{\partial P_p}{\partial t} + \mathbf{U} \cdot \nabla P_p + \gamma_p P_p \nabla \cdot \mathbf{U} = \frac{1}{3}\nabla \cdot (K_p \cdot \nabla P_p) - (\gamma_p - 1)\mathbf{\Pi_p} : (\nabla \mathbf{U}); \tag{14}$$

$$\begin{aligned} \frac{\partial}{\partial t}\left(\frac{1}{2}\rho \mathbf{U^2} + \frac{P_p}{\gamma_p - 1}P_p + \frac{P_g}{\gamma_g - 1} + \frac{1}{2\mu_0}B^2\right) + \nabla \cdot (\frac{1}{2}\rho \mathbf{U}U^2 + \frac{\gamma_p}{\gamma_p - 1}P_p\mathbf{U} + \frac{\gamma_g}{\gamma_g - 1}P_g\mathbf{U} \\ + \frac{1}{\mu_0}B^2\mathbf{U} - \frac{1}{\mu_0}\mathbf{U} \cdot \mathbf{BB} + \mathbf{\Pi_p} \cdot \mathbf{U} - \frac{1}{\gamma_p - 1}\frac{\mathbf{K_p}}{3} \cdot \nabla P_p) = 0; \end{aligned} \tag{15}$$

$$\mathbf{E} = -\mathbf{U} \times \mathbf{B}; \ \frac{\partial \mathbf{B}}{\partial t} = -\nabla \times \mathbf{E}; \ \mu_0 \mathbf{J} = \nabla \times \mathbf{B}; \ \nabla \cdot \mathbf{B} = 0, \tag{16}$$

where ρ, U, and P_g are the fluid density, velocity, and thermal gas pressure, respectively. We denote all quantities pertaining to the background thermal gas with the subscript "g" and those related to the energetic particles, such as PUIs, with the subscript "p".

3. Pickup Ion Mediated Shock Structure

The PUI mediated plasma model derived in the last section has many applications, one of which is determining the structure of shocks mediated by PUIs. The hydrodynamic form of the one-dimensional single-fluid-like PUI mediated plasma model is studied here. The viscosity term in one-dimensional form may be expressed as $\Pi_p = -\frac{4}{3}\eta_p \frac{dU}{dx}$. The η_p is the coefficient of the PUI collisionless viscosity and is typically assumed to be constant. The thermal gas is isotropic and PUIs are non-thermal particles. The continuity, momentum and pressure equations governing the PUIs and the thermal background gas are given by

$$\frac{\partial \rho}{\partial t} + \frac{\partial (\rho U)}{\partial x} = 0; \tag{17}$$

$$\rho\left(\frac{\partial U}{\partial t} + U\frac{\partial U}{\partial x}\right) = -\frac{\partial P_g}{\partial x} - \frac{\partial P_p}{\partial x} + \frac{4}{3}\eta_p \frac{\partial^2 U}{\partial x^2}; \tag{18}$$

$$\frac{\partial P_g}{\partial t} + U\frac{\partial P_g}{\partial x} + \gamma_g P_g \frac{\partial U}{\partial x} = 0; \tag{19}$$

$$\frac{\partial P_p}{\partial t} + U\frac{\partial P_p}{\partial x} + \gamma_p P_p \frac{\partial U}{\partial x} = \frac{1}{3}\frac{\partial}{\partial x}\left(K_p \frac{dP_p}{dx}\right) + \frac{4}{3}\eta_p(\gamma_p - 1)(\frac{\partial U}{\partial x})^2. \tag{20}$$

The shock structure equation can be derived by combining the steady-state form of equations (17)-(20), which yields

$$\frac{d^2 y}{dx^2} + \left[\frac{3}{4\eta_p}\rho_1 U_1\left(\left(\frac{y_s}{y}\right)^{\gamma_g + 1} - 1\right) - \frac{3}{K_p}U_1 y\right]\frac{dy}{dx} = \frac{9\gamma_p}{4K_p\eta_p}\frac{\rho_1 U_1^2}{\gamma_g M_{s1}^2}(1 - y)P(y), \tag{21}$$

where y is the inverse compression ratio and

$$P(y) \equiv \gamma_g M_{s1}^2 \frac{\gamma_p+1}{2\gamma_p}(y-\mu_p^2) - \left(1 + \frac{P_{p1}}{P_{g1}} - \frac{(\gamma_g-\gamma_p)}{\gamma_p(\gamma_g-1)} \frac{(1-y^{1-\gamma_g})}{(1-y)}\right), \tag{22}$$

$\mu_p^2 \equiv \frac{(\gamma_p-1)}{(\gamma_p+1)}$ and $y_s^{\gamma_g+1} \equiv \frac{1}{M_{s1}^2}$. The normalized form of the shock structure equation may be written as

$$\frac{d^2y}{d\overline{x}^2} + \frac{1}{Sc_p}\left[\frac{3}{4}\left(\left(\frac{y_s}{y}\right)^{\gamma_g+1} - 1\right) - 3Sc_p y\right]\frac{dy}{d\overline{x}} = \frac{1}{Sc_p}\frac{9}{4}\frac{\gamma_p}{\gamma_g M_{s1}^2}(1-y)P(y), \tag{23}$$

where $\overline{x}$ is x normalized to the diffusion length scale, i.e. $\overline{x} \equiv x/L$, $L \equiv (K_p/U_1)$, and $Sc_p \equiv (\eta_p/\rho_1 K_p)$ is the Schmidt number. Equation (23) determines the shock structure in the presence of the PUI heat flux and PUI viscosity. Further discussion related to the shock structure was presented in Mostafavi et al. (2016, 2017) [14, 15].

Axford et al. (1982) [9] investigated cosmic ray mediated shock structure in the presence of only the dissipation term associated with energetic particles (i.e., the heat flux). In this case, the shock structure equation exhibits two kinds of solutions. One solution yields a smooth transition connecting an upstream state to a downstream state. This solution describes an energetic particle fully mediated shock. An example of a smooth transition is presented in Fig. (1a). The second kind of solution exhibits double-valued solutions and is not physical. In this case, there is no smooth transition from an upstream state to a downstream state (Fig. (1b)). To solve this issue, Drury & Völk (1981) and Axford et al. (1982) [9, 10] introduced a gas sub-shock to reach a downstream state. Since, in this regime, the energetic particle heat flux was insufficient to smooth the shock transition, they assumed that there should be a dissipation mechanism associated with the thermal gas (a gas sub-shock) to determine the shock transition. This assumption is unlikely to be correct in all cases. For example, the HTS converts most of the kinetic energy to PUI heating and the thermal gas remains relatively cold and is only heated adiabatically [11]. Voyager 2 observed the thermal background gas to only 180,000 K and not 10^6 K [2]. We will discuss the HTS in greater detail below.

The structure of shock with the inclusion of PUI viscosity is given by equation (23), which is a second-order differential equation. This equation does not possess singular points that introduce double-valued solutions. Therefore, all transitions between upstream and downstream states are smooth and there is no need to add a gas sub-shock (see Fig. (2a)-(2d)). Figures (2b) and (2d) use the same parameters as figures (1b) and (1d), for which both needed gas sub-shocks. However, figures (2b) and (2d) show a smooth transition from the upstream to the downstream state because of the inclusion of PUI viscosity. Fig. (1b) shows an example for relativistic particles such as ACRs. In this case, the downstream state determined by the inclusion of a gas sub-shock is completely different that the downstream state determined from the inclusion of energetic particle viscosity. By introducing gas sub-shock, it is assumed than an unspecified thermal dissipation mechanism acts only on the thermal gas. However, Fig. (2d) shows that when non-relativistic particles such as PUI are considering then the downstream state with either PUI viscosity or a gas sub-shock included is the same. The reason for that is for both PUIs and the thermal gas have the same adiabatic index $\gamma = 5/3$. Figures (3a)-(3d) plot particle pressure, showing that as the Mach number increases, the PUI pressure become dominate. Shocks with a very small Mach number and small initial PUI pressure for upstream remain dominated by the thermal gas (Figs. (3b) and (3d)).

4. Comparison of the Pickup Ion Mediated Plasma Model with Observations

Voyager 2 and 1 are measuring the properties of the heliosphere boundaries and the VLISM. We numerically simulate the HTS and a shock in the VLISM using a PUI mediated plasma model and compare them with the observations.

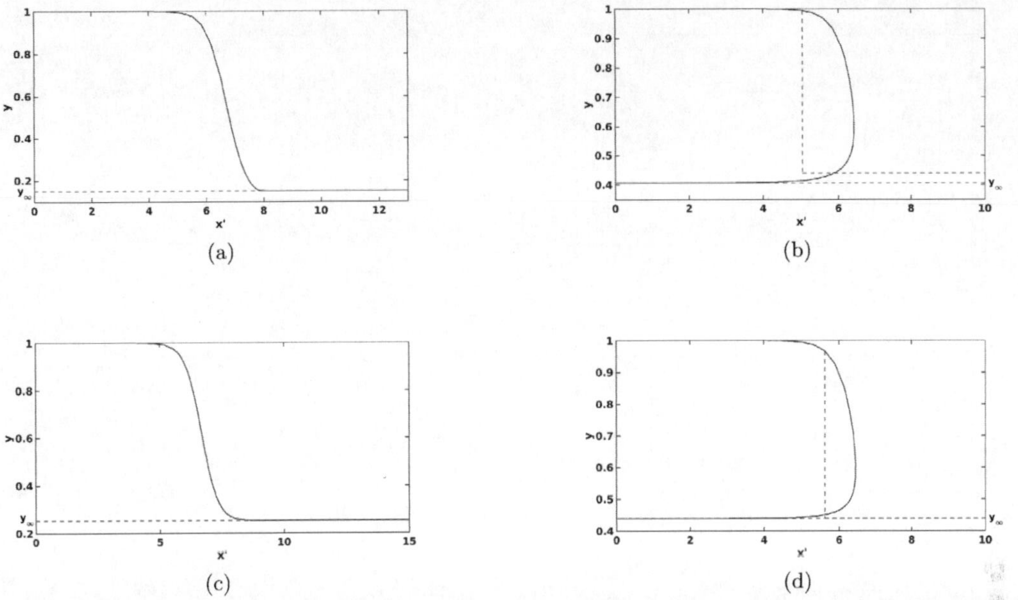

Figure 1: Shock structure in the absence of the collisionless PUI or cosmic ray viscosity. Here $\gamma_g = 5/3$, and $P_{p1}/P_1 = 0.01$. Following [9] a gas sub-shock is typically added to connect the downstream and upstream states, as illustrated by the jump in y to the downstream solution (green dashes). Red dashes correspond to the Rankine-Hugoniot condition downstream state, y_∞. a: $M_{s1} = 14$ and $\gamma_p = 4/3$. A smooth transition exists between the upstream and downstream states. b: $M_{s1} = 2$ and $\gamma_p = 4/3$. Solutions are doubled-valued and a smooth transition is not possible. c: Same as Fig. (1a), except here $\gamma_p = 5/3$. d: Same as Fig. (1b), except here $\gamma_p = 5/3$.

4.1. The Heliospheric Termination Shock

In this section, we describe the structure of the HTS using our PUI mediated plasma model and compare the structure with Voyager 2 observations. The upstream temperature of the HTS was predicted by global MHD models to be 10^6 K, but the Voyager 2 plasma instrument measured the temperature to be about 150,000-180,000 K. The PUI temperature upstream of the HTS is inferred by [16] and [17] to be 1.56×10^6 K. Although the PUI number density is 20% of the total proton densities, the PUI pressure is the dominant term and is much greater than the thermal gas pressure (i.e., $P_{g1} = 2.76 \times 10^{-16}$ Pa and $P_{p1} = 5.38 \times 10^{-15}$ Pa). Voyager 2 crossed the HTS multiple times and the properties of these crossing are determined by the observations ([2]). Here, we model the second HTS crossing, TS-2, which moved with a speed of about 90 km/s in the outward direction. Figures (4a) and (4b) are plots of the inverse compression ratio and the normalized PUI and thermal gas pressures as a function of normalized distance. From Fig. (4a), the HTS compression ratio is $r = \rho_2/\rho_1 \simeq 3$, which is a little greater than the value $r = B_2/B_1 \approx 2.38$, measured by [3] based on the observation that the shock was highly perpendicular. Fig. (4b) shows that the thermal gas experiences almost no heating at the shock (being heated adiabatically), with the incident ram pressure being converted almost entirely to PUI pressure.

Unnormalized plots of the PUI and thermal gas temperature profiles as a function of unnormalized distance are shown in Figures (4c) and (4d). The thermal gas downstream

IOP Conf. Series: Journal of Physics: Conf. Series **900** (2017) 012016 doi:10.1088/1742-6596/900/1/012016

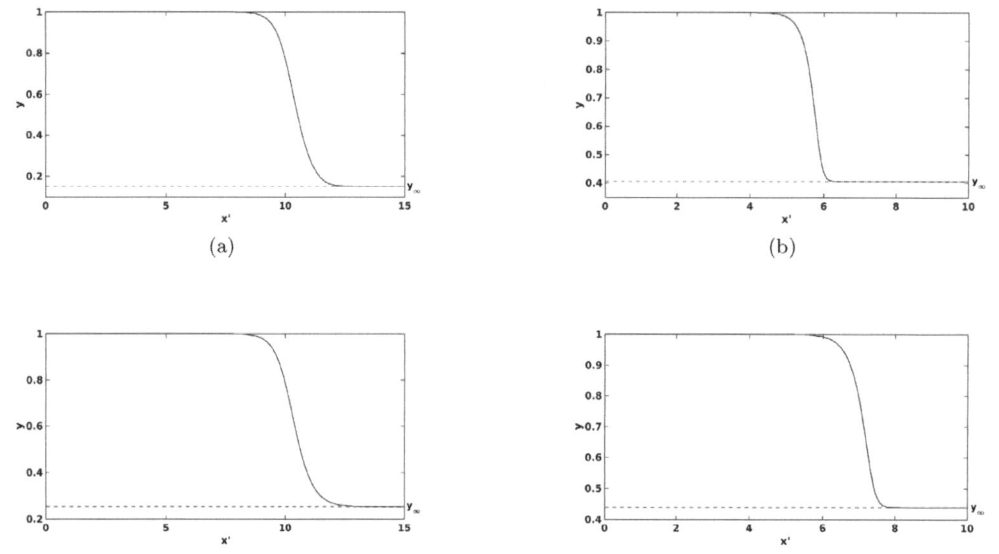

(a) (b) (c) (d)

Figure 2: Smoothed shock transitions when the collisionless cosmic ray or PUI viscosity is present. Here $\gamma_g = 5/3$, and $P_{p1}/P_1 = 0.01$. Red dashes correspond to the Rankine-Hugoniot condition downstream state, y_∞. a: $M_{s1} = 14$ and $\gamma_p = 4/3$ (all parameters are the same as Fig. 1a). b: $M_{s1} = 2$ and $\gamma_p = 4/3$ (all parameters are the same as Fig. (1b). c: Same as Fig. (2a), except here $\gamma_p = 5/3$. d: Same as Fig. (2b), except here $\gamma_p = 5/3$.

temperature is about 1.27×10^5 K and the PUI temperature is about 1.82×10^7K. These temperatures are consistent with the temperatures determined by [16] and from Voyager 2 observations ([2]). The total temperature downstream of the HTS, on the assumption that $n_p = 0.25 n_g$, is calculated to be about 3.69×10^6 K. This is consistent with the estimated total temperature downstream of the HTS by [16], which was about 3.4×10^6 K. The partition of the thermal gas internal energy in the heliosheath is about 2.5% and the PUIs form the dominant component with about 97.5% of the total internal energy. To determine the width of the HTS, we calculate the scattering time for PUIs in the vicinity of the HTS since the distance is normalized to the diffusion length scale. Voyager 2 observations of the HTS showed that the HTS upstream magnetic field was about 0.05 nT ([3]). The scattering time scale is taken to be $2 \times \Omega_g$, giving 4.17×10^3 s, and the diffusion coefficient is calculated to be 2×10^{13} m^2/s. Thus the shock thickness of the PUI mediated HTS is about 3×10^5 km, consistent with observations of TS-2 that showed the shock width was 3×10^5 km ([2]).

The thermal gas Mach number through the shock (Fig. (5a)) shows that the flow remains supersonic with respect to the thermal gas downstream of the HTS. This is a consequence of the thermal gas remaining cold through the shock. Observations of the HTS showed that the small increase in the downstream thermal gas temperature did not correspond to a transition from a supersonic to subsonic flow through the HTS (see Fig. (4) in [2]). Figure (5b) shows the combined thermal gas and PUI Mach number through the HTS in which the sound speed is based on the long wavelength sound speed ($= \sqrt{a_g^2 + a_p^2}$, [5]). Here, because of the inclusion of PUIs, the transition from upstream to downstream does result in a supersonic to subsonic flow transition.

IOP Conf. Series: Journal of Physics: Conf. Series **900** (2017) 012016 doi:10.1088/1742-6596/900/1/012016

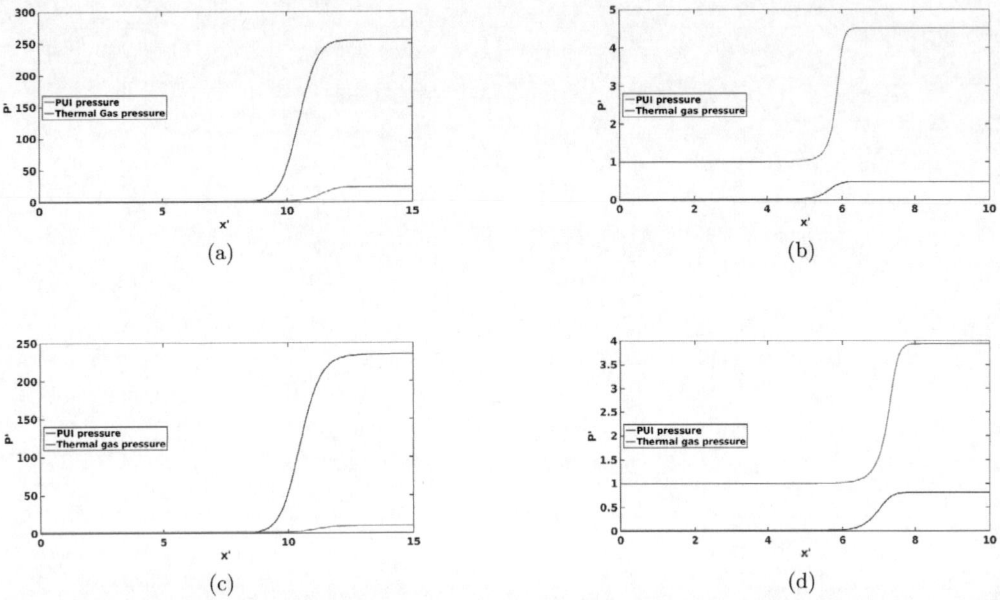

Figure 3: PUI and thermal gas pressure normalized to the thermal gas pressure far upstream. Here $\gamma_g = 5/3$, and $P_{p1}/P_1 = 0.01$. a: $M_{s1} = 13$ and $\gamma_p = 4/3$ (all the parameters are the same as Fig. 2a). b: $M_{s1} = 2$ and $\gamma_p = 4/3$ (all parameters are the same as Fig. (2b)). c: Same as Fig. (3a), except here $\gamma_p = 5/3$. d: Same as Fig. (3b), except here $\gamma_p = 5/3$.

4.2. A shock in the VLISM

Voyager 1 is now measuring the properties of the VLISM. Burlaga et al. (2013) [6] described the first in situ observations of a shock in the VLISM. The observed collisionless shock was very broad, weak, laminar, low beta, and quasi-perpendicular. Since this shock is 10^4 times broader than shocks with the similar properties at 1 AU, there is no theoretical explanation to describe its structure. We use [6, 18] observations to numerically simulate a shock in the VLISM. Observations show that in the VLISM, the Alfvén speed dominates the sound speed and the shock wave appears to be very weak with a Mach number of about 1.4. Because the shock was highly perpendicular, we can assume that the total "thermal pressure" is given by $P_g + B^2/2\mu_0$. As is well known (e.g., [19]) if the magnetic field configuration is perpendicular to the flow, the choice of $\gamma_g = 2$ reduces the MHD equations to a form identical to the hydrodynamic equations except that $P_T = P_g + B^2/2\mu_0$ and B/ρ is constant. In view of the Voyager 1 observations ([6, 18]), we set $P_{T1} = P_{g1} + B_1^2/2\mu_0 = 8.28 \times 10^{-14}$ Pa in our gas dynamic equations and compute the structure and downstream properties of the interstellar shock. The PUI pressure in the VLISM was considered by [5] based on a multi-component treatment of the VLISM ([20]). This yields the ratio $P_{p1}/P_{T1} = 8.4 \times 10^{-3}$. Fig. (6) shows the properties of a VLISM shock which is weak. From Fig. (6a), we show that the compression ratio ($= 1/y$) is about 2.3, whereas [6] observed B_2/B_1 to be about 1.4. Fig (6b) shows that the shock is mediated by the combined thermal gas and magnetic field pressure, the total of which is the dominant component downstream of the shock. The fast magnetosonic Mach number through the shock shows that the flow changes from supersonic to subsonic (Fig. (6c)). Figure (6d) shows that the combined total thermal gas and PUI Mach number through the VLISM shock is a supersonic-subsonic transition. Here, one can see because PUIs do not contribute a large pressure through the

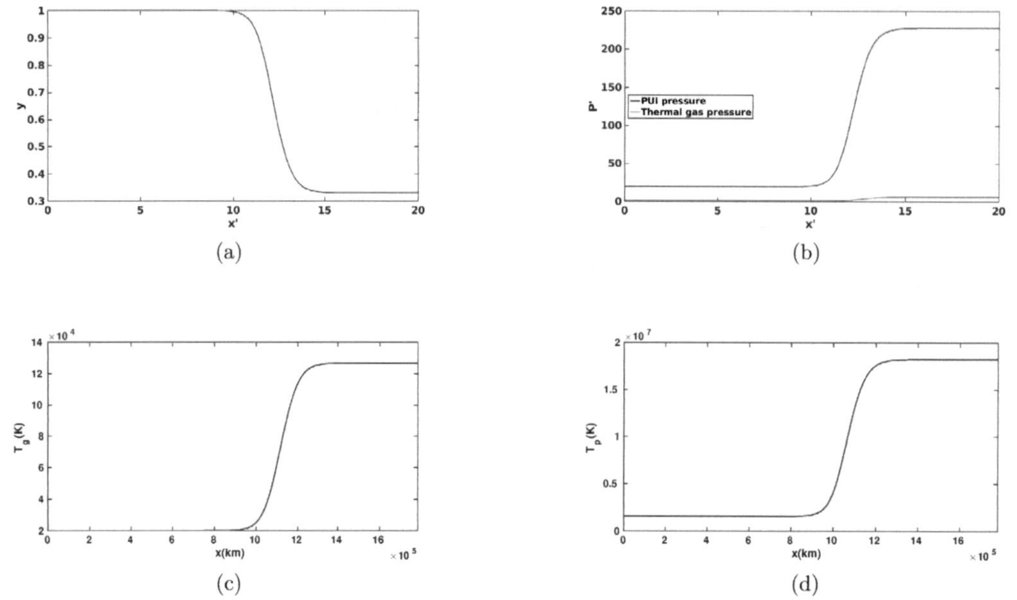

Figure 4: Smoothed shock transition corresponding to HTS parameters when both the collisionless PUI heat flux and viscosity are included. Here $\gamma_g = 5/3$, $\gamma_p = 5/3$, $P_{p1}/P_1 = 19.5$, and $M_{s1} = 13.8$. a: Inverse compression ratio showing the smoothed shock. b: PUI and thermal gas pressure normalized to the thermal gas pressure far upstream. The HTS is mediated by energetic particles with almost all the upstream ram energy being converted to downstream PUI internal energy. c: The unnormalized thermal gas temperature through the shock as a function of unnormalized distance shows that thermal gas remains relatively cold. d: The unnormalized PUI temperature through the shock as a function of unnormalized distance shows that PUI are strongly heated at the HTS.

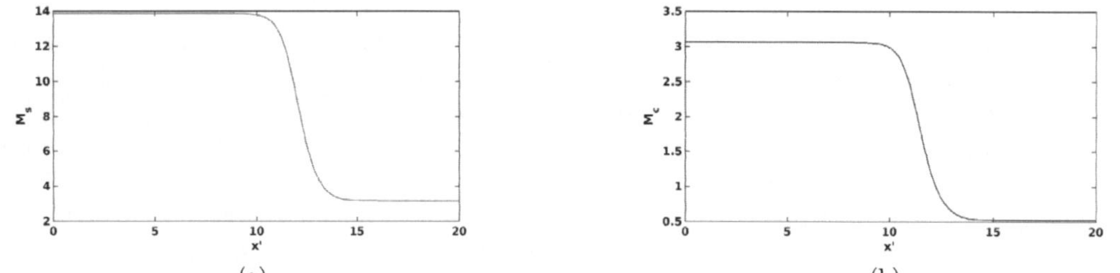

Figure 5: a: The thermal gas Mach number, $M_s \equiv u/a_g$, through the HTS. The flow remains supersonic with respect to the thermal gas sound speed even downstream of the HTS. b: The combined thermal gas and PUI Mach number, $M_c \equiv u/\sqrt{a_g^2 + a_p^2}$, through the shock. Relative to the combined sound speed, the flow is subsonic downstream of the HTS.

shock, the combined Mach number is roughly the same as Fig.(6c).

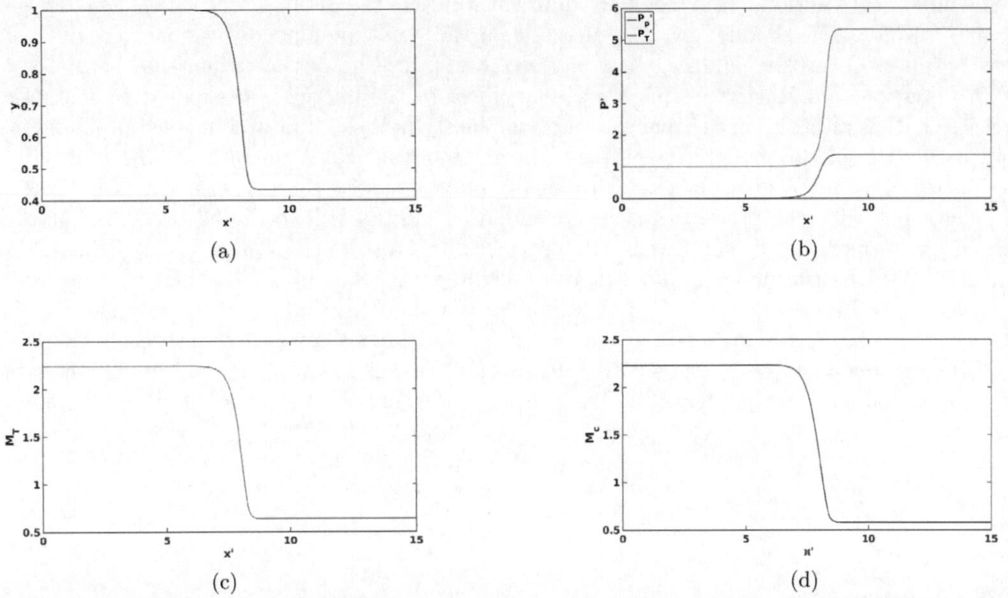

(a) (b) (c) (d)

Figure 6: A smoothed shock transition corresponding to the VLISM parameters when both the collisionless PUI heat flux and viscosity is included. Here $\gamma_g = 2$, $\gamma_p = 5/3$, $P_{p1}/P_{T1} = 0.0084$, $M_{s1} = 2.24$, and $P_{T1} = P_{g1} + B_1^2/2\mu_0$. a: The inverse compression ratio showing the smoothed shock as a function of normalized position. b: The PUI pressure and the combined thermal gas and magnetic field pressure (i.e., total thermal gas pressure) normalized to P_T far upstream. c: The total thermal gas Mach number through the shock shows the flow changes from supersonic to subsonic through the shock. d: The combined effective total thermal gas and and PUI Mach number through the shock, showing a supersonic-subsonic transition.

5. General Case

In this section, we consider the most general case in which we have included the dissipation terms (i.e., collisionless viscosity and heat flux) associated with both thermal gas and PUIs. Assuming the thermal gas is slightly non-Maxwellian, its distribution function introduces higher order corrections that yield both heat conduction and viscosity in the thermal gas system.

The 1D continuity, momentum and pressure equations governing the background thermal gas and non-thermal PUIs in this case are given now by

$$\frac{\partial \rho}{\partial t} + \frac{\partial (\rho U)}{\partial x} = 0; \tag{24}$$

$$\rho(\frac{\partial U}{\partial t} + U\frac{\partial U}{\partial x}) = -\frac{\partial P_g}{\partial x} - \frac{\partial P_p}{\partial x} + \frac{4}{3}\eta_p \frac{\partial^2 U}{\partial x^2} + \frac{4}{3}\eta_g \frac{\partial^2 U}{\partial x^2}; \tag{25}$$

$$\frac{\partial P_g}{\partial t} + U\frac{\partial P_g}{\partial x} + \gamma_g P_g \frac{\partial U}{\partial x} = \frac{1}{3}\frac{\partial}{\partial x}\left(K_g \frac{dP_g}{dx}\right) + \frac{4}{3}(\gamma_g - 1)\eta_g (\frac{\partial U}{\partial x})^2; \tag{26}$$

$$\frac{\partial P_p}{\partial t} + U\frac{\partial P_p}{\partial x} + \gamma_p P_p \frac{\partial U}{\partial x} = \frac{1}{3}\frac{\partial}{\partial x}\left(K_p \frac{dP_p}{dx}\right) + \frac{4}{3}(\gamma_p - 1)\eta_p (\frac{\partial U}{\partial x})^2. \tag{27}$$

η_g and K_g represents the collisionless thermal gas viscosity and heat flux, respectively. The thermal gas does not obey the adiabatic gas law and its entropy increases through the shock due to the dissipation terms associated with thermal gas in the system.

IOP Conf. Series: Journal of Physics: Conf. Series **900** (2017) 012016 doi:10.1088/1742-6596/900/1/012016

The numerical solution of the model for different values of the dissipation coefficients is shown in Fig. (7a)-(8d). The thermal gas and energetic particle heating depends on the choice of their specific values of viscosity and heat conduction. Fig. (7a) shows that when the thermal gas viscosity and heat conduction terms are absent, most of the energy is dissipated to energetic particles or PUIs rather than thermal gas and thus the shock is essentially controlled by energetic particles or PUIs for this moderately high gas Mach number shock. When $\eta_p \gg \eta_g$ and $K_p \gg K_g$, the solution almost looks like the case without η_g and K_g but the thermal gas is slightly hotter. A model shock with the properties $\eta_p \approx \eta_g$ and $K_p \gg K_g$ is still mediated by PUIs but the thermal gas is much more heated than the previous cases and $P_{g2} \simeq P_{p2}$ (Fig. 8c). The shock is mediated by the thermal gas and only very slightly mediated by PUIs when $\eta_p \ll \eta_g$ and $K_p \approx K_g$. The Earth's bow shock is an example of a shock mediated by thermal gas. As is well-known, in the quasi-perpendicular regime, the dissipation process is due to reflected thermal ions. [2] compared one example of a planetary shock (Neptune's bow shock) with the HTS data and showed that most of the energy was transferred to the thermal plasma at Neptune's inbound bow shock.

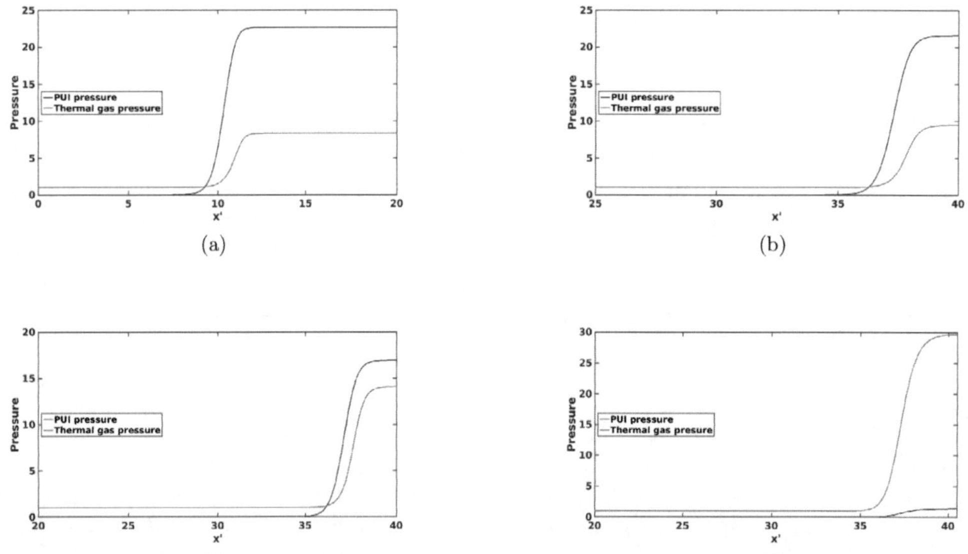

(a) (b) (c) (d)

Figure 7: PUI and thermal gas pressure normalized to the thermal gas pressure far upstream. Here $K_g, K_p > \eta_g, \eta_p$; $\gamma_g = 5/3$; $\gamma_p = 5/3$; $P_{p1}/P_{g1} = 0.01$; and $M_{s1} = 5$. a: $\eta_g = K_g = 0$. b: $\eta_p = 10\eta_g, K_p = 10K_g$. c: $\eta_p = \eta_g, K_p = 10K_g$. d: $K_g = K_p, \eta_g = 10\eta_p$.

5.1. Numerical method

A relaxation method is used to solve the system of equations numerically. The general model includes thermal gas dissipation terms (i.e., collisionless heat conduction and viscosity). The boundary quantities can be calculated by solving the Rankine-Hugoniat conditions. The thermal gas does not obey the adiabatic law and $P_g y^{\gamma_g}$ is not constant, which means the entropy increases

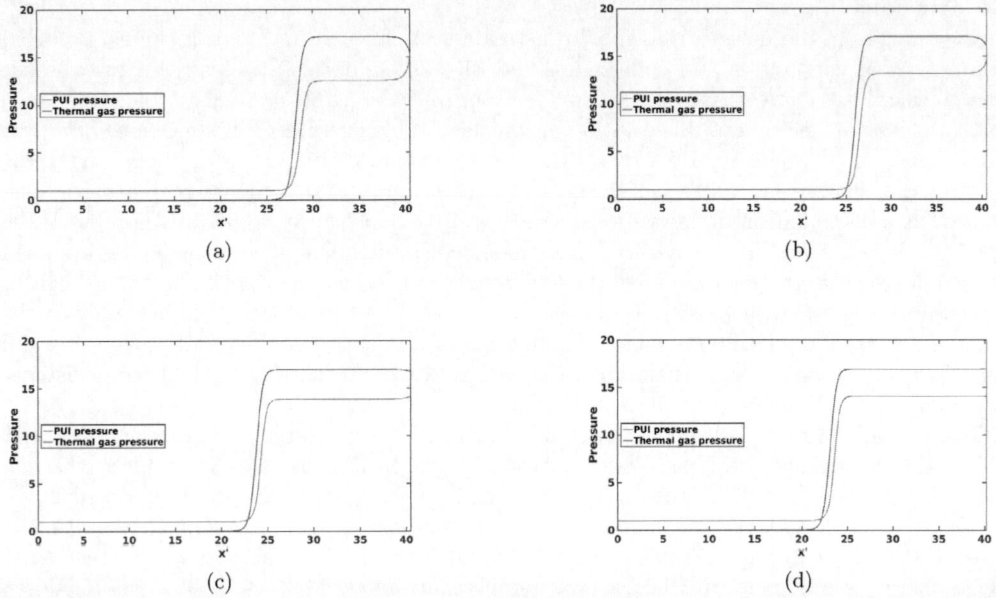

Figure 8: PUI and thermal gas pressure normalized to the thermal gas pressure far upstream. Here $K_g, K_p > \eta_g, \eta_p$; $\gamma_g = 5/3$; $\gamma_p = 5/3$; $P_{p1}/P_{g1} = 0.01$, $M_{s1} = 5$ $\eta_p = \eta_g$, and $K_p = 10K_g$. (a)-(d) show the intermediate steps in the changing downstream pressures as the numerical solution converges to a stable shock structure.

through the shock. The mass, momentum, and energy flux equations in conservative form are

$$\rho U = constant; \tag{28}$$

$$\rho U^2 + P_g + P_p - \frac{4}{3}\eta_p \frac{dU}{dx} - \frac{4}{3}\eta_g \frac{dU}{dx} = constant; \tag{29}$$

$$\begin{aligned} \frac{1}{2}\rho U^3 + \frac{\gamma_g}{\gamma_g - 1} P_g U + \frac{\gamma_p}{\gamma_p - 1} P_p U - \frac{1}{3}\frac{K_p}{\gamma_p - 1}\frac{dP_p}{dx} - \frac{1}{3}\frac{K_g}{\gamma_g - 1}\frac{dP_g}{dx} - \frac{4}{3}\eta_p U \frac{dU}{dx} \\ - \frac{4}{3}\eta_g U \frac{dU}{dx} = constant. \end{aligned} \tag{30}$$

Equations (28)-(30) show that two equations and three unknowns (i.e., the inverse compression ratio, thermal gas and PUI pressure) exist. The conservative form of equations can only specify the inverse compression ratio and total pressure $(P_g + P_p)$, but it can not determine the ratio of PUI pressure and thermal gas pressure at the downstream state separately. Since the relaxation method needs the boundary conditions on the both sides, a numerical code has been written to determine P_g and P_p downstream (i.e., P_{p2} and P_{g2}). In our numerical code, an initial guess is made for both P_{p2} or P_{g2} and then the system is iterated. The pressure profile is not stable for the downstream state because of the wrong guess, and for example in Fig. (8a), the downstream PUI pressure is greater than the initial guess. The downstream pressures are changed in every step subject to the constraint that $P_{g2} + P_{p2}$ satisfies equations (28)-(30) and the system is iterated until it reaches a stable downstream state. Figs. (8b) and (8c) show two intermediate steps as the downstream pressures changes, and Fig. (8d) is the final step where shock structure is stable.

6. Conclusions

All transitions from the upstream to the downstream state for a system that includes both PUI collisionless heat conduction and collisionless viscosity are smooth. Classically, in the absence of the collisionless energetic particle viscosity, a double-valued solution exists for some Mach number regimes. To eliminate this solution and reach a downstream state, a gas sub-shock was added to the shock waves [9, 10]. In fitting a gas sub-shock certain assumptions about the physics of the thermal gas and the sub-shock are made. In our PUI-mediated plasma model, we find that a combination of energetic particle viscosity and heat conduction smooths all the shocks and there is no need to introduce a gas sub-shock. In fitting a gas sub-shock, it is assumed that for relatively large Mach numbers the energetic particle heat conduction is not enough to smooth the shock and so there is an unspecified thermal dissipative process that acts only on the thermal gas. The downstream state of a combination of energetic particles heat conduction and a gas sub-shock can be different than the downstream state of an energetic particles viscous-heat conduction smoothed shock.

The observed HTS by Voyager 2 is mediated by PUIs, and despite being a perpendicular shock, reflected thermal protons do not provide a dissipation mechanism and the thermal gas remains relatively cold in crossing the HTS and they are only heated adiabatically. We numerically simulate the TS-2 and show that the thermal gas experiences very little heating at the shock (the downstream thermal plasma temperature $\approx 1.27 \times 10^5$ K) and almost all the internal energy is converted to PUIs, whose temperature increases to $\approx 1.82 \times 10^7$ K. When measured using the thermal gas sound speed, the flow of the model HTS remains supersonic downstream of the shock. However, when using the combined sound speed, accounting for both PUIs and thermal gas, the downstream Mach number is indeed subsonic. Based on our choice of the heat flux coefficient, the HTS thickness is about 3×10^5 km. All these results are consistent with Voyager 2 observations ([2, 16]). The HTS compression ratio is found to be about 3, which is a little stronger than inferred by [3]. Voyager 1 observed an interstellar shock ([6]) which was weak, laminar, low beta, and quasi-perpendicular. We use the observed shock data to numerically simulate a shock in the VLISM. The shock wave appears to be very weak with a Mach number at about 2.24 and a compression ratio about 2.3. The flow through the shock becomes subsonic downstream relative to the combined thermal gas and magnetic field pressure.

The general case considered different values of the dissipation coefficients associated with both the thermal gas and energetic particles (i.e., η_p, η_g, K_p, and K_g). The choice of different values of the dissipation term yields either a thermal gas or energetic particle mediated shock. A large viscous thermal gas term yields a gas dynamic shock that is partially mediated by energetic particles. One example of this kind of shock is Neptune's inbound bow shock ([21]). An example of the shock mediated by energetic particles is the HTS.

References

[1] Lario D, Decker R B, Roelof E C and Viñas A F 2015 *Journal of Physics Conference Series* **642** 012014

[2] Richardson J D, Kasper J C, Wang C, Belcher J W and Lazarus A J 2008 **454** 63–66

[3] Burlaga L F, Ness N F, Acuña M H, Lepping R P, Connerney J E P and Richardson J D 2008 **454** 75–77

[4] Florinski V, Decker R B, le Roux J A and Zank G P 2009 **36** L12101

[5] Zank G P, Hunana P, Mostafavi P and Goldstein M L 2014 **797** 87

[6] Burlaga L F, Ness N F, Gurnett D A and Kurth W S 2013 **778** L3

[7] Zank G P, Mostafavi P and Hunana P 2016 *Journal of Physics Conference Series* **719** 012014

[8] Zank G P 2016 *Geoscience Letters* **3** 22

[9] Axford W I, Leer E and McKenzie J F 1982 **111** 317–325

[10] Drury L O and Völk J H 1981 **248** 344–351

[11] Zank G P, Pauls H L, Cairns I H and Webb G M 1996 **101** 457–478

[12] Jokipii J R and Williams L L 1992 **394** 184–187

[13] Richardson J D 2008 **35** L23104

[14] Mostafavi P, Zank G P and Webb G M 2016 *American Institute of Physics Conference Series* (*American Institute of Physics Conference Series* vol 1720) p 090003

[15] Mostafavi P, Zank G P and Webb G M 2017 *under press.*

[16] Zank G P, Heerikhuisen J, Pogorelov N V, Burrows R and McComas D 2010 **708** 1092–1106

[17] Burrows R H, Zank G P, Webb G M, Burlaga L F and Ness N F 2010 **715** 1109–1116

[18] Burlaga L F, Florinski V and Ness N F 2015 **804** L31

[19] Zank G P, Shaikh D and Ao X 2006 *Physics of the Inner Heliosheath* (*American Institute of Physics Conference Series* vol 858) ed Heerikhuisen J, Florinski V, Zank G P and Pogorelov N V pp 308–313

[20] Zirnstein E J, Heerikhuisen J, Zank G P, Pogorelov N V, McComas D J and Desai M I 2014 **783** 129

[21] Richardson J D, Paularena K I, Lazarus A J and Belcher J W 1995 **22** 325

IOP Conf. Series: Journal of Physics: Conf. Series **900** (2017) 012017 doi:10.1088/1742-6596/900/1/012017

Pressure Waves in the Heliosheath?

John D. Richardson

Kavli Center for Astrophysics and Space Sciences, Massachusetts Institute of Technology, Cambridge, MA 02139 USA

jdr@space.mit.edu

Abstract. Models predict that solar transients drive pressure waves through the heliosphere. Pressure pulses are observed near solar maximum upstream of the termination shock and in the heliosphere. These pressure pulses may generate the plasma oscillations observed in the local interstellar medium. We investigate whether the observed plasma, particle, and magnetic field observations are consistent with the presence of pressure waves. The results are mixed. The plasma density, temperature, keV, and MeV particle intensities vary in phase with similar amplitude as expected for pressure waves. The galactic cosmic rays are correlated with the plasma and particles with a ~30-day lag. However, the magnetic field and velocity show only a weak correlation with the plasma and particles.

1. Introduction

The Voyager spacecraft are making the first observations of the heliosheath and local interstellar medium (LISM). Voyager 1 (V1) was at 134 AU in March 2017 in the LISM and Voyager 2 (V2) was at 114 AU in the heliosheath. This paper provides an update of recent V2 observations and compares them with V1 observations. We also investigate whether fluctuations observed in the heliosheath are pressure waves.

2. Recent Voyager 2 Observations

Figure 1 shows 25-day running averages of V2 plasma data through early 2017 and magnetic field data through 2015 [1]. The speed at V2 continues to fluctuate between 120 and 160 km/s but has averaged about 145 km/s throughout the heliosheath; the expected slowdown is not observed [2]. The flow in the heliosheath initially rotated tailward, with the RT angle (using the standard RTN coordinate system) going from 15° after the termination shock to almost 60° in 2014, but since then it has not turned further. The RN angle increased from 5° near the termination shock to about 30° in 2013 but since then has decreased slightly. The plasma flows do not suggest that V2 is nearing the heliopause. At V1 the flow speed was near zero throughout an 8 AU stagnation region before the heliopause crossing; no signs of a stagnation region are observed at V2.

The plasma density increased in 2011, probably associated with the arrival of solar maximum at V2, and has on average remained at about 0.002 cm^{-3} with significant increases in a merged interaction region (MIR) in 2012 [3] and in a probably MIR in 2015. The temperature of the thermal protons is correlated with the density and averages about 50,000 K. The magnetic field data are also on average constant, with increases just after the termination shock and in the MIR in 2012. At V1 the magnetic field B started to increase about two years before the heliopause crossing; no evidence of such an increase is observed at V2 through 2015.

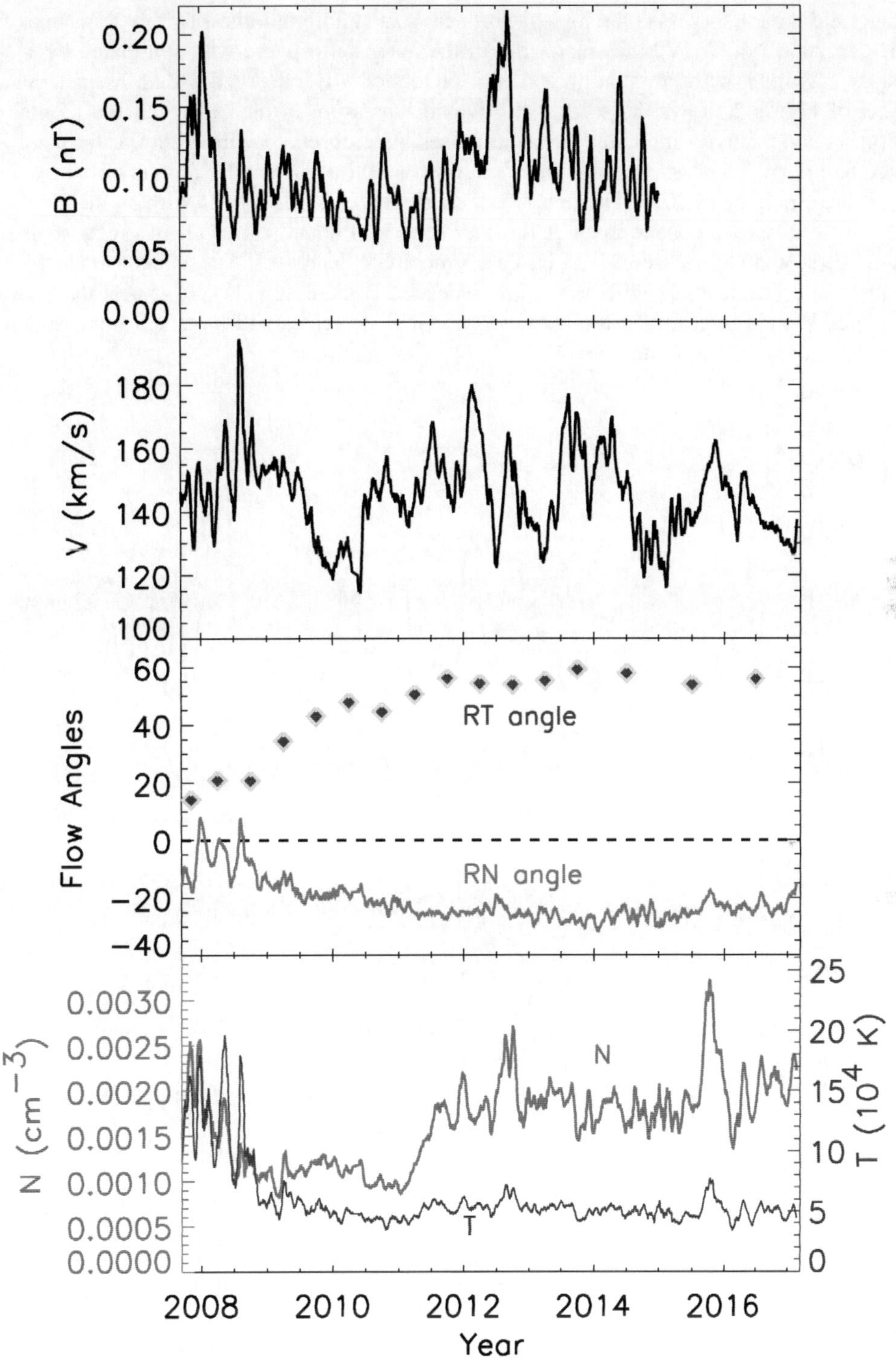

Figure 1. The magnetic field magnitude B, plasma speed, plasma flow angles, plasma density, and plasma temperature measured by V2 in the heliosheath.

Figure 2 extends the pressure pulse observations of Richardson et al. [4] to include 2 more years of magnetic field data. V1 crossed the heliosheath near solar minimum and saw few transients, with only one MIR in 2006 [5]. At V2 the solar wind in the outer heliosphere was dominated by MIRs in the 2001 solar maximum with pressure increases of factors of 4-10 observed 1-2 times each year [4]. The top panel of Figure 2 shows the solar wind dynamic pressure in the HSH and the monthly sunspot numbers. As solar activity increased pressure pulses were observed with a similar frequency to those observed before the TS crossing but with much lower amplitudes, usually factors of 0.5 except for one factor of three increase in 2015. The pressure increase in 2012 was a MIR with an increase in B and a decrease in the galactic cosmic ray (GCR) intensity [5]. Richardson et al. [4] suggest that pressure pulses D, E, F, and G could be MIRs based on the GCR decreases. Subsequent magnetic field data show that while D is a small MIR, E has no associated B increase [1]. We expect that that the 2015 event, which has the largest pressure increase and GCR decrease, will have a B increase but analysis of these data is not yet complete.

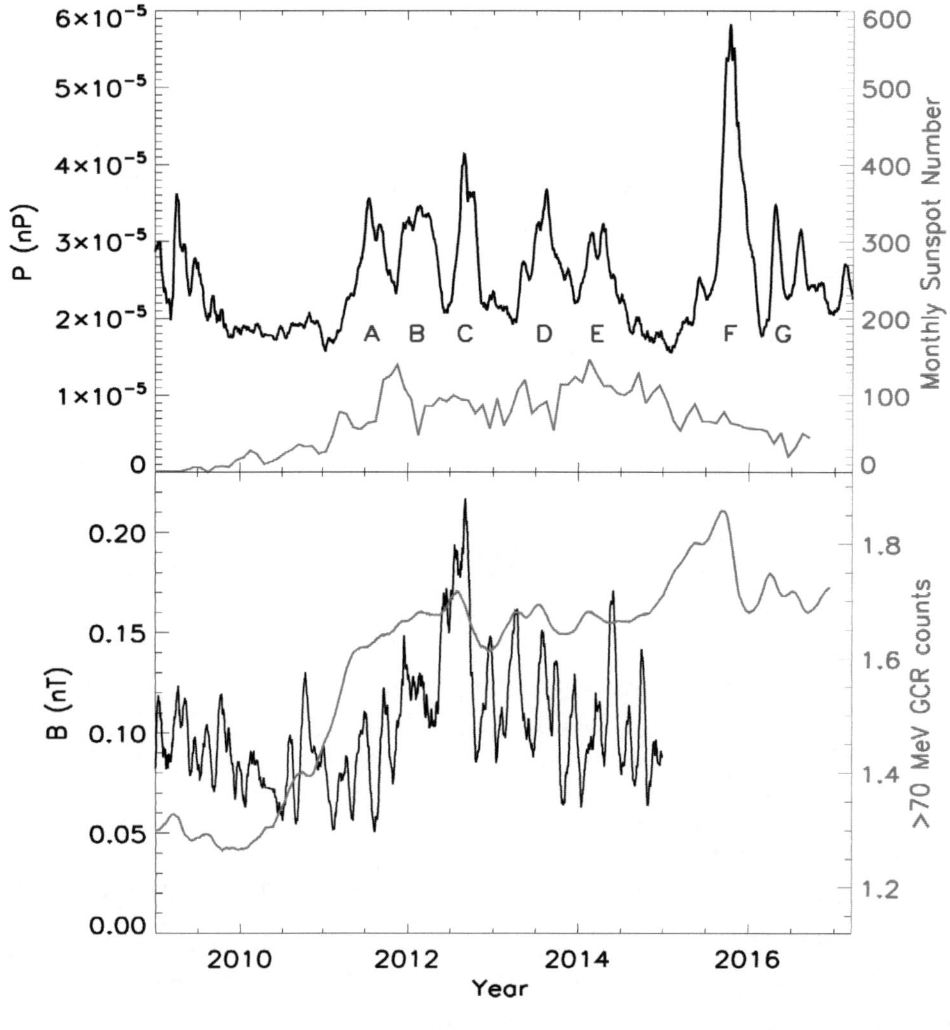

Figure 2. The top panel shows the dynamic pressure in the heliosheath at V2 and the monthly sunspot number. The bottom panel shows the magnetic field magnitude and the CRS >70 MeV GCR count rates.

3. Pressure waves in the heliosheath

Several studies predict that solar transients drive pressure waves in the heliosheath [6, 7, 8, 9, 10, 11, 12]. These pressure pulses are partially transmitted and partially reflected near the heliopause. The reflected waves again encounter the termination shock, moving it inward [9, 10]. The transmitted waves drive weak shock waves into the LISM [8].

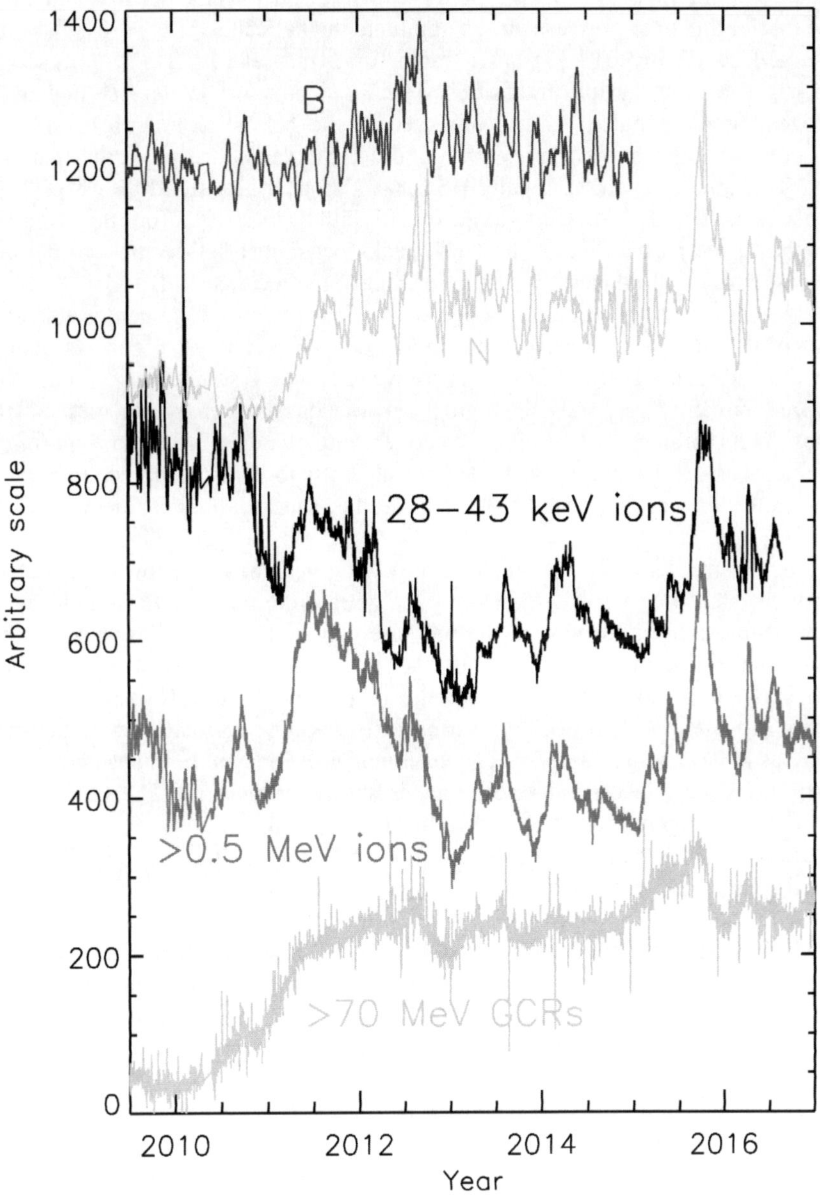

Figure 3. The profiles of B, N, keV ions, MeV ions, and GCRs in the heliosheath

IOP Conf. Series: Journal of Physics: Conf. Series **900** (2017) 012017 doi:10.1088/1742-6596/900/1/012017

We investigate the properties of the pressure pulses in Figure 3. If these events are pressure waves the thermal plasma density and temperature, particle intensities, speed, and magnetic field should all vary in phase. Figure 3 shows the V2 magnetic field, thermal plasma density, 28-43 keV ion intensity, >0.5 MeV ion intensity, and GCR intensity. This comparison is qualitative, each quantity is scaled to fit in the plot. The density N and B are 11-day running averages and the particle intensities are one-day averages. The plot covers 2009.5 through early 2017; from the termination shock crossing at 2007.7 to 2009.5 solar minimum conditions are prevalent and pressure pulses are not observed [2]. The transition from solar minimum at V2 occurs in 2011 where the plasma density and average magnetic field increase. After this time pressure pulses dominate the heliosheath.

The first MIR observed at V2 in 2012 [3] has a factor of 2 increase in B and corresponding increases in the density and all the particle intensities. The large increase in density in late 2015 corresponds to the biggest pressure pulse V2 has observed in the heliosheath, with a factor of 3 increase in pressure, increases in the particle intensities, and a GCR decrease suggesting that B also increases in this event. The smaller event at about 2013.6 has also been identified as a MIR [1] and has the same characteristics as the 2012 MIR. Increases in early 2014, in 2015 before the large event, and two in 2016 occur in all the particle data. The GCR peaks occur just before the particle peaks, suggesting they are at least partially produced by pile up in front of the increase in B.

The panels of Figure 4 compare the fluctuation amplitudes; we plot for each parameter x 25-day (1 solar rotation) averages of (x - <x>)/<x> where we average <x> over 301 days. Figure 4a compares the thermal proton density and temperature. For a pressure wave N and T should vary in phase, with Tperp (to which V2 predominantly responds) changing due to conservation of the first adiabatic invariant as B increases. The correlation coefficient is between the changes in N and T is 0.88 and peaks at zero lag. We note that N and T are derived from the same fits to PLS spectra so are not totally independent quantities, but this correlation is not seen in the supersonic solar wind, only the heliosheath.

Figure 4b compares the plasma densities and the 28-43 keV ion intensities. The low energy charged particle (LECP) experiment intensities are expected to vary slightly more than the PLS densities since the LECP ions are both compressed and heated in the pressure waves. The data show that this effect occurs before 2014; after that the fluctuations are roughly the same. The correlation coefficient is 0.60 with highest values at small lags, so these values change in phase. Figure 4c compares plasma densities with cosmic ray subsystem (CRS) >0.5 MeV ion intensities. The correlation with between N and the >0.5 MeV particles is 0.66 with a lag of 0. The keV and MeV ions are both accelerated in the heliosheath so their variations are expected to be similar, as shown in the data. The keV and MeV intensity changes have a correlation coefficient of 0.85 in this time period.

Figure 4d compares the density and CRS >70 MeV GCR intensity fluctuations. The GCRs originate outside the heliosphere and diffuse through the heliosphere. Previous work shows that GCR intensities peak before MIRs since the increased magnetic fields reduce diffusion and show Forbush decreases after the MIRs [13]. The data show that the GCR intensities increase before the density and lower-energy particle peaks. The best correlation with density is 0.61 with a 33 day lag (GCR features are observed before density features). The correlation of GCR and >0.5 MeV intensities is .85 with a 25 day lag. The magnitude of the GCR fluctuations is less than that for other particle intensities by about a factor of ten.

Figure 4e compares magnetic field and density variations. For pressure waves we expect these fluctuations to be similar. However, the plot shows that B variations are larger and more frequent than those of the plasma and particles. The correlation between B and N variations is weak, only 0.34. We have tested other correlations with B, such as dynamic pressure NV^2 and VR, and these correlations are also weak. We note that magnetic field data during the large 2015-2016 variations are not yet available.

IOP Conf. Series: Journal of Physics: Conf. Series **900** (2017) 012017 doi:10.1088/1742-6596/900/1/012017

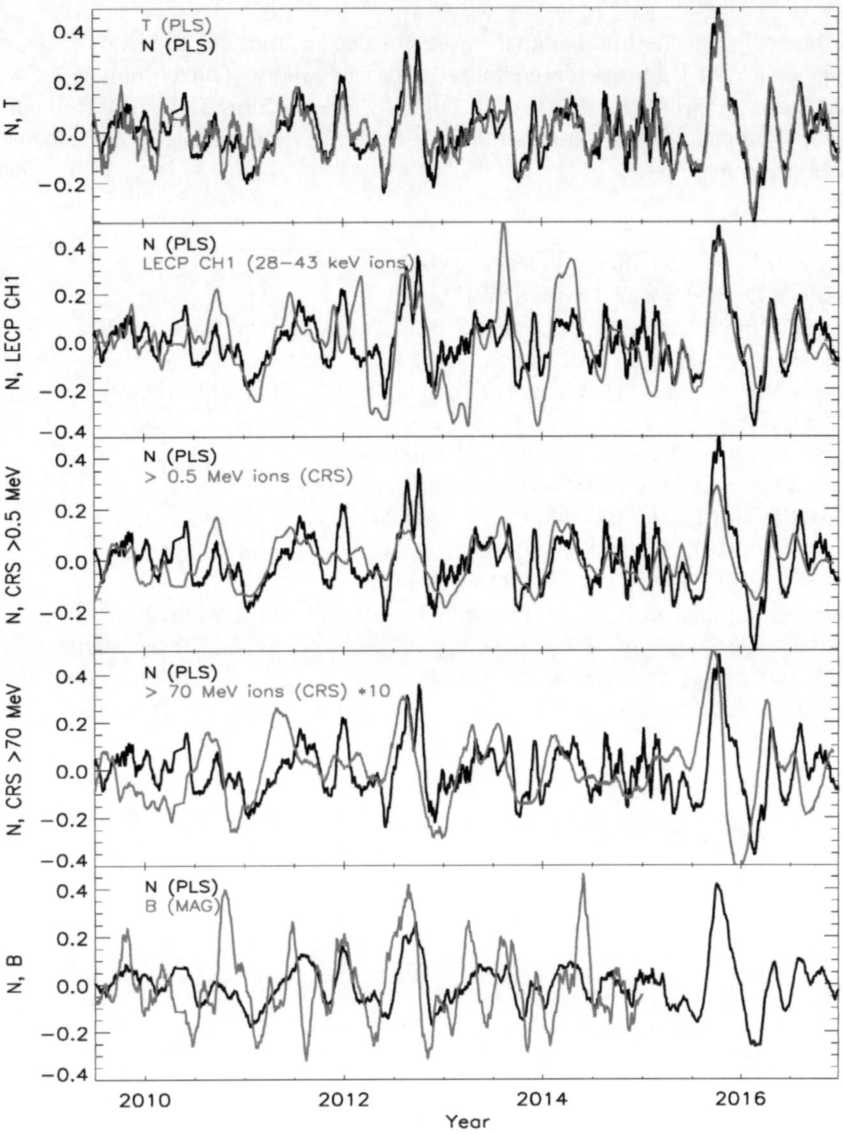

Figure 4. Comparison of changes in density (black curves) compared to changes in other particle intensities and B (red curves) . The plots of each parameter x show (x-<x>)/<x> where <x> is averaged over 301 days.

4. Discussion and Summary

We have tested the conjecture that pressure waves are important components of the structure in the heliosheath. The results are mixed. Particle intensities changes of the thermal plasma, keV energy, and MeV energy plasma are correlated, in phase, and have similar amplitudes. The thermal plasma density and temperature are highly correlated. These results are consistent with pressure waves. The GCRs are correlated with lower energy ion intensities but with a ~25 day lag, consistent with these particles being influenced by magnetic fields in pressure waves. The magnetic field, however, is only weakly correlated with the particle intensities and plasma properties. This result is not consistent with pressure waves and is not understood. We will continue these studies as more data become available.

IOP Conf. Series: Journal of Physics: Conf. Series **900** (2017) 012017 doi:10.1088/1742-6596/900/1/012017

Acknowledgments

We acknowledge support from the International Space Science Institute for the team ``Facing the Most Pressing Challenges to Our Understanding of the Heliosheath and its Outer Boundaries'' which spurred this work. We thank SPDF for the MAG data and CRS for providing GCR data through their web site. JDR was supported under NASA contract 959203 from the Jet Propulsion Laboratory to the Massachusetts Institute of Technology.

5. References

[1] Burlaga L F, Ness N F and Richardson J D 2017 *ApJ*, in press
[2] Richardson, J D and Decker R B 2015 *ApJ* **792** 126
[3] Burlaga L F, Ness N F, Richardson J D, Decker R B and Krimigis S M 2016 *ApJ* **818** 147
[4] Richardson J D, Wang C, Liu Y D, Šafránková J, Němeček Z and Kurth W S 2017 *ApJ* **834** 190
[5] Burlaga L F, Ness N F, Acuna M H, Wang Y-M and Sheeley Jr. N R 2008 *J. Geophys. Res.* **114** A06106.
[6] Story T R and Zank G P 1997 *J. Geophys. Res.* **102** 17381
[7] Steinolfson R S and Gurnett D A 1995 *Geophys. Res. Lett.* **22** 651
[8] Zank G P and Muller H-R 2003 *J. Geophys. Res.* **108** 1240
[9] Washimi H, Zank G P, Hu Q, Tanaka T and Munakata, K. 2007 *ApJ* **670** L139A
[10] Washimi H, Zank G P and Hu, Q. 2011 *MNRAS* **416** 1475
[11] Washimi H, Webber W, Zank G P, Hu Q, Florinski V, Adams J and Kubo Y 2012 *ApJ Lett.* **757** L2
[12] Zank G. P. 2015 *Ann. Rev. Astron. Astrophys.* 53 **449**
[13] McDonald F B, Burlaga L F, McGuire R E and Ness N F 2000 *J. Geophys. Res.* **105** 20997

IOP Conf. Series: Journal of Physics: Conf. Series **900** (2017) 012018 doi:10.1088/1742-6596/900/1/012018

Observations of Low-Frequency Magnetic Waves due to Newborn Interstellar Pickup Ions Using ACE, Ulysses, and Voyager Data

Charles W. Smith[1], Poornima Aggarwal[2], Matthew R. Argall[1], Leonard F. Burlaga[3], Maciej Bzowski[4], Bradford E. Cannon[5], S. Peter Gary[6], Meghan K. Fisher[1], Jason A. Gilbert[7], Sophia J. Hollick[1], Philip A. Isenberg[1], Colin J. Joyce[1], Neil Murphy[8], Raquel G. Nuno[9], Zackary B. Pine[1], John D. Richardson[10], Nathan A. Schwadron[1], Ruth M. Skoug[11], Justyna M. Sokół[4], David K. Taylor[12], and Bernard J. Vasquez[1]

[1] Space Science Center, Institute for the Study of Earth, Oceans and Space, University of New Hampshire, Durham, NH 03824
[2] Electrical Engineering Department, Cooper Union, New York, New York, USA
[3] NASA Goddard Space Flight Center, Code 673, Greenbelt, MD, USA
[4] Space Research Centre, Polish Academy of Sciences, (CBK PAN), Warsaw, Bartycka 18A, Poland
[5] Physics Department, Florida State University, Tallahassee, Florida, USA
[6] Space Science Institute, Boulder, Colorado, USA
[7] Department of Climate and Space Sciences and Engineering, University of Michigan, Ann Arbor, USA
[8] Jet Propulsion Laboratory, Mail Stop 180-600, 4800 Oak Grove Drive, Pasadena, California, USA
[9] School of Earth and Space Exploration, Arizona State University, Tempe, AZ, USA
[10] Kavli Institute for Astrophysics and Space Research, Massachusetts Institute of Technology, 37-655, Cambridge, Massachusetts, USA
[11] Space and Atmospheric Sciences, Los Alamos National Laboratory, Los Alamos, NM, USA
[12] Rensselaer Polytechnic Institute, Troy, New York, USA

E-mail: Charles.Smith@unh.edu

Abstract. Wave excitation by newborn interstellar pickup ions (PUIs) plays a significant role in theories that attempt to describe IBEX and Voyager observations in the solar wind and heliosheath. The same dynamic processes can be far-reaching and extend into the inner heliosphere to at least 1AU and likely to smaller heliocentric distances. While the high-resolution magnetic field measurements required to study these waves are not yet available in the heliosheath, we have studied a range of available observations and found evidence of waves due to interstellar PUIs using ACE (1998–2015 at 1 AU), Ulysses (1996–2006 at 2 to 5 AU, high and low latitudes) and Voyager (1978–1979 and 2 to 6 AU) observations. Efforts to extend the Voyager observations to 35 AU are ongoing. We have examined these data sets and report on observations of low-frequency waves that result from newborn interstellar pickup H^+ and He^+ ions. Although not as common as theory originally predicted, we presently have identified 524 independent occurrences. Our conclusion from studying these waves is that they are seen only when the ambient turbulence is sufficiently weak. The instability that generates these waves requires a slow accumulation of wave energy over several to tens of hours to achieve observable wave amplitudes. In regions where the turbulence is moderate to strong, the turbulence absorbs the wave energy before it can reach observable levels and transports the energy to the dissipation scales where it heats the background thermal particles. Only intervals with the weakest turbulence will permit energy accumulation over this time scale. These conditions are most often, but not exclusively, achieved in solar wind rarefaction regions.

1. Introduction

The solar wind (SW) and embedded heliospheric magnetic field (HMF) form a system of diverse internal dynamics. They also form a barrier to the entry of interstellar charged particles into the heliosphere. In contrast, neutral atoms enter and transit the heliosphere freely at $\sim 23\,\mathrm{km\,s^{-1}}$ [1, 2, 3, 4] until they are ionized, primarily by collisional charge exchange with solar wind ions or photoionization by solar EUV flux [5, 6, 7, 8, 9, 10, 11]. Electron impact ionization may also be important for some particle species. The newly ionized atoms then become embedded in the flow as "interstellar pickup ions" (PUIs) that gyrate about the HMF at the solar wind speed at the point of ionization. As such, they constitute an energetic ion population and a source of free energy for wave excitation.

The theory of wave generation by interstellar PUIs is an outgrowth of the theory of foreshock dynamics and the theory of wave excitation at comets [12]. It predicts that waves excited by interstellar PUIs should be commonly seen throughout the heliosphere as a regular feature of the HMF spectrum. They are not. Turbulence transport theory argues that the wave energy excited by interstellar PUIs is absorbed by the turbulent energy cascade and used to heat the background plasma [13, 14, 15, 16, 17, 18, 19, 20, 21, 22, 23, 24, 25, 26, 27, 28, 29, 30, 31, 32, 33, 34]. Thermal proton temperatures are approximately constant beyond ~ 20 AU as a result of turbulent driving by interstellar PUIs, thereby offsetting expansive cooling. The turbulent reprocessing of the accumulating wave energy makes observations of waves due to interstellar PUIs rare and difficult to find.

In this paper we review observations of magnetic waves due to interstellar PUIs as seen by the Advanced Composition Explorer (ACE), Ulysses and Voyager spacecraft. We present their observational characteristics and the circumstances under which they are seen.

2. Observations

Because PUIs move at the solar wind speed in the plasma frame, the resonant instability produces right-hand polarized, sunward propagating fast mode waves in the plasma frame. The Doppler shift results in a spacecraft-frame wave signature that is left-hand polarized and seen at spacecraft-frame frequencies $f_{sc} \geq f_{i,c}$ where $f_{i,c} = e_i B/(m_i c 2\pi)$ is the ion cyclotron frequency of the source ion [12]. Here e_i is the charge of the pickup ion, B is the mean magnetic field intensity, m_i is the mass of the pickup ion, and c is the speed of light. The waves at $f_{sc} > f_{i,c}$ result from pitch-angle scattering of the source ion. Waves are not seen at $f_{sc} < f_{i,c}$ because this would require energization of the source ion which is not generally seen outside of shocks. This gives these waves a unique signature that is distinct from many solar wind processes.

2.1. Early Ulysses Observations

The first observations of waves due to ionized hydrogen (H), pickup protons, were reported using Ulysses magnetic field data [35] recorded from early 1992 until late 1993 [36]. Thirty-one distinct intervals of waves were reported. Figure 1 reproduces one of them. Waves are seen in the power spectrum at $f_{sc} > f_{p,c}$ and most of the energy lies in the components perpendicular to the mean magnetic field. Their propagation direction was unresolved due to the insufficient data rate of the thermal ion instrument and this has remained a common problem in the study of PUI waves. It was reported that the observation of the waves did not depend on the strength

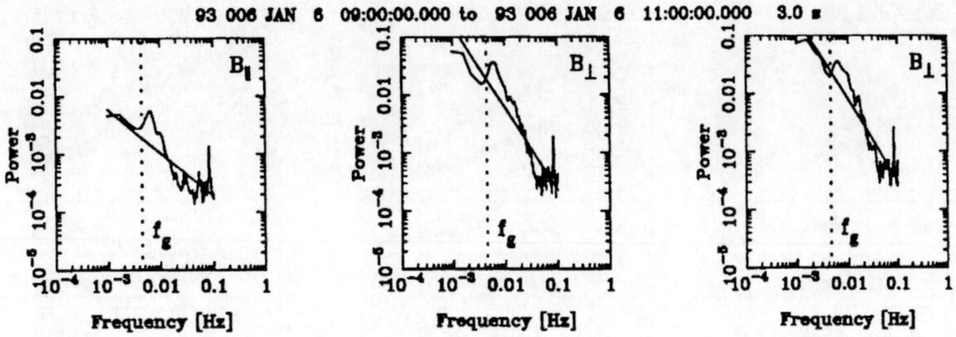

Figure 1. Power spectra for first reported observations of magnetic waves due to newborn interstellar H^+ recorded by the Ulysses spacecraft. Data interval is January 6, 1993 from 09:00 to 11:00 UT. (left to right) Spectra of the fluctuation component parallel to the mean magnetic field and two components perpendicular to the mean magnetic field. Note that the fluctuation vector is largely transverse to the mean field. Reproduced from [36].

of the magnetic field $|\mathbf{B}|$, but the waves were most often seen when the orientation of the field was quasi-radial.

The Ulysses spacecraft moved from 5.4 to 4.6 AU and from $-1°$ to $-40°$ heliographic latitude during the time that the events were gathered [36]. H ionization becomes most efficient at this distance from the Sun whereas inside ~ 4 AU the neutral H density falls precipitously so that neutral H density at 1 AU is only $\sim 1\%$ that of the interstellar medium [37]. The above wave observations occured where the source was strongest.

2.2. Early Voyager Observations
While developing a turbulence data base of magnetic field spectra from the Voyager 2 spacecraft [38], a wave event was found that could be attributed to newborn interstellar PUIs [39]. The event was recorded on January 7, 1979 (Day-of-Year or DOY 7) when the spacecraft was at 4.5 AU and $-5.4°$ heliographic latitude. Unlike the previously reported Ulysses events, this interval contained waves due to both H^+ and He^+. Neutral He is the second most abundant interstellar atom. No shock or other likely sources of the waves were observed. The waves were observed in the middle of a seven-day rarefaction interval lasting from DOY 4 through DOY 10. There was a strong forward shock seen midway on DOY 12, but no evidence of waves between the above wave interval and the shock crossing five days later. It is unlikely that the observed waves were due to shock acceleration.

The mean field intensity was $|\mathbf{B}| \simeq 0.5$ nT. As a result, $f_{i,c}$ for protons was only marginally resolved by the 96 s thermal ion instrument, PLS [40], whereas $f_{i,c}$ for He^+ was readily observed. Cross-correlation of the magnetic and velocity fluctuations showed that the waves were Sunward propagating as expected.

Realizing that low-frequency waves due to interstellar PUIs had been found in rarefaction regions, a second survey of Voyager 1 and 2 observations from DOY 329 of 1977 through DOY 288 of 1979 (2 to 6.2 AU) was undertaken [42]. Figure 2 shows three examples of spectra computed, but not shown, as part of that study. The left stack shows an example of waves due to newborn interstellar pickup He^+. Note the abrupt rise in the power at the He^+ cyclotron frequency $f_{He,c}$, along with the moderate rise in the degree of polarization and coherence and the fact that the same frequencies are left-hand circularly polarized in the spacecraft frame as shown by the ellipticity. The middle panel shows typical solar wind spectra without observable waves. It is

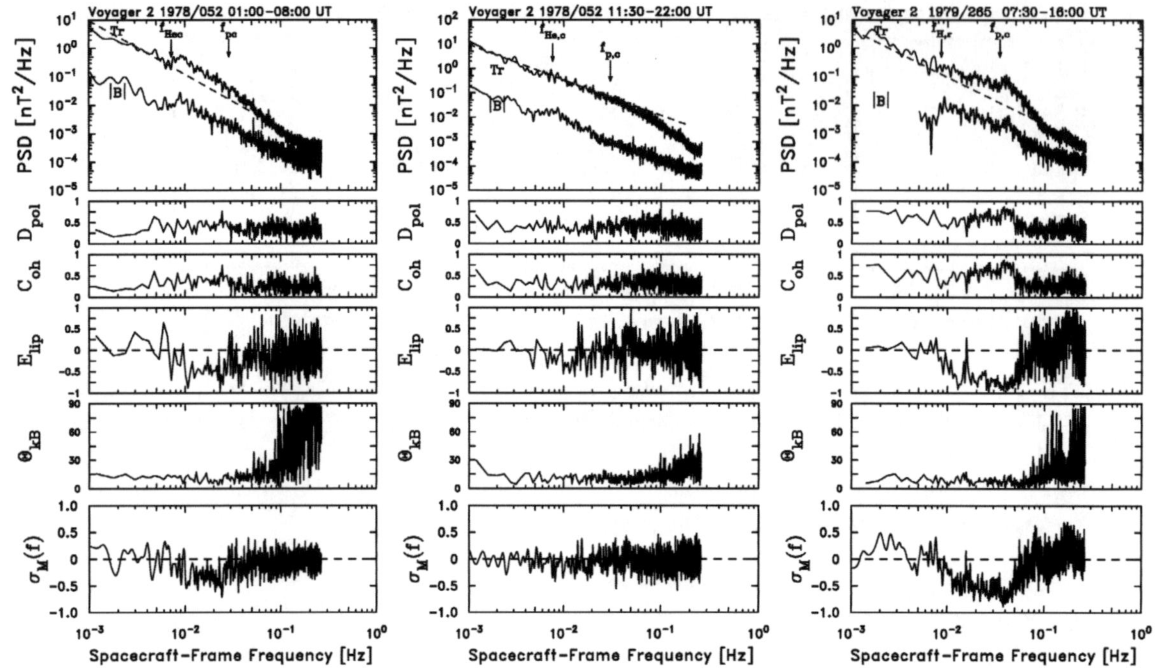

Figure 2. Our analysis of three intervals of Voyager 2 magnetic field data with times marked at the top of each stack. Each stack contains (top to bottom) the spectrum of the total power and power in $|\mathbf{B}|$, degree of polarization, coherence, ellipticity, angle between the minimum variance direction and the mean magnetic field, and the normalized mangetic helicity. The left stack shows our analysis of a time when waves due to newborn interstellar pickup He^+ are present. The middle stack shows our analysis of a control interval without evidence of waves due to PUIs. The right stack shows our analysis of a time when waves due to newborn interstellar pickup H^+ are present.

a broken power law without significant spectral features at either $f_{He,c}$ or $f_{p,c}$. The right stack shows an example of waves due to newborn interstellar H^+. There is an abrupt rise in the power spectrum at $f_{p,c}$ and a moderate enhancement in the degree of polarization and coherence. The ellipticity $\simeq -1$, which extends back to $f_{He,c}$, suggests that newborn He^+ may also be involved, but not enough to significantly alter the power spectrum. The magnetic helicity agrees closely with the ellipticity.

Eleven events were found using data from both Voyager 1 and 2 magnetic field data [42] including the event previously studied [39]. Ten events showed waves due to He^+, seven showed waves due to H^+, and six showed signatures of both sources. One event that showed a power spectral feature indicative of a H^+ source was right-hand polarized in the spacecraft frame. In all other regards, the observation appears to be excited by newborn interstellar pickup H^+. There is presently no good explanation for the observed polarization.

2.3. Later Ulysses Observations

Since the observations were proving to be scarce and there are no reliable large-scale signatures of likely wave intervals, unlike foreshock dynamics which are readily found by the presence of shocks, a better search method was needed. A spectrogram code was written for the Ulysses magnetic field data to provide continuous output of spectra including polarization parameters.

Data from 1996 through 1999 and 2003 through 2006 was analyzed [43, 44, 45, 46]. The gap from 2000 through 2002 represents the fast latitude scan and was omitted from this study. A total of 502 wave intervals were uncovered with all but two seen beyond 4 AU. Those two seen inside 4 AU were beyond 3.5 AU. There was a discernable peak in the observation rate at 5 AU once the distribution of heliocentric distances across the spacecraft trajectory was taken into account. Wave events tend to last less than 5 hrs in the spacecraft data. This is not a measure of their lifetime in the plasma, but rather a measure of the spatial extent of the waves. The waves are seen in $\sim 3.6\%$ of the data. Quasi-radial mean magnetic fields are favored, but not required. A strong solar cycle dependence is not seen.

The observed wave events consistently demonstrated high degrees of polarization and coherence at the relevant frequencies. They had minimum variance directions that were quasi-parallel to the mean magnetic field. Most observations were left-hand polarized in the spacecraft frame, but $\sim 10\%$ were right-hand polarized.

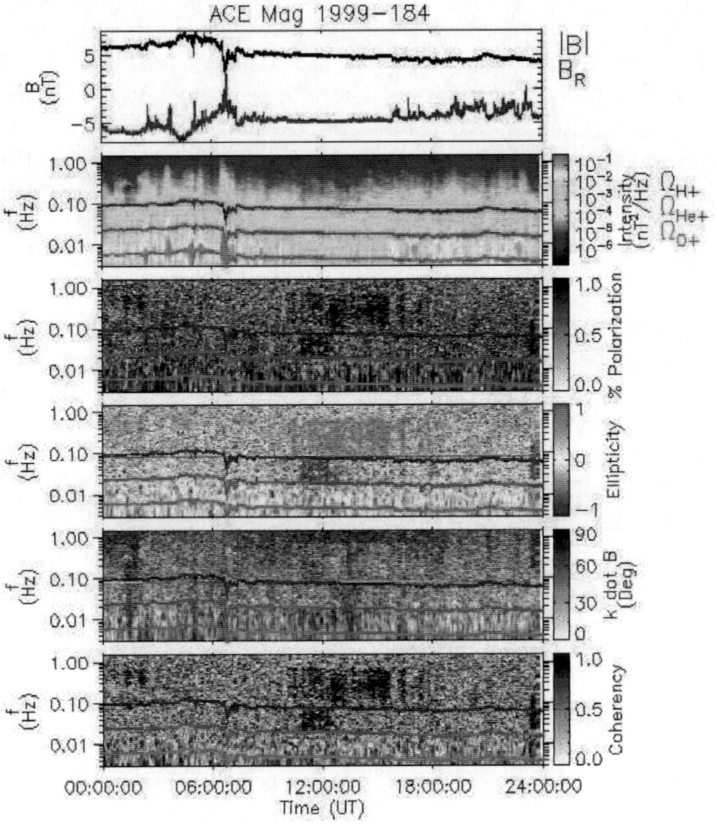

Figure 3. Daily spectrogram of ACE magnetic field data for DOY 184 of 1999. Top to bottom, the panels are the total magnetic field intensity $|\mathbf{B}|$ and radial component B_R; power spectrum; percent polarization; ellipticity; angle between the minimum variance direction and the mean magnetic field; and the coherence. The spacecraft-frame gyrofrequencies of H^+, He^+, and O^+ are indicated in the spectrograms by the blue, red, and magenta traces, respectively. Note the blue patch around noon in the ellipticity at frequencies $f_{He,c} < f_{sc} < f_{p,c}$. These fluctuations are highly coherent and strongly polarized with nearly circular polarization. These are left-hand polarized waves in the spacecraft frame that arise from newborn interstellar pickup He^+. The right-hand polarized fluctuations at higher frequencies have not been studied.

IOP Conf. Series: Journal of Physics: Conf. Series **900** (2017) 012018 doi:10.1088/1742-6596/900/1/012018

2.4. ACE Observations

Taking a lesson from the Ulysses analysis, a daily spectrogram code was written for the ACE magnetic field data [47, 48]. Figure 3 shows one such daily spectrogram that contains waves due to pickup He^+. Those spectrograms are now available to anyone at the ACE Science Center [49] as a routine data product and are very useful in preliminary examinations for a range of investigations at kinetic physics scales. Eighteen years of data were surveyed from launch in late 1997 into 2015. Twenty-five wave events were found to result from newborn interstellar He^+ [50, 51]. We did not investigate the ACE data for waves due to pickup H^+ since the density of interstellar H is very low at 1 AU and it is difficult to discount foreshock sources in this case.

Data rates for the thermal ion instrument on ACE preclude cross-field correlation in an effort to determine wave propagation directions at this time scale [52]. All twenty-five wave events show high degrees of polarization and coherence with minimum variance directions there were quasi-parallel to the mean magnetic field. All but two of the wave events were left-hand polarized in the spacecraft frame. This has become a recurring theme in this analysis. Approximately 10% of all waves seen in association with interstellar pickup ion sources have the "wrong" polarization. This is true for both H^+ and He^+ sources.

Each December, ACE passes through the He focusing cone [53], a region with $\sim 3\times$ the normal density of interstellar neutral He and as a result, we expect to see a higher abundance of wave events during this period. However, waves are seen throughout the year and at any time in the solar cycle. The relative strength of the wave source does not seem to provide the sole explanation for when waves are seen. This appears to be true for Ulysses and Voyager observations as well.

3. Observability

In our more extensive survey efforts [42, 46, 51] we also analyzed an approximately equal number of control intervals selected from times near the wave intervals when the ambient conditions appeared similar but no waves were observed. The Ulysses wave events were shown to exist when the background power spectrum is lower than in control intervals [46]. However, this effect alone could not explain absence of the intense wave features which were originally predicted [12]. Our studies have indicated the importance of an additional process.

We take the expression for the time-asymptotic wave power spectrum [12]:

$$I_{\pm}(k,\infty) = \frac{1}{2}\left\{[C(k)]^2 + 4I_+(k,0)I_-(k,0)\right\}^{1/2} \pm \frac{1}{2}C(k) \tag{1}$$

where

$$\begin{aligned} C(k) &= I_+(k,0) - I_-(k,0) \\ &+ 2\pi N V_A m|\Omega_{i,c}|k^{-2}\left[\Omega_{i,c}k^{-1}v_0^{-1} - \left(\Omega_{i,c}k^{-1}v_0^{-1} - \mu_0\right)|\Omega_{i,c}k^{-1}v_0^{-1} - \mu_0|^{-1}\right] \end{aligned} \tag{2}$$

and $I_+(k,0)$ ($I_-(k,0)$) is the background spectrum for the anti-sunward (sunward) propagating fluctuations at wavenumber k. All k are assumed to be parallel to the mean magnetic field. For sunward propagating waves, the wave vectors $k < 0$ ($k > 0$) denote the right-hand polarized fast-mode (left-hand polarized Alfvén) waves. Here m_i is ion mass, N_i is the PUI number density, $\Omega_{i,c} = e_i B/(m_i c)$ is the ion cyclotron frequency $= 2\pi f_{i,c}$, V_A is the Alfvén speed, v_0 is the PUI speed in the plasma frame $\simeq V_{SW}$ where V_{SW} is the solar wind speed and μ_0 is the pitch angle of the newborn ions.

The time required to scatter the PUI is on the order of an hour and the time required to accumulate the observed wave energy is on the order of 20 hrs. This suggests that the PUI distribution remains a scattered distribution throughout the wave energy accumulation process.

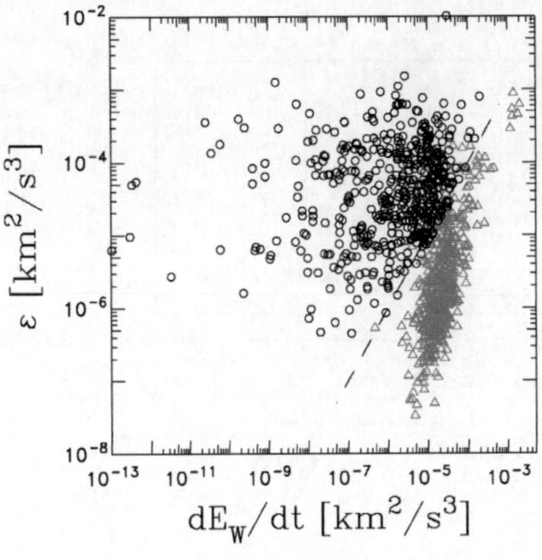

Figure 4. Scatter plot for wave growth rate dE_W/dt vs. turbulent transport rate ϵ for the Ulysses wave observations (red triangles) and control intervals (black circles). The diagonal line is unity. Wave observations favor growth rates that are greater than turbulence rates. Control intervals where waves are not seen favor turbulence rates that are greater than wave growth rates. Reproduced from [46]. ACE and Voyager observations show similar results for both He$^+$ and H$^+$ pickup ions [42, 51].

Therefore, we do not use linear theory based on the initial PUI ring distribution to compute the rate of wave energy excitation over the lifetime of the observation. For our purposes we will take the derivative of eq. 1–2 holding all background variables constant so that the accumulation rate, dE_W/dt is controlled by the PUI production rate, dN_i/dt.

We can also compute the rate at which the wave energy is transported through the inertial range via turbulent cascade (according to MHD extensions of the Kolmogorov theory [54, 55, 56]):

$$\epsilon = \frac{f^{5/2}[E(f)]^{3/2} \cdot 21.8^3}{V_{SW} N_P^{3/2}} \tag{3}$$

where $E(f)$ is the measured magnetic field power spectral density in units of $\mathrm{nT^2\,Hz^{-1}}$, which is assumed to vary as $f^{-5/3}$. Equipartition of magnetic and kinetic energy is assumed, $R_A = 1$. N_P and 21.8^3 are part of the conversion of the magnetic field to Alfvén units. This choice is based on the assumption that the turbulent cascade of the background fluctuations is essentially two-dimensional and this controls the rate at which the wave energy is absorbed by the turbulent cascade. Eq. 3 disagrees with that used by [55] by a factor of $2\pi\,[(5/3)(1+R_A)/C_K]^{3/2} \sim 6\pi$ where the Komogorov constant $C_K = 1.6$ [57], but this expression is more nearly in agreement with the observed heating rates at 1 AU [56].

Figure 4 shows a scatter plot for the computed wave excitation and turbulent cascade rates for wave events and control intervals used in the Ulysses analysis [46]. The dashed line is unity. Although not perfect, wave events tend to show wave growth rates in excess of the turbulent cascade rates while the control intervals show computed turbulence rates that exceed the wave growth rates. Thus we find that the waves are seen when the background turbulence is sufficiently weak to allow the wave energy to accumulate to observable levels, thereby explaining why these waves are so rarely observed. Similar results are found for ACE [51] and Voyager [42] observations and for waves due to He$^+$ and H$^+$ PUIs.

4. Later Voyager Analyses

We have modified the daily spectrogram code used in the ACE analysis to plot Voyager magnetic field data and applied the code to the data currently available in the National Space Science Data Center (NSSDC). The data extends from launch in September 1977 through 1990 and includes data resolution 1.92 s, 9.6 s and 48 s. We have scanned the data, found $\sim$ 400 wave events, and are in the process of analyzing the results.

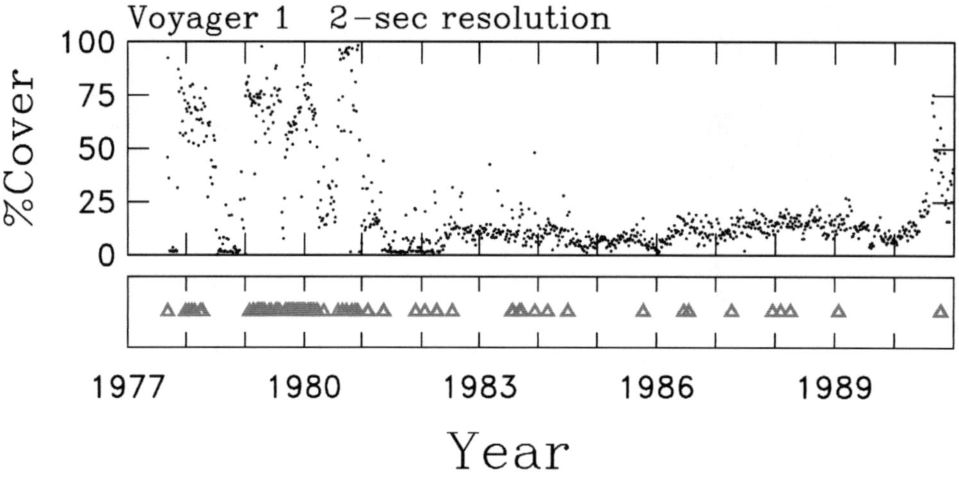

Figure 5. (top) Daily data coverage of Voyager 1 2 s magnetic field data. Note abrupt decline in coverage beginning 1981. (bottom) Times of waves due to newborn interstellar ions (either H^+ or He^+ or both) currently under study. Approximately 200 intervals showing waves due to PUIs are currently known. The Voyager 2 data set shows $\sim 3\times$ as many wave intervals in the same time interval.

Figure 5 shows the daily coverage of Voyager 1 magnetic field 1.92 s data (top) and the times when waves due to pickup ions are seen (bottom). This list continues to be refined. An incomplete search of the Voyager 2 data set reveals $\sim 3\times$ as many wave events for the same years. Wave observations extend out to include observations in 1990 when Voyager 1 was at $\sim$ 40 AU and Voyager 2 was at $\sim$ 35 AU. We find waves due to He^+ and H^+ source ions that include the most distant observations. Reduced data coverage starting in 1981 and the shorter data intervals due to reduced spacecraft tracking work in tandem with the reduced $|\mathbf{B}|$ and $f_{i,c}$ to make resolution of the waves increasingly difficult as the spacecraft move to greater heliocentric distances. This reduces our ability to study the waves at the greater heliocentric distances, but they are still found.

5. Bernstein Waves

Although distinct from the above observations, an additional wave event was discovered in the Voyager 2 magnetic field data [41]. This event was measured on DOY 350 of 1978, 199 days before the Jovian encounter. Strong aliasing in the 1.92 s data suggested a signal at greater than the Nyquist frequency and so the highest resolution (0.32 s) MAG data was obtained to study this event. The waves were seen to evolve from $\sim$ 0.18 Hz to $\sim$ 0.67 Hz in the spacecraft frame with the magnetic fluctuations transverse to the mean magnetic field. The magnetic field intensity was $|\mathbf{B}| \simeq 0.3$ nT and perpendicular to the radial (solar wind flow) direction. The solar wind speed was a nominal 423 km s^{-1}. The thermal proton density and temperature were unusually low at 0.07 cm^{-3} and 0.57 eV, respectively. A careful analysis ruled out the likelihood of whistler waves and it was concluded that these signals were Bernstein waves. Gyrating newborn interstellar PUIs were inferred to be the source of the waves. No other such event has since been found in the Voyager data.

6. Discussion and Summary

The papers we have reviewed here present many additional features of the waves due to newborn interstellar pickup ions. They show the frequency of occurence, duration of wave events in the data set, relative rate of observation according to solar cycle and season, the properties of the waves, and how those properties sometimes differ from current theoretical predictions, and the time required to reach the observed wave energy levels. In the process, we have demonstrated the utility of computing automated daily spectrograms in an effort to aid in the search for candidate events.

We have used a prediction for the turbulent cascade that derives from the Kolmogorov theory for hydrodynamics. Other theories exist [13, 58, 59, 60, 61, 62, 63, 64, 65, 66]. Third-moment expressions for the turbulent cascade suggest a high degree of variability between similar samples that lies outside what was considered here [67, 68], although the long growth time also suggests that the wave growth persists over a sufficiently extended period of time to allow the local cascade rate to vary significantly and possibly average to what the steady-state theories suggest.

In many applications of plasma instabilities that operate in the solar wind at kinetic scales, such as in planetary, interplanetary and cometary foreshocks, wave excitation and particle scattering are rapid processes due in large part to the greater density of energetic particles. Little to no concern is given to whether the background turbulence can act to limit wave growth. However, the long wave growth times associated with interstellar PUIs provide an exception to the rule. This does not mean that waves are not excited when they are not observed. Every indication from theory and observations suggests that the wave excitation process is ongoing at all times. When the turbulence is too strong to permit the waves to grow to observable levels, the wave energy is still being absorbed by the turbulence and used to heat the background ions. Only the ability to accumulate wave energy to observable levels is absent.

Acknowledgments

This work was supported in part by NASA grants NNX07AH75G, NNX13AF97G, NNX11AJ37G, NNX17AB86G and NNX13AH66G, NASA contract NNG14PN24P, and by NSF grants ATM0635863, AGS0962506, AGS1358103 and AGS1357893. Additional support was provided by Caltech subcontract 44A1085631 to the University of New Hampshire and the Polish National Science Center grant 2015/18/M/ST9/00036. Portions of this research were carried out at the Jet Propulsion Laboratory, California Institute of Technology, under a contract with the National Aeronautics and Space Administration. Data used in these studies are available from the NSSDC and ACE Science Center.

References

[1] Witte M, Rosenbauer H, Banaskiewicz M, and Fahr H 1993 Adv. Space Res. **13** 121-126
[2] Geiss J, and Witte M 1996, in Proc. First ISSI Workshop, edited by von Steiger R, Lallement R, and Lee M A, Space Sci. Rev. **78** 229–238
[3] Gloeckler G, Fisk L A, and Geiss J 1997 Nature **386** 374–377
[4] McComas D J, Alexashov D, Bzowski M, Fahr H, Heerikhuisen J, Izmodenov V, Lee M A, Möbius E, Pogorelov N, Schwadron N A, and Zank G P 2012 Science **336** 1291–1293
[5] Blum P W, and Fahr H J 1970 Astron. Astrophys. **4** 280–290
[6] Axford W I 1972, in *Solar Wind*, edited by C P Sonett, P J. Coleman, and J M Wilcox, p. 609, NASA Spec. Pub. SP-308 Washington, D.C.
[7] Holzer T E 1972 J. Geophys. Res. **77** 5407–5431
[8] Vasyliunas V M, and Siscoe G L 1976 J. Geophys. Res. **81** 1247–1252
[9] Möbius E, Hovestadt D, Klecker B, Scholer M, Gloeckler G, and Ipavich F M 1985 Nature **318** 426
[10] Gloeckler G, Geiss J, Balsiger H, Fisk L A, Galvin A B, Ipavich F M, Ogilvie K W, von Steiger R, and Wilken B 1993 Science **261** 70–73
[11] Geiss J, Gloeckler G, Mall U, von Steiger R, Galvin A B, and Ogilvie K W 1994 Astron. Astrophys. **282** 924

[12] Lee M A, and Ip W H 1987 J. Geophys. Res. **92** 11,041–11,052
[13] Zhou Y, and Matthaeus W H 1990 J. Geophys. Res. **95** 14,881–14,892
[14] Williams L L, and Zank G P 1994 J. Geophys. Res. **99** 19229–19244
[15] Williams L L, Zank G P, and Matthaeus W H 1995 **100** 17059–17067
[16] Zank G P, Matthaeus W H, and Smith C W 1996 J. Geophys. Res. **101** 17,093–17,107
[17] Matthaeus W H, Zank G P, Smith C W, and Oughton S 1999 Phys. Rev. Lett. **82**(17) 3444–3447
[18] Smith C W, Matthaeus W H, Zank G P, Ness N F, Oughton S, and Richardson J D 2001 J. Geophys. Res. **106** 8253–8272
[19] Isenberg P A, Smith C W, Matthaeus W H 2003 Astrophys. J. **592** 564–573
[20] Isenberg P A Astrophys. J. **623** 502–510
[21] Breech B, Matthaeus W H, Mininie J, Oughton S, Parhi S, Bieber J W, and Bavassano B 2005 Geophys. Res. Lett. **32** L06103
[22] Smith C W, Isenberg P A, Matthaeus W H, and Richardson J D 2006 Astrophys. J. **638** 508–517
[23] Breech B, Matthaeus W H, Mininie J, Bieber J W, Oughton S, Smith C W, and Isenberg P A 2008 J. Geophys. Res. **113** A08105
[24] Cranmer S R, Matthaeus W H, Breech B A, and Casper J C 2009 Astrophys. J. **702** 1604–1614
[25] Ng C S, Bhattacharjee A, Munsi D, Isenberg P A, and Smith C W 2010 J. Geophys. Res. **115** A02101
[26] Isenberg P A, Smith C W, Matthaeus W H, and Richardson J D 2010 Astrophys. J. **719** 716–721
[27] Oughton S, Matthaeus W H, Smith C W, Breech B, and Isenberg P A 2011 J. Geophys. Res. **116** A08105
[28] Usmanov A V, Matthaeus W H, Breech B A, and Goldstein M L 2011 Astrophys. J. **727** 84
[29] Usmanov A V, Goldstein M L, and Matthaeus W H 2012 Astrophys. J. **754** 40
[30] Zank G P, Dosch A, Hunana P, Florinski V, Matthaeus W H, and Webb G M 2012 Astrophys. J. **745** 35
[31] Usmanov A V, Goldstein M L, and Matthaeus W H 2014 Astrophys. J. **788** 43
[32] Adhikari L, Zank G P, Bruno R, Telloni D, Hunana P, Dosch A, Marino R, and Hu Q 2015 Astrophys. J. **805** 63
[33] Adhikari L, Zank G P, Bruno R, Telloni D, Hunana P, Dosch A, Marino R, and Hu Q 2015 J. Phys.: Conference Series **642** 012001
[34] Usmanov A V, Goldstein M L, and Matthaeus W H 2016 Astrophys. J. **820** 17
[35] Balogh, A, Beek T J, Forsyth R J, Hedgecock P C, Marquedant R J, Smith E J, Southwood D J, and Tsurutani B T 1992 *Astron. Astrophys. Supp.* **92** 221–236
[36] Murphy N, Smith E J, Tsurutani B T, Balogh A, and Southwood D J 1995 Space Sci. Rev. **72**(1-2) 447–453
[37] Zank G P 1999 Space Sci. Rev. **89** 413–688
[38] Behannon, K W, Acuña M H, Burlaga L F, Lepping R P, Ness N F, and Neubauer F M 1977 Space Sci. Rev. **21** 235–257
[39] Joyce C J, Smith C W, Isenberg P A, Murphy N, and Schwadron N A 2010 Astrophys. J. **724** 1256–1261
[40] Bridge, H S, Belcher J W, Butler R J, Lazarus A J, Mavretic A M, Sullivan J D, Siscoe G L, and Vasyliunas V M 1977 Space Sci. Rev. **21** 259–287
[41] Joyce C J, Smith C W, Isenberg P A, Gary S P, Murphy N, Gray P C, and Burlaga L F 2012 Astrophys. J. **745** 112
[42] Aggarwal P, Taylor D K, Smith C W, Joyce C J, Fisher M K, Isenberg P A, Vasquez B J, Schwadron N A, Cannon B E, and Richardson J D 2016 Astrophys. J. **822** 94
[43] Smith C W, Isenberg P A, Joyce C J, Cannon B E, Murphy N, Nuno R G, and Schwadron N A 2010 in Proceedings of Ninth Annual International Astrophysics Conference: Pickup Ions Throughout the Heliosphere and Beyond, AIP Conf. Proc. **1302** 186-191
[44] Cannon B E, Smith C W, Isenberg P A, Vasquez B J, Joyce C J, Murphy N, and Nuno R G 2013 Solar Wind 13, AIP Conf. Proc. **1539** 334–337
[45] Cannon B E, Smith C W, Isenberg P A, Vasquez B J, Murphy N, and Nuno R G 2014 Astrophys. J. **784** 150
[46] Cannon B E, Smith C W, Isenberg P A, Vasquez B J, Joyce C J, Murphy N, and Nuno R G 2014 Astrophys. J. **787** 133
[47] Stone E C, Frandsen A M, Mewaldt R A, Christian E R, Margolies D, Ormes J F, and Snow F 1998 Space Sci. Rev. **86**(1-4) 1–22
[48] Smith C W, Acuña W H, Burlaga L F, L'Heureux J, Ness N F, and Scheifele J 1998 Space Sci. Rev. **86**(1-4) 613–632
[49] Garrard T L, Davis A J, Hammond J S, and Sears S R 1998 Space Sci. Rev. **86** 649–663
[50] Argall M R, Fisher M K, Joyce C J, Smith C W, Isenberg P A, Vasquez B J, Schwadron N A, and Skoug R M 2015 Geophys. Res. Lett. **42** 9617–9623
[51] Fisher M K, Argall M R, Joyce C J, Smith C W, Isenberg P A, Vasquez B J, Schwadron N A, Skoug R M, Sokół J M, Bzowski M, Zurbuchen T H, and Gilbert J A, 2016 Astrophys. J. **830** 47

[52] McComas, D J, Bame S J, Barker P, Feldman W C, Phillips J L, Riley P, and Griffee J W 1998 Space Sci. Rev. **86**(1-4) 563–612
[53] Gloeckler G, Möbius E, Geiss J, Bzowski M, Chalov S, Fahr H, McMullin D R, Noda H, Oka M, Rucínski D, Skoug R, Terasawa T, von Steiger R, Yamazaki A, and Zurbuchen T 2004 Astron. Astrophys. **426** 845–854
[54] Kolmogorov A N 1941 Dokl. Akad. Nauk SSSR **30** 301–305 (Reprinted in Proc. R. Soc. London A **434** 9–13, 1991)
[55] Leamon R J, Smith C W, Ness N F, and Wong H K 1999 J. Geophys. Res. **104** 22,331–22,344
[56] Vasquez B J, Smith C W, Hamilton K, MacBride B T, and Leamon R J 2007 J. Geophys. Res. **112** A07101
[57] Batchelor G K 1953 *The Theory of Homogeneous Turbulence*, Cambridge Univ. Press, New York
[58] Iroshnikov P S 1964 Sov. Astron. **7** 566
[59] Kraichnan R H Phys. Fluids 1965 **8** 1385–1387
[60] Dobrowolny M, Mangeney A, and Veltri P 1980 Phys. Rev. Lett. **45** 144–147
[61] Higdon J C 1984 Astrophys. J. **285** 109–123
[62] Goldreich P, and Sridhar S 1997 Astrophys. J. **485** 680–688
[63] Boldyrev S 2005 Astrophys. J. Lett. **626** L37–L40
[64] Boldyrev S 2006 Phys. Rev. Lett. **96** 115002
[65] Politano H, and Pouquet A 1998a Phys. Rev. E **57**(1) R21–R24
[66] Politano H, and Pouquet A 1998b Geophys. Res. Lett. **25** 273–276
[67] Coburn J T, Smith C W, Vasquez B J, Forman M A, and Stawarz J E 2014 Astrophys. J. **786** 52
[68] Coburn J T, Forman M A, Smith C W, Vasquez B J, and Stawarz J E 2015 Phil. Trans. Royal Soc. A **373** 20140150

IOP Conf. Series: Journal of Physics: Conf. Series **900** (2017) 012019 doi:10.1088/1742-6596/900/1/012019

Critical Fluctuations in Beam-Plasma Systems and Solar Type III Radio Bursts

G. Thejappa[1], R. J. MacDowall[2]

[1] Department of Astronomy, University of Maryland, College Park, MD 20742, USA
[2] NASA/Goddard Space Flight Center, Greenbelt MD 20771, USA

Abstract. It is shown that the Langmuir waves are excited similar to critical fluctuations during phase transitions when the negative absorption due to electron beam traveling radially outward in the solar atmosphere is balanced by the positive absorption due to collisions in the corona and due to scattering on electron density inhomogeneities in the interplanetary medium. The effective temperature of the Langmuir fluctuations range from 10^{11} to 10^{13} K, explaining the majority of the type III bursts. The Rayleigh scattering and direct coupling due to density gradient as well as due to density inhomogeneities are discussed in the context of fundamental radiation and the combination scattering for second harmonic. The number density of electrons in type III beams is estimated and compared with observations. It is also shown that the stabilization of type III beams is achieved automatically since the instability does not develop in the case of critical fluctuations.

1. Introduction

For four decades the solar type III radio burst phenomenon has attracted the attention of both solar radio astronomers and space plasma physicists because of the intriguing complexity revealed by observations and theory. The direct detection of electron beams [17, 18] and Langmuir waves [10, 11] in association with type III bursts together with tracking of the type III burst sources in the interplanetary medium [4, 5] confirmed that electrons and Langmuir waves are associated with type III bursts [7, 8]. However the dynamics of type III electron beams and their interaction with the ambient plasma through the excited Langmuir waves is an unsolved problem. In order that the electron beams don't loose their energy by resonantly interacting with Langmuir waves, various nonlinear processes such as the induced scattering of Langmuir waves off the background ions in weak turbulence regime [14] and oscillating two-stream instability in strong turbulence regime [22] were invoked. However the observed electron density inhomogeneities [2, 3] do not allow Langmuir waves to grow to very high intensities as pointed out by several authors [20]. As far as the conversion of Langmuir wave energy into electromagnetic energy is concerned, there is no consensus either for the fundamental or for the second harmonic emission.

In the present paper we investigate the phenomenon of critical fluctuations near the boundary of the instability which is analogous to the anomalously growing fluctuations near the phase transition points called critical points [12, 23]. We show that the problems concerning the beam stabilization [25] and the estimation of the number density of beam electrons are automatically solved if type III bursts are assumed as manifestations of critical fluctuations. We also examine different physical processes involved in conversion of Langmuir into electromagnetic waves at the fundamental and second harmonic plasma frequencies.

2. Critical Fluctuations in the Beam-Plasma Systems

The effective temperature T_{eff} of Langmuir waves excited by an electron beam propagating radially outward in the solar atmosphere is determined in the linear approximation by the transfer equation:

$$\frac{dT_{eff}}{dl} = \alpha - \mu T_{eff}, \tag{1}$$

where α and μ are the emission and absorption coefficients respectively in a layer of thickness dl. The emission coefficient α is connected to the emissivity α_ω as:

$$\alpha = \frac{(2\pi)^3 \alpha_\omega}{\kappa k^2}, \tag{2}$$

where κ is the Boltzmann constant and k is the wave number. The emissivity α_ω is given by [1]:

$$\alpha_\omega = \frac{\omega_{pe}^2 \omega v_{ph}^2 m_e N_b}{6(2\pi)^{5/2} \Delta v_b v_{Te}^2 N_e} \exp(-\frac{(v_{ph} - v_b cos\theta)^2}{2\Delta v_b^2}, \tag{3}$$

where m_e is the electron mass; ω_{pe}, N_e and v_{Te} are the plasma frequency, density and thermal velocity of electrons in the ambient plasma; N_b, v_b and Δv_b are the density, velocity and velocity spread of the electron beam; ω and v_{ph} are the frequency and the phase velocity of the excited Langmuir waves. The absorption coefficient μ is determined by the absorption due to collisions (μ_c), effective absorption due to scattering on electron density inhomogeneities (μ_{sca}), Landau damping by the ambient electrons (μ_L) and negative absorption due to electron beam (μ_b):

$$\mu = \mu_c + \mu_{sca} + \mu_L + \mu_b. \tag{4}$$

The absorption due to collisions is given by:

$$\mu_c = \frac{2 v_{ph} \nu_c}{3 v_{Te}^2}, \tag{5}$$

where ν_c is the effective electron-ion collision frequency determined by the temperature (T_e) and density (N_e) as:

$$\nu_c = \frac{5.5 N_e}{T_e^{3/2}} \ln(220 \frac{T_e}{N_e^{1/3}}). \tag{6}$$

IOP Conf. Series: Journal of Physics: Conf. Series **900** (2017) 012019 doi:10.1088/1742-6596/900/1/012019

The effective absorption coefficient due to scattering by density inhomogeneities (μ_{sca}) is given by:

$$\mu_{sca} = \frac{2v_{ph}}{3v_{Te}^2}\nu_{sca}, \tag{7}$$

where the effective damping ν_{sca} is given by the diffusion coefficient D_θ :

$$\nu_{sca} = D_\theta(\Delta\theta)^2, \tag{8}$$

where $(\Delta\theta)$ is the mean angle of deflection given approximately as:

$$(\Delta\theta)^2 \simeq (\frac{\Delta v_b}{v_b})^2 \tag{9}$$

and D_θ is given by [21, 20]:

$$D_\theta = \frac{\pi}{12}\frac{\omega_{pe}}{(k\lambda_D)^2}\frac{\bar{q}}{k}\langle\frac{\delta n^2}{n^2}\rangle. \tag{10}$$

Here λ_D is the Debye length $\bar{q}$ is the typical wave number associated with the fluctuations and $\langle\frac{\delta n^2}{n^2}\rangle$ is the relative level of density fluctuations. The absorption due to Landau damping by the ambient electrons is:

$$\mu_L = \sqrt{\frac{\pi}{18}}\frac{\omega_{pe}^2\omega^3}{k^4 v_{Te}^5}\exp(-\frac{\omega^2}{2v_{Te}^2 k^2}). \tag{11}$$

In the corona, the absorption due to Landau damping can also be important due to smoothly varying nature of the density distribution as shown by [28, 27, 29]. However in the interplanetary medium it is negligibly small. In the present study we neglect it both in the corona as well as in the interplanetary medium. And also the relative level of density fluctuations increases with the radial distance in the solar atmosphere and hence we consider that in the corona the main contribution to the absorption is due to collisional damping whereas the absorption due to scattering of Langmuir waves by the electron density inhomogeneities plays the dominant role in the interplanetary medium. The absorption due to beam (μ_b) is given by:

$$\mu_b = \sqrt{\frac{\pi}{18}}\frac{\omega_{pe}^2 v_{ph}^3}{\omega v_{Te}^2 \Delta v_b^2}\frac{N_b}{N_e}(v_{ph} - v_b cos\theta)\exp(-\frac{(v_{ph} - v_b cos\theta)^2}{2\Delta v_b^2}). \tag{12}$$

The value μ_b is positive for $v_{ph} > v_b \cos\theta$, whereas it is negative in the region where $v_{ph} < v_b \cos\theta$. Therefore for $(v_{ph} - v_b \cos\theta) = -\Delta v_b$,and for marginally stable condition, where the absorption μ tends to zero, we can obtain the ratio $\frac{N_b}{N_e}$ in the corona by equating μ_b to μ_c and in the interplanetary medium by equating μ_{sca} with μ_b, which can be written as:

$$\frac{N_b}{N_e} = \sqrt{\frac{13.2}{\pi}}\frac{\omega\Delta v_b^2}{\omega_{pe}^2 v_{ph}^2}\nu, \tag{13}$$

where $\nu = \nu_c$ in the corona and $\nu = \nu_{sca}$ in the interplanetary medium. The solution of equation (1) in the marginally stable condition i.e., when the negative absorption due to electron beam is approximately balanced by the damping due either to collisions in the corona or to scattering on electron density fluctuations in the interplanetary medium., i.e., optically thin case $\mu L \leq 1$ can be written as:

$$T_{eff} = \alpha L. \tag{14}$$

Here L is the thickness of the layer in which the Langmuir waves exist which is determined by:

$$L = \frac{N_e}{2|\nabla N_e|} \frac{v_{Ti}}{v_{ph}}, \tag{15}$$

where v_{Ti} is the ion thermal velocity. In the corona approximately for 100 MHz plasma frequency layer the thickness L can be approximated as $L \simeq 10^9$ cm whereas in the interplanetary medium for plasma frequency layer of 13 kHz it can be approximated as $L \simeq 4.5 \times 10^{10}$ cm.

From equation (13) for $\Delta v_b \simeq v_b \simeq 10^{10}$ cm, $\omega \simeq \omega_{pe}$ and $\nu = \nu_c$, we obtain $\frac{N_b}{N_e} \simeq 2.63 \frac{\nu_c}{\omega_{pe}}$. In the interplanetary medium, since it is difficult to measure the effective damping due to scattering on density fluctuations, we evaluate it by assuming that the values for the beam density measured by [18] simultaneously with type III bursts as the critical density. Therefore for the values reported by [18], where $\frac{N_b}{N_e} \simeq 3.5 \times 10^{-5}$, $v_b \simeq 3.5 \times 10^9$ cms^{-1} and $\frac{\Delta v_b}{v_b} \simeq 0.15$, the effective damping rate (ν_{sca}) is calculated as $\nu_{sca} \simeq 0.6 \times 10^{-3} \omega_{pe}$, i.e., for $f_{pe} \simeq 13$ kHz, we obtain $\nu_{sca} \simeq 48.4$ s^{-1}. Therefore we obtain the ratio of critical beam density to electron density of ambient plasma $(\frac{N_b}{N_e})_{cri} = 5.23 \times 10^{-8}$ for $f_{pe} \simeq 100$ MHz and $\nu_c \simeq 12.5$ s^{-1}. The effective temperature of Langmuir waves emitted spontaneously by such beams as critical fluctuations can be obtained by using equations (2), (3) and (14). We obtain $T_{eff} \simeq 10^{12}$ K for $(v_{ph} - v_b cos\theta \simeq -\Delta v_b$, $v_{ph} \simeq v_b \simeq \Delta v_b \simeq c/3$, $v_{Te} = 3.89 \times 10^8$ cms^{-1} $f \simeq f_{pe} \simeq 100$ MHz and $L \simeq 10^9$ cm. In the case of interplanetary type III bursts, for $\frac{N_b}{N_e} \simeq 3.5 \times 10^{-5}$; $v_{ph} - v_b \cos\theta \simeq -\Delta v_b$; $v_{ph} \simeq 3.5 \times 10^9$ cms^{-1}; $\frac{\Delta v_b}{v_b} \simeq 0.15$; $v_{Te} \simeq 1.74 \times 10^8$ cms^{-1}; $f \simeq f_{pe} \simeq 13$ kHz and $L \simeq 4.5 \times 10^{10}$ cm, we obtain $T_{eff} \simeq 2.1 \times 10^{12}$ K.

[13] estimated the maximum value of the energy density of Langmuir waves due to critical fluctuations when the damping approaches zero as $\sqrt{g}$ by including the damping due to nonlinear interactions, where g is the plasma parameter defined as $g = \frac{1}{N_e \lambda_D^3}$. Thus approximated effective temperature is much higher than the effective temperature obtained by assuming that critical fluctuations are manifestations of spontaneous emission.

3. Conversion of Langmuir Fluctuations into Electromagnetic Waves

Various nonlinear mechanisms were proposed for conversion of Langmuir waves into electromagnetic waves at the fundamental plasma frequency f_{pe}. These are (1) spontaneous and induced scattering of Langmuir waves by thermal ions [7, 24, 19, 30, 31], (2) scattering of Langmuir waves by low frequency waves, such as ion-acoustic and whistler waves [26, 19] (3) strong turbulence processes [9, 16] and (4) gradient coupling of Langmuir waves with electromagnetic waves [6, 34, 35, 15]. In the present case, the energy density of critical Langmuir fluctuations is much less than the threshold

IOP Conf. Series: Journal of Physics: Conf. Series **900** (2017) 012019 doi:10.1088/1742-6596/900/1/012019

energy densities required for the induced scattering on thermal ions, decay of Langmuir waves into electromagnetic and low frequency waves and strong turbulence. Therefore only spontaneous scattering of Langmuir waves on thermal ions and direct coupling of Langmuir waves with electromagnetic waves may be important for conversion process at fundamental plasma frequency.

The brightness temperature of electromagnetic radiation T_B emitted by Langmuir waves of effective temperature T_{eff} is given by [32]:

$$T_B \simeq \frac{c^2\Omega_L}{3v_{Te}^2\Omega_T} T_L Q \exp(-\tau), \tag{16}$$

where Ω_L and Ω_T are the solid angles of Langmuir and electromagnetic waves respectively, τ is the optical depth from the source to the observer and Q is the efficiency of conversion from Langmuir to electromagnetic waves. The efficiency of conversion due to Rayleigh scattering is [7]: $Q = \frac{\omega_{pe}^4 L}{12\pi N_e c^3 v_{Te}}$ giving 3.2×10^{-6} and 5.7×10^{-12} respectively at 100 MHz and 13 kHz. The efficiency of conversion due to gradient coupling between Langmuir and electromagnetic waves is defined by [32]: $Q = \frac{\pi v_{ph}^2}{4\Omega_L c^2}(\frac{3c}{2\omega L_n})^{\frac{2}{3}}$, where L_n is the scale length of the density variation which is $\simeq 1.7\times10^{10}$ cm and 1.5×10^{13} cm respectively making Q as 5.7×10^{-7} and 6.24×10^{-7} at 100 MHz and 13 kHz respectively. Therefore we believe that the direct coupling due to the gradient in the electron density in the ambient plasma may be the appropriate conversion mechanism for generation of electromagnetic waves from the Langmuir waves. By using the equation (16) for the parameters adopted in the present paper for 100 MHz and 13 kHz, we obtain 1.72×10^8 K and $7.\times10^3 K$ through Rayleigh scattering and 3.06×10^7 K and 7.66×10^8 K through direct coupling respectively at 100 MHz and 13 kHz for the effective temperatures of the Langmuir critical fluctuations.

As far as the electromagnetic emission at $2\omega_{pe}$ is concerned, if the generation mechanism is assumed to be the merging of beam excited Langmuir waves with back-scattered secondary Langmuir spectrum i.e., the modified idea of [7], the absorption due to decay of electromagnetic wave into two plasma waves is the most dominant absorption process. The optical depth due to such a decay is [33]:

$$\mu L = \frac{2e^2\omega_{pe}^2\kappa T_{eff}}{15\sqrt{3}m_e^2c^3v_{Te}^2v_{ph}}L. \tag{17}$$

For the parameters assumed for 100 MHz and 13 kHz $\mu L \ll 1$ and the brightness temperature T_B is given by:

$$T_B \simeq \alpha_1 L, \tag{18}$$

where the spontaneous scattering coefficient in this case is given by [33]:

$$\alpha_1 \simeq \frac{2}{15\sqrt{3}}\frac{e^2\omega_{pe}^2\kappa T_{eff}^2}{m^2c^3v_{Te}^2v_{ph}}L. \tag{19}$$

Therefore for $T_{eff} \simeq 10^{12} K$ we obtain $T_B \simeq 3.\times10^{10}$ K and 3.1×10^5 K at 100 MHz and 13 kHz respectively.

4. Discussion and Conclusions

Since the velocity of the electron beams responsible for type III bursts is known directly by measuring the drift rate of type III bursts in the dynamic spectrum, the ratio of beam to ambient electron density is an unknown quantity in type III phenomena. The critical beam density derived in the present paper where the phenomenon of critical fluctuations starts playing the dominant role in generation of Langmuir waves may give the required beam densities. Higher densities where the instability is playing the role in excitation of Langmuir fluctuations may be necessary to explain the brightest type III bursts. In the case of the interplanetary medium, the exact knowledge of electron density fluctuations is necessary to estimate the effective damping due to scattering of Langmuir waves on fluctuations. The problem of beam stabilization is not important in the case of critical fluctuations since the effective growth is almost negligible.

As far as conversion mechanism for the fundamental emission is concerned, the direct coupling between Langmuir and electromagnetic waves due density gradient in the ambient electron distribution appears to be the most favorable mechanism both in the corona as well as in interplanetary medium. The efficiency of conversion due to Rayleigh scattering is comparable to that due to direct coupling in the corona whereas it is much smaller at lower frequencies. But the efficiency of conversion at 100 MHz due to Rayleigh scattering is comparable to direct coupling. Therefore we believe that direct coupling may be the dominant process in converting the Langmuir waves into electromagnetic waves both in the corona as well as in the interplanetary medium.

In the case of second harmonic emission, the combination scattering may account for the emission at higher frequencies whereas it is too inefficient at lower frequencies thus requiring to invoke other mechanisms. Therefore we conclude that: (1) the phenomenon of critical fluctuations in the beam-plasma system may be very important for type III bursts, (2) the ratio N_b/N_e in weak type III bursts can be determined by the collisional damping in the corona, and by damping due to scattering on density fluctuations in the interplanetary medium, (3) since the instability does not develop in the case of critical fluctuations, the question of beam stabilization does not arise in the present model, (4) the direct coupling between Langmuir and electromagnetic waves appears to play a dominant role in converting the langmuir energy into electromagnetic energy at fundamental plasma frequency and (5) even though the combination scattering is capable of explaining the emission at twice the plasma frequency, it is too inefficient to explain the type III bursts at lower frequencies.

4.1. Acknowledgments

G. T. acknowledges the NASA STEREO and Wind projects for the financial support

5. References

[1] Andronov A A 1961 *Izv. Vuzov, Radiofiz* **4** 861 (in Russian)
[2] Celnikier L M, Harvey C C, Jegou R, Kemp M, and Moricet P 1983 *Astron. Astrophys* **181** 138
[3] Celnikier L M, Muschietti L, and Goldman M V 1987 181, 138.
[4] Fainberg J, Evans L G, and Stone R G 1972 *Science* **178** 743
[5] Fainberg J, and Stone R G 1974 *Space Sci. Rev.* **16** 145
[6] Field G B 1956 *Astrophys. J.* **124** 555
[7] Ginzburg V L, and Zheleznyakov V V 1958 *Soviet Astron.* **2** 653

[8] Ginzburg V L and Zheleznyakov V V 1959 *Soviet Astron.* **3** 235
[9] Goldman M V, Reiter F G, and Nicholson D R 1980 *Phys. Fluids* **23** 388
[10] Gurnett D A, and Anderson R R 1976 *Science* **194** 1159
[11] Gurnett D A and Anderson R R 1977 *J Geophys. Res.* **82** 632
[12] Ichimaru S 1962 *Ann. Phys.* **20** 78
[13] Ichimaru S 1970 *Phys. Fluids* **13** 1560
[14] Kaplan S A, and Tsytovich V N 1968 *Soviet Astr.* **11** 834
[15] Kellogg P J 1986 *Astron. Astrophys.* **169** 329
[16] Kruchina E N, Sagdeev R Z, and Shapiro V D 1980 *Sov. Phys. JETP letters* 32
[17] Lin R P, Potter D W, Gurnett D A, and Scarf F L 1981 *Astrophys. J.* **251** 364
[18] Lin R P, Levedahl W K, Lotko W, Gurnett D A, and Scarf F L 1986 *Astrophys. J.* bf 308 954
[19] Melrose D B 1980 *Space Sci. Rev.* **26** 3
[20] Muschietti L, Goldman M V, and Newman D 1985 *Solar Phys.* **96** 181
[21] Nishikawa K and Ryutov D 1976 *J. Phys. Soc. Japan* **41** 1757
[22] Papadopoulos, K., Goldstein, M. L. and Smith, R. A.: 1974, Astrophys. J., 190, 175.
[23] Rosenbluth M and Rostoker N 1962 *Phys. Fluids* **5** 776
[24] Smith D F 1977 *Astrophys. J.* **216** L53.
[25] Sturrock P A 1964 (in) *The Physic of Solar Flares, AAS- NASA Symp., NASA Sp-50* 357
[26] Takakura T 1979 Solar Phys. **62** 375
[27] Thejappa G 1991 *Solar Phys.* **132** 173
[28] Thejappa G, Gopalswamy N, and Kundu M R 1990 *Solar Phys.* **127** 165
[29] Thejappa, G. and Kundu, M. R.: 1991, Solar Phys., 132, 155.
[30] Zaitsev V V 1975 *Sov. Astron. Lett.* **5**, 206
[31] Zaitsev V V 1977 *Radio Phys. Quantum Electron.* 20
[32] Zheleznyakov V V 1970 *Radio Emission of the Sun and Planets*, Pergamon, Newyork.
[33] Zheleznyakov V V 1977 *Electromagnetic Waves in Space Plasmas*, Nauka, Moscow
[34] Zheleznyakov V V and Zlotnik E Ya 1962, *Radio Phys. Quantum Electron.* **5**, 644 (in Russian).
[35] Zheleznyakov V V and Zlotnik E Ya 1963, *Radio Phys. Quantum Electron.* **6**, 634 (in Russian).

IOP Conf. Series: Journal of Physics: Conf. Series **900** (2017) 012020 doi:10.1088/1742-6596/900/1/012020

Towards a parallel collisionless shock in LAPD

M S Weidl[1,*], P Heuer[1,*], D Schaeffer[1], R Dorst[1], D Winske[2], C Constantin[1], C Niemann[1]

[1] UCLA Physics & Astronomy, 475 Portola Plaza, Los Angeles, California 90095, USA
[2] Los Alamos National Laboratory, Los Alamos, New Mexico 87545, USA

E-mail: mwei@physics.ucla.edu

Abstract. Using a high-energy laser to produce a super-Alfvénic carbon-ion beam in a strongly magnetized helium plasma, we expect to be able to observe the formation of a collisionless parallel shock inside the Large Plasma Device. We compare early magnetic-field measurements of the resonant right-hand instability with analytical predictions and find excellent agreement. Hybrid simulations show that the carbon ions couple to the background plasma and compress it, although so far the background ions are mainly accelerated perpendicular to the mean-field direction.

1. Introduction

Collisionless shocks are a virtually ubiquitous phenomenon in space, from planetary bow shocks in our solar system [1, 2, 3, 4, 5] to the shocks occurring in supernova remnants [6, 7] and presumably the jets of active galactic nuclei [8]. Our understanding of the complex process of shock formation has greatly improved since Sagdeev's seminal paper [9], thanks to both satellite observations [10, 11] and numerical simulations [12, 13]. Yet the complicated three-dimensional structure that one would predict for a parallel shock on theoretical grounds [14] is difficult to confirm, either because of the nature of one- or few-point satellite measurements or because three-dimensional kinetic simulations are computationally too expensive.

To avoid these particular difficulties, our group is working on producing a parallel collisionless shock in a laser-plasma experiment at the Large Plasma Device (LAPD) [15]. Simulations assuming a quasi-one-dimensional carbon front [16] showed that the available energy density is sufficient to observe the early stages of collisionless-shock formation. In this proceeding, we present measurements of the resonant right-hand instability in our experiment and compare them with predictions of shock-formation theory [17]. This research continues previous work on the generation and investigation of a perpendicular collisionless shock [18, 19].

2. Theory and definitions

For an externally magnetized plasma which consists of two ion species and charge-compensating electrons with a relative drift velocity along the magnetic-field direction much larger than either thermal velocity, the dispersion relation for parallel-propagating waves is [17]

$$\omega^2 - k^2c^2 = \omega_{p,0}^2 \frac{\omega - kV_0}{\omega - kV_0 \pm \Omega_0} + \omega_{p,b}^2 \frac{\omega - kV_b}{\omega - kV_b \pm \Omega_b} + \omega_{e,0}^2 \frac{\omega - kv_0}{\omega - kv_0 \mp \Omega_e} + \omega_{e,b}^2 \frac{\omega - kv_b}{\omega - kv_b \mp \Omega_e}, \quad (1)$$

where ω and k are frequency and wavenumber along the magnetic field, while ω_p and ω_e are the plasma frequencies of the proton and electron populations for background (subscript 0) and beam (subscript b) species. V and v denote the drift velocities of ions and electrons, respectively; Ω_0, Ω_b, and Ω_e are the positive gyrofrequencies of background ions, beam ions, and electrons.

We define right-handed circular polarization with respect to the magnetic field as having the same sense of rotation as the Larmor motion of electrons with no parallel velocity component in a given frame of reference. It follows that the upper signs in the denominators of eqn. (1) correspond to a right-hand circularly polarized wave, the lower signs to left-handed polarization.

The electron terms can be simplified if we assume that the frequency range of interest is much smaller than the Doppler-shifted electron gyrofrequency, $|\omega - kv_{0/b}| \ll \Omega_e$. Moreover, we will choose as our frame of reference a frame in which the net electron current vanishes, such that $\omega_{e,0}^2\, v_0 + \omega_{e,b}^2\, v_b = 0$, and define $\omega_e^2 = (\omega_{e,0}^2 + \omega_{e,b}^2)\, \Omega_0/\Omega_e$ to obtain

$$\omega^2 - k^2c^2 = \omega_{p,0}^2\, \frac{\omega - kV_0}{\omega - kV_0 \pm \Omega_0} + \omega_{p,b}^2\, \frac{\omega - kV_b}{\omega - kV_b \pm \Omega_b} - \omega_e^2\, \frac{\omega}{\pm\Omega_0}. \tag{2}$$

Although this choice of frame simplifies the solution of the dispersion relation and saves us from having to consider the electric field created by a net current, this solution is only valid in the zero-current frame. In the experiment, on the other hand, wave measurements are taken in the lab frame in which the background ions are at rest; consequently, a Doppler shift becomes necessary for comparing the results.

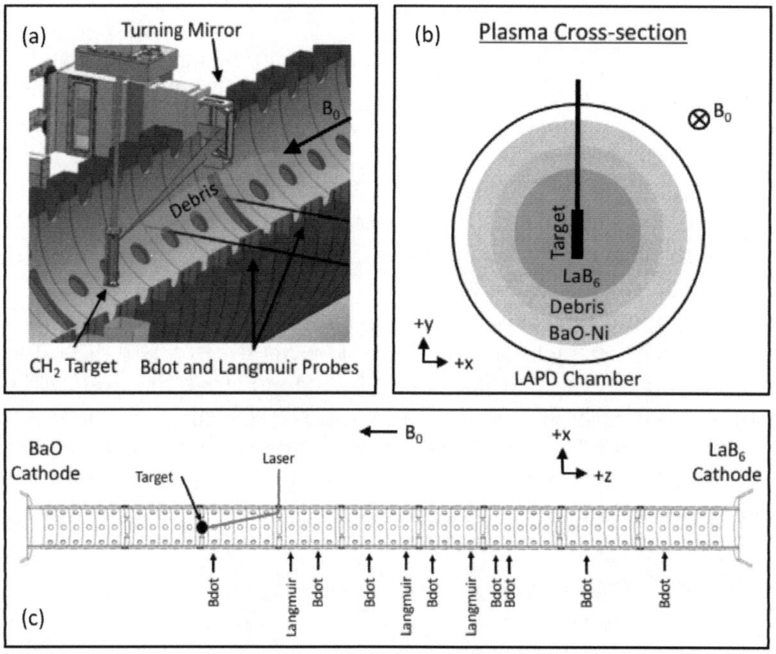

Figure 1: *a*) Schematic of turning mirror reflecting the laser beam on the polyethylene (CH_2) target inside LAPD. *b*) Cross section of LAPD with relative sizes of target, high-density helium plasma created by the LaB_6 cathode, debris/blow-off ion cloud, and lower-density helium plasma created by a BaO cathode. *c*) Cut along LAPD with positions of target, cathodes, and sample configuration of Bdot and Langmuir probes

IOP Conf. Series: Journal of Physics: Conf. Series **900** (2017) 012020 doi:10.1088/1742-6596/900/1/012020

3. Experimental setup

The background plasma in our experiments is confined in the 17-metre-long LAPD. A discharge current is created with a lanthanum hexaboride (LaB_6) cathode and ionizes helium to He^+ with a density of 10^{13} cm^{-3} at an ion temperature of $T_i = 1$ eV. In future experiments, significantly higher densities will be facilitated by a second LaB_6 cathode. Pancake magnets surrounding the plasma maintain a magnetic field of 300 gauss. Thus the nominal Alfvén velocity evaluates to $v_A = 106$ km s^{-1}, the gyroperiod of a singly charged helium ion is $\tau_{\mathrm{He}^+} = 8.7$ μs. An overview of the experiment is shown in Figure 1.

In order to create the beam ions, the high-energy Raptor laser ($E \leq 250$ J, $\lambda = 1053$ nm) is focused on a high-density polyethylene target and various charge states of carbon are ablated at super-Alfvénic velocities, predominantly C^{3+} and C^{4+} [20]. By adjusting the alignment of the target, we control the angle between the bulk velocity of the blow-off ions and the external magnetic field. After a previous campaign of experiments at LAPD studied the generation of collisionless shocks perpendicular to the magnetic-field direction [21], we will focus in this report on our recent progress towards exciting a parallel collisionless shock.

4. Measurements

As the ablated carbon ions stream through LAPD, they cannot couple to the background plasma collisionally because their mean-free-path is far longer than the available distance. Creating a shock requires instead that the carbon ions excite electromagnetic waves of such an amplitude that the background plasma is accelerated and compressed efficiently by the steepening wave fronts which propagate along the external magnetic field.

We quantify the carbon ions free-streaming down the central axis of LAPD with two types of probes. Langmuir probes detect the carbon-ion saturation current at various distances from the target. In addition, a fibre probe positioned 30 centimetres from the target transmits light through a monochromator to an avalanche photodiode (APD) to measure the spontaneously emitted fluorescence of C^{4+} ions. Time-of-flight analysis yields a bulk velocity of $v_C \approx 2.5\ v_A$ after 2.0 metres that is slowed down to 2.0 v_A after another 4.0 metres, although some ions appear to stream ahead with a velocity of 4.5 v_A. It is presumably these latter fast 'vanguard' ions that are the predominant cause of the observed magnetic instabilities.

All three components of the magnetic field on the central axis are measured with several

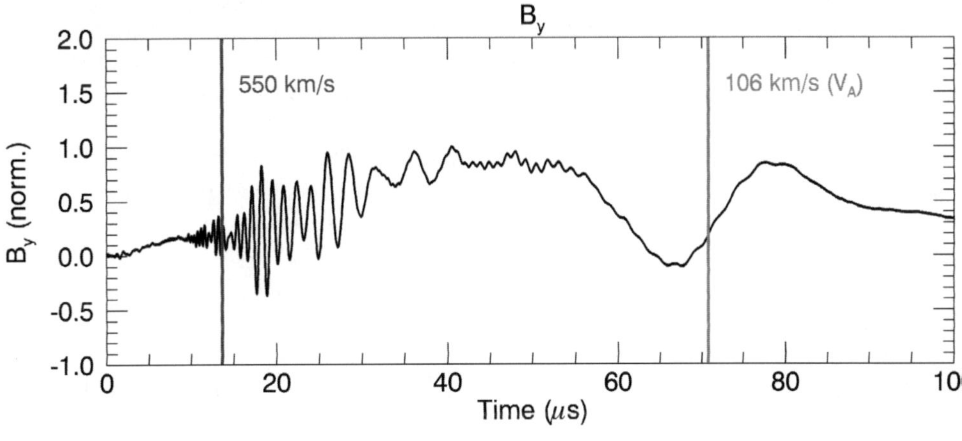

Figure 2: Sample time trace of the radial magnetic-field component measured 7.5 metres from the target on the central axis of the plasma column

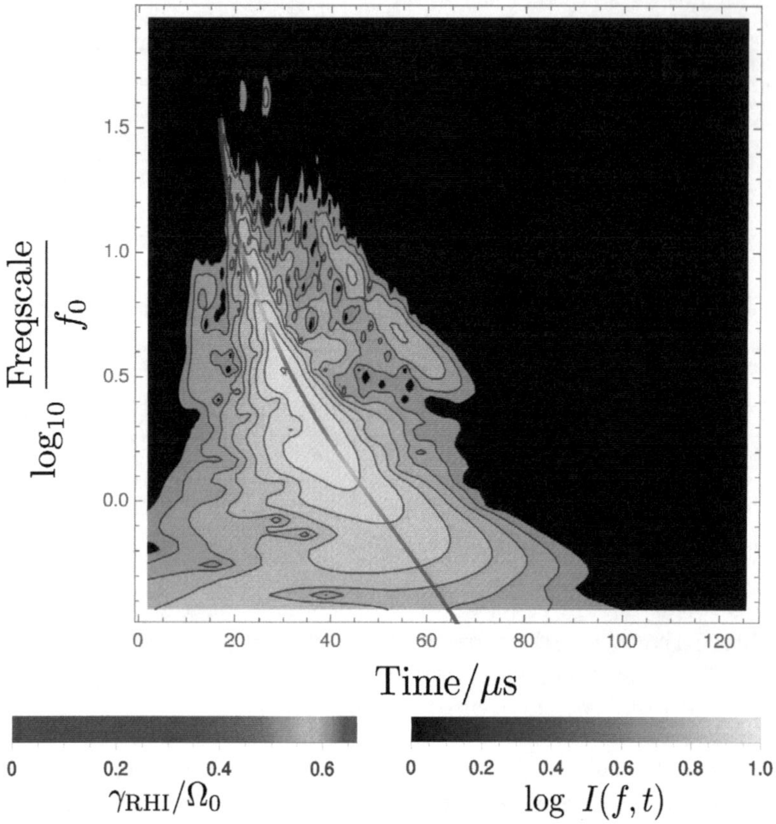

Figure 3: Wavelet transform of the 7.5 m–Bdot signal (grey contours, arb. units), ETA and growth rate (coloured line)

magnetic-flux or 'Bdot' probes. The amplitude of the waves in the radial component decreases from 6 G after 1.5 metres to about 3 G after 11.5 metres. A sample time trace for the radial component, taken at a distance of 7.5 metres from the target, is shown in figure 2. As that plot demonstrates, the parallel-propagating waves excited by the carbon ions can be distinguished unambiguously into two groups: a fast high-frequency chirp propagating with a group velocity of about 550 km s^{-1} or 5.2 v_A, and a larger-amplitude wave packet traveling at about the Alfvén velocity. The latter component is clearly dominated by modes with oscillation periods at least twice the gyroperiod of background ions.

To investigate the wave dispersion of the high-frequency chirp in closer detail, we show in figure 3 iso-intensity contours of a Morlet-wavelet transform of the signal measured 7.5 metres from the target. The highest frequencies at about $10^{1.5}$ f_0 are picked up by the probe after about 15 seconds ($f_0 = \Omega_0/(2\pi)$), followed by a smooth transition towards lower frequencies all the way down to the sub-cyclotron range. During this chirp, the spectral intensity of the signal passes through two maxima, one at $10^{0.9}$ f_0 after 24 seconds, the other one at $10^{0.5}$ f_0 after 30 seconds. A closer look at the phase relation of the two perpendicular magnetic-field components shows that the waves during this period are almost exclusively right-hand polarized.

We can thus compare the timing of this signal to the dispersion that we expect for waves in the helium plasma on the fast-magnetosonic/whistler branch with right-hand polarization, which corresponds to taking the upper signs in the denominators of equation (1). We neglect the beam terms in the dispersion relation because the carbon is too dilute to affect the group velocity

IOP Conf. Series: Journal of Physics: Conf. Series **900** (2017) 012020 doi:10.1088/1742-6596/900/1/012020

over most of the distance covered by the waves. Using the plasma frequency and Alfvén velocity given above, we calculate the group velocity, and thus the expected time of arrival (ETA) at the probe location, as a function of the Doppler-shifted frequency, plot the result over the measured intensity contours in figure 3, and find close to perfect agreement in the regime above Ω_0. At lower frequencies, left-handed waves on the Alfvénic branch of the dispersion relation contribute more and more to the measured signal and arrive later than predicted for waves of right-handed polarization.

Close to the target, where the carbon ions excite the waves, the beam terms must be retained. Making several simplifying assumptions, such as neglecting the finite temperature of the beam and the presence of blow-off ions other than C^{4+}, we can derive an exact analytical expression for the growth rate γ_{RHI} of the right-hand instability from the dispersion relation (2). Since this growth rate is far more sensitive to the beam velocity than to its density (at moderately super-Alfvénic velocities, [22]), we expect that the fastest ions detected by the Langmuir probe at 2 metres make the dominant contribution to the instability and assume a beam velocity of 4.2 v_A. At a relative number density $n_b/n_0 = 5.8$ %, the growth rate as a function of frequency exhibits two easily separated peaks. The colour coding of the ETA line in figure 3, which represents that growth rate, shows that these two fastest growing modes are located extremely close to the maxima of the measured spectral intensity.

All these data are thus described by the cold-plasma dispersion relation to a surprisingly high level of accuracy. Determining the effect of the carbon-ion beam on the helium background, however, in particular the amount of compression and acceleration, requires numerical simulation.

5. 2D-hybrid simulations

In order to capture the important beam-resonant instabilities, we use a hybrid code that treats ions with the particle-in-cell method and approximates the electrons as an inertia-less magnetohydrodynamic fluid [23]. The magnetic field is computed on a two-dimensional grid 20 inertial lengths ($= 20\ \delta_i$) wide and extending 256 inertial lengths along the mean-field direction, which is the x axis. Resolutions in time and space are $0.001/\Omega_0$ and $0.125\ \delta_i$, respectively; each grid cell is initially populated with 100 background-ion quasiparticles.

We model the ablated carbon-ion beam as a 50:50 mixture of C^{3+} and C^{4+} of a total initial density 300 times the helium-ion density n_0, concentrated in a volume measuring $0.5 \times 0.03\ \delta_i^2$ around the point $(x, y) = (2\ \delta_i, 10\ \delta_i)$. The angular distribution of velocity and density is proportional to $\cos^2\ \varphi$ and to $\cos^4\ \varphi$, respectively [24], such that the resulting mean x-velocity is $\langle v_x \rangle = 1.3\ v_A$ for C^{3+} and 3.9 v_A for C^{4+}. To match the Langmuir-probe measurements, we add an isotropic thermal-velocity component corresponding to $\beta = 10$. The resulting initial velocity distribution has a mean $\langle v_x \rangle = 2.6\ v_A$ averaged over both carbon species, as in the experiment.

Qualitatively as well as quantitatively, we find strong agreement in both frequency and amplitude between the right-handed waves as they appear in time traces of the magnetic field in the simulation and the right-handed waves as measured by the Bdot probes (see figure 4). This strong agreement for the right-hand instability, which is mainly driven by the fastest beam ions, is to be expected as the fastest C^{4+} quasiparticles in the simulation have a velocity of around $v_x \approx 5.0\ v_A$, similar to the Langmuir-probe measurements of the experiment. A detailed comparison of the left-hand circularly polarized low-frequency pulse between simulation and experiment is currently in progress.

The effect of the propagating carbon ions on the background-plasma density is shown in figure 5. While the two clouds of ablated ions spread and re-converge because of their gyromotion in the magnetic field, the helium ions are compressed in a thin (about one inertial length in diameter) channel along the central axis of LAPD. Although the density in this channel

IOP Conf. Series: Journal of Physics: Conf. Series **900** (2017) 012020 doi:10.1088/1742-6596/900/1/012020

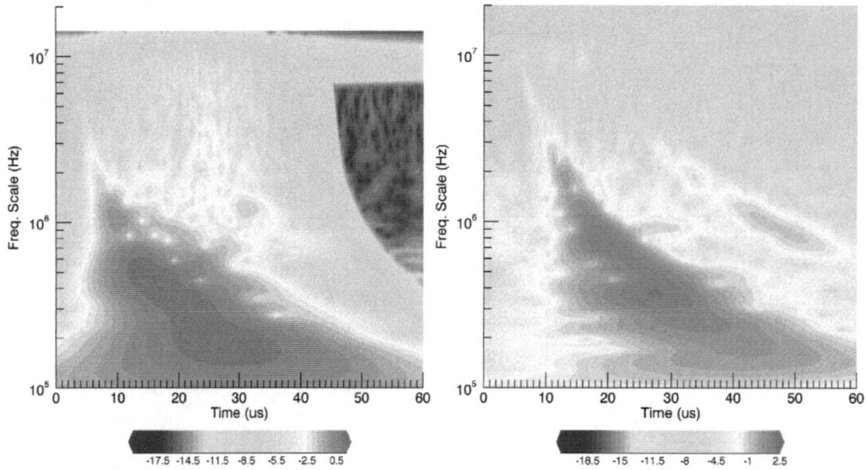

Figure 4: Comparison of Morlet-wavelet transforms of a magnetic-field time trace from the hybrid simulation (left) and the experimentally measured Bdot signal (right) 7.5 metres from the target (gyrofrequencies are $T_g^{-1} = 1.1 \cdot 10^5$ Hz for He^+ and $1.5 \cdot 10^5$ Hz for C^{4+})

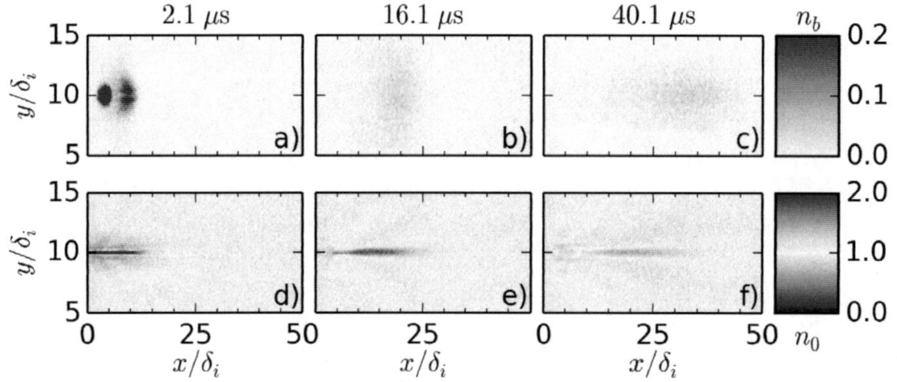

Figure 5: Simulated density of carbon beam ions (*a–c*) and helium background ions (*d–f*) as the beam, which consists of slow C^{3+} and fast C^{4+}, propagates to the right

doubles, this compression does not result in a shock since the acceleration of the helium ions is mainly perpendicular to the central axis. Energy transport in the parallel direction is too small to sustain the compression, which ceases after only about 50 inertial lengths from the target. Further work will focus on improving the collimation of blow-off ions and thus the conversion of beam energy into parallel kinetic energy of helium ions.

6. Summary

We have ablated carbon ions from a plastic target in a highly magnetized plasma with a high-energy laser. Measurements of the current have confirmed that the ions propagate at several Alfvén velocities, and we have shown that the waves that the ions produce as they propagate along the field direction are excited by the right-hand instability, one of the beam instabilities associated with the formation of the parallel bow shock.

However, the amplitude of the waves is not yet high enough to accelerate the background ions efficiently in the parallel direction. Instead, hybrid simulations suggest that the background plasma within a distance of 4 metres from the target is compressed by a factor of 2 perpendicularly to the mean-field direction. In order to form a parallel collisionless shock, we are currently investigating methods to decrease the perpendicular spread of the carbon blow-off, which will result in improved coupling between the beam ions and the parallel velocity of the background ions.

Acknowledgments

We would like to thank the staff of the Large Plasma Device for their help in carrying out this experiment. This work was performed at the Basic Plasma Science Facility at UCLA, funded by the National Science Foundation (NSF) and the Department of Energy (DOE), and was supported by the Defense Threat Reduction Agency (DTRA) under contract number HDTRA1-12-1-0024, the DOE under contract numbers DE-SC0006538:0003 and DE-NA0001995, and NSF award 1414591.

References

[1] Fairfield D H 1971 *J. Geophys. Res.* **76** 6700
[2] Scarf F L, Gurnett D A, Kurth W S and Poynter R L 1979 *Nature* **280** 796
[3] Szego K *et al.* 2003 *J. Geophys. Res. (Space Physics)* **108** 1287
[4] Acuna M H and Ness N F 1980 *Science* **207** 444–446
[5] Achilleos N *et al.* 2006 *J. Geophys. Res. (Space Physics)* **111** A03201
[6] Reynolds S P and Ellison D C 1992 *Astrophysical Journal Letters* **399** L75–L78
[7] Morlino G and Caprioli D 2012 *Astronomy & Astrophysics* **538** A81
[8] Ros E, Zensus J A and Lobanov A P 2000 *Astronomy & Astrophysics* **354** 55–66
[9] Sagdeev R Z 1966 *Reviews of Plasma Physics* **4** 23
[10] Schwartz S J, Burgess D, Wilkinson W P, Kessel R L, Dunlop M and Luehr H 1992 *J. Geophys. Res.* **97** 4209–4227
[11] Kucharek H, Möbius E, Scholer M, Mouikis C, Kistler L, Horbury T, Balogh A, Réme H and Bosqued J 2004 *Annales Geophysicae* **22** 2301–2308
[12] Winske D and Leroy M M 1984 *J. Geophys. Res.* **89** 2673–2688
[13] Lembege B and Dawson J M 1989 *Physical Review Letters* **62** 2683–2686
[14] Schwartz S J and Burgess D 1991 *Geophys. Res. Letters* **18** 373–376
[15] Gekelman W *et al.* 2016 *Review of Scientific Instruments* **87** 025105
[16] Weidl M S, Winske D, Jenko F and Niemann C 2016 *Physics of Plasmas* **23** 122102
[17] Gary S P 1991 *Space Science Reviews* **56** 373–415
[18] Niemann C *et al.* 2014 *Geophys. Res. Letters* **41** 7413–7418
[19] Bondarenko A S *et al.* 2017 *Nature Physics* **13**
[20] Schaeffer D B, Bondarenko A S, Everson E T, Clark S E, Constantin C G and Niemann C 2016 *Journal of Applied Physics* **120** 043301
[21] Schaeffer D B, Winske D, Larson D J, Cowee M M, Constantin C G, Bondarenko A S, Clark S E and Niemann C 2017 *Physics of Plasmas* **24** 041405
[22] Gary S P, Foosland D W, Smith C W, Lee M A and Goldstein M L 1984 *Physics of Fluids* **27** 1852–1862
[23] Winske D and Gary S P 2007 *J. Geophys. Res. (Space Physics)* **112** A10303
[24] Lewis C L S, Cunningham P F, Pina L, Roy A K and Ward J M 1982 *Journal of Physics D: Applied Physics* **15** 69

IOP Conf. Series: Journal of Physics: Conf. Series **900** (2017) 012021 doi:10.1088/1742-6596/900/1/012021

An Improved Parametrized Representation of the Secondary He Neutral Flow in its Source Region

Brian E. Wood[1]

[1] Naval Research Laboratory, Space Science Division, Washington, DC 20375, USA

E-mail: brian.wood@nrl.navy.mil

Abstract. Analysis of data from the *Interstellar Boundary EXplorer* (IBEX) has revealed the presence of a flow of neutral helium through the inner solar system that most likely emanates from the outer heliosheath, where a distinct population of neutral He is produced by charge exchange processes. This secondary He flow has been modeled using codes designed to study interstellar flows through the heliosphere, but a laminar flow is not a good approximation for the outer heliosheath. I present a simple parametrization for a more appropriate divergent flow, and demonstrate how the secondary He particles might provide a means to remotely measure the divergence of the ISM flow around the heliopause.

1. Introduction

Launched in 2008, the *Interstellar Boundary EXplorer* (IBEX) mission is designed to study various populations of neutral particles flowing through the inner solar system [1]. One such population is that of the local interstellar medium, which is partly neutral. The plasma component of the ISM will be deflected around the heliopause, but the neutral component can penetrate into the inner heliosphere, where these particles can be observed and provide useful diagnostics on the characteristics of the undisturbed ISM just outside the heliosphere. Of particular interest are neutral He atoms. Unlike neutral H, for example, neutral He has a very low charge exchange cross section and will therefore reach the inner solar system with its flow relatively unaffected by particle interactions. Observations of neutral He have therefore been crucial for measuring the local ISM flow vector, based on both IBEX data and similar observations from the older *Ulysses* mission [2, 3, 4, 5, 6].

However, in the process of studying IBEX observations of the ISM neutral He flow, a second component of neutral He was discovered, which was dubbed the "Warm Breeze" component [7]. A signature of this secondary component may also be present in *Ulysses* data [8]. Based on continuing analysis of these data, it is becoming increasingly certain that the second component is due to a population of neutral He created by charge exchange outside the heliopause (i.e., the outer heliosheath) [9]. Whether a bow shock exists or not, the ISM plasma outside the heliopause will be heated, compressed, and deflected [10]. The secondary He component observed by IBEX is presumed to be created by charge exchange with this heated, compressed, and deflected ISM plasma flow, predominantly by the $He^0 + He^+ \rightarrow He^+ + He^0$ reaction, which is important due to the significant abundance of both He^0 and He^+ in the ISM [3, 11].

Initial analyses of the secondary He component have typically used the same techniques used to analyze the primary ISM component, assuming a Maxwellian laminar flow from infinity, affected primarily by the Sun's gravity and photoionization as the neutrals approach the inner solar system. The most recent such analysis found a secondary flow of $V_\infty = 11.3$ km s^{-1} towards ecliptic coordinates $(\lambda,\beta)=(71.6°,-12.0°)$, with a temperature of $T = 9500$ K, and an abundance of 5.7% of the primary ISM component [9].

However, the reliability of these measurements are questionable, because although the "laminar flow from infinity" assumption is appropriate for the primary component, it is a poor approximation for the outer heliosheath source region of the secondary component, where the flow could more accurately be described as a "divergent flow from about 150 AU." The inferred temperature of the secondary neutrals is particularly suspect, as $T = 9500$ K is much lower than expected for the outer heliosheath, where $T \approx 20,000$ K is more likely [12]. The goal of this article is to present an alternative flow parametrization that is more appropriate for the secondary component, and to show how the divergence of that flow affects the observed He atoms at 1 AU.

2. Parametrizing a Divergent Flow

We define our flow in a coordinate system with the Sun at the origin and the x-axis pointing in the upwind direction of the ISM flow (i.e., the stagnation axis), as in Figure 1. For our flow from the outer heliosheath, we define an outer boundary at a distance of 150 AU, as this is roughly the distance to the heliopause. The most important difference between the outer heliosheath flow and that of the ISM is the divergence of the flow around the heliopause. If ψ is the angle of the flow from the x-axis, then we define a divergence parameter D such that $\psi = D \times \theta$, with θ being the angle defined in Figure 1. In practice, we assume a maximum value of $\psi = 80°$ regardless of D. Figure 1 shows the flow patterns associated with $D = 0.5$ and $D = 1.0$.

In order to illustrate how D affects the shape of a He beam observed at 1 AU, we use a simple flow model in two-dimensions, where the flow from the outer boundary is assumed to be affected only by solar gravity, and the flow is modeled using a routine that computes particle trajectories using simple numerical integration [13]. We assume the velocity distribution function (VDF) at the outer boundary (at $r = 150$ AU) is Maxwellian. We then compute the resulting particle distribution observed at (X, Y)=(1, 0) in the coordinate system shown in Figure 1, where for simplicity we are assuming the observer is on the stagnation axis at 1 AU.

The black line in Figure 2 shows the observed distribution for a flow with $V = 20$ km s^{-1} and $T = 10^4$ K, assuming no divergence (e.g., $D = 0$). Limitations in the sampling resolution of the VDF in the simulation lead to some noise in the observed He

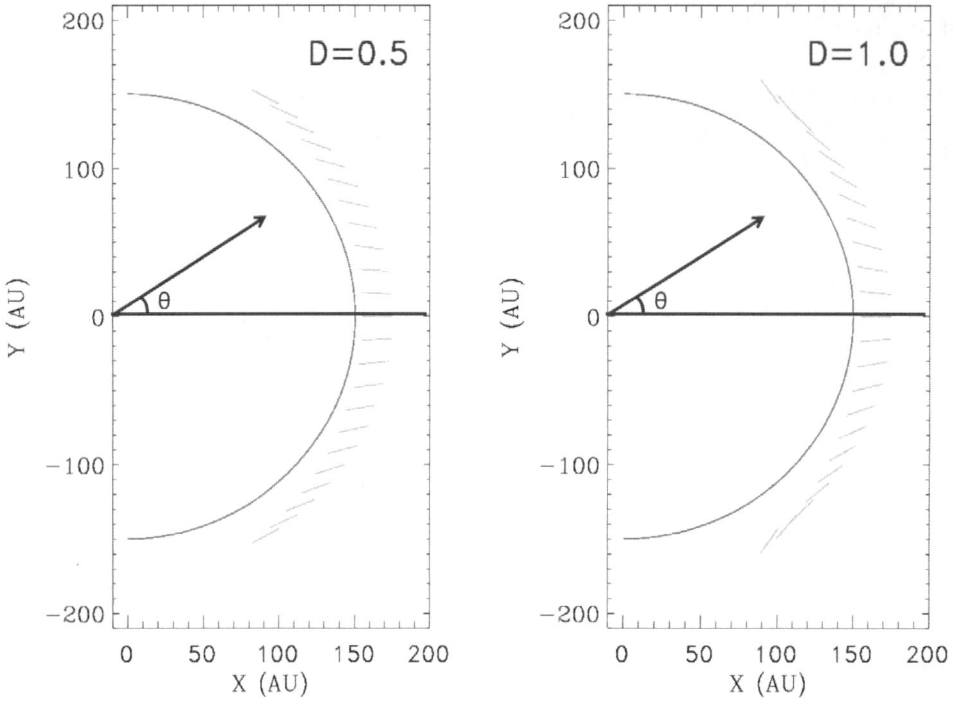

Figure 1. Representations of divergent flows from a distance of 150 AU (red line) for divergence parameters D=0.5 and D=1.0.

distribution. Colored lines in Figure 2 show how the observed distribution narrows as D is increased from 0 to 1.2. Beam widths, W, can be quantified as the full-width-at-half-maxima of Gaussians fitted to the distributions. In Figure 2, increasing D from 0 to 1.2 leads to a decrease in width from $W = 24.3°$ to $W = 11.4°$.

The decrease of W with increasing D happens because as the observer looks farther from the stagnation axis (i.e, as θ increases), the flow at the boundary becomes directed further away from the line of sight relative to a laminar flow, the result being a faster decrease in observed flux with θ compared to a laminar flow, and therefore a narrower He beam. Decreasing T would also naturally narrow the observed distributions, as the VDF at the source is narrower. Thus, if D is increased, one can preserve the same W by increasing T, illlustrating how assuming a divergent flow would increase the $T = 9500$ K measurement of [9] towards a more reasonable value. The width will also be affected by the velocity V, because as V is decreased there is more dispersion introduced into the observed distribution by the Sun's gravitational effects.

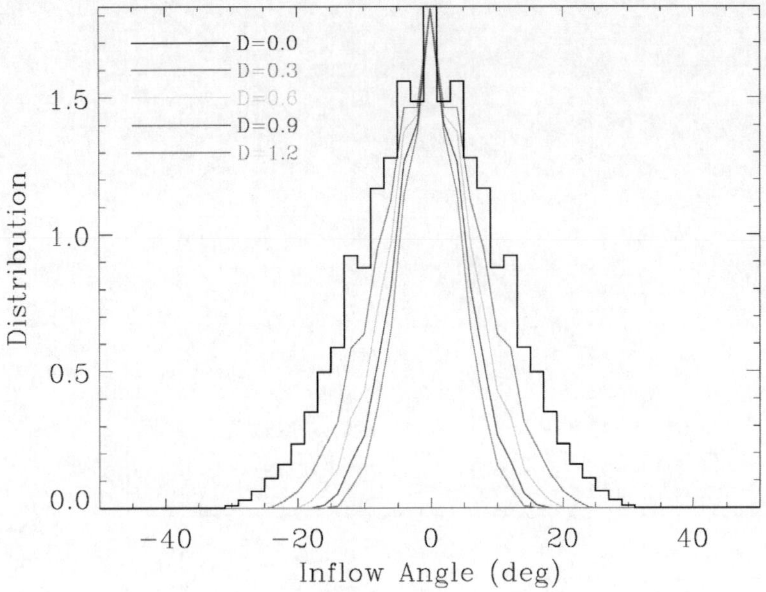

Figure 2. Models of the He distribution observed at (X, Y)=(1, 0), assuming a divergent flow from 150 AU with $V = 20$ km s^{-1}, $T = 10^4$ K, and five values for the divergence parameter, in the range $D = 0.0 - 1.2$.

3. Remotely Measuring Flow Divergence

In order to further explore the dependence of W on D, T, and V; a number of trials like those in Figure 2 are computed, assuming a range of D, T, and V values, and measuring the W values associated with them. The observed (D, T, V, W) points are fitted with the following power law relation:

$$W = C\left(\frac{V}{20}\right)^{\alpha}\left(\frac{T}{10^4}\right)^{\beta}(D+1)^{\gamma}, \tag{1}$$

with V in km s^{-1} and T in K. A least squares analysis is used to determine the values of C, α, β, and γ that lead to the best fit to the measurements. The best fit is shown in Figure 3, with $C = 24.1°$, $\alpha = -0.84$, $\beta = 0.52$, and $\gamma = -0.91$. The quality of the fit demonstrates that the power law relation provides a reasonable approximation for how W depends on D, T, and V.

Based on equation (1), the [9] fit to the secondaries, (D, T, V)=(0, 9500, 11.3), suggests a He beam width of $W = 37.9°$ in the context of our simple 2-D model geometry. Global heliospheric models suggest $V \approx 9$ km s^{-1} and $T \approx 21,000$ K for the outer heliosheath [7, 12]. Assuming these values for V and T, we can use equation (1) to compute the value of D necessary to recover the $W = 37.9°$ value that is crudely representative of the IBEX measurements. The resulting divergence is $D = 0.94$. We

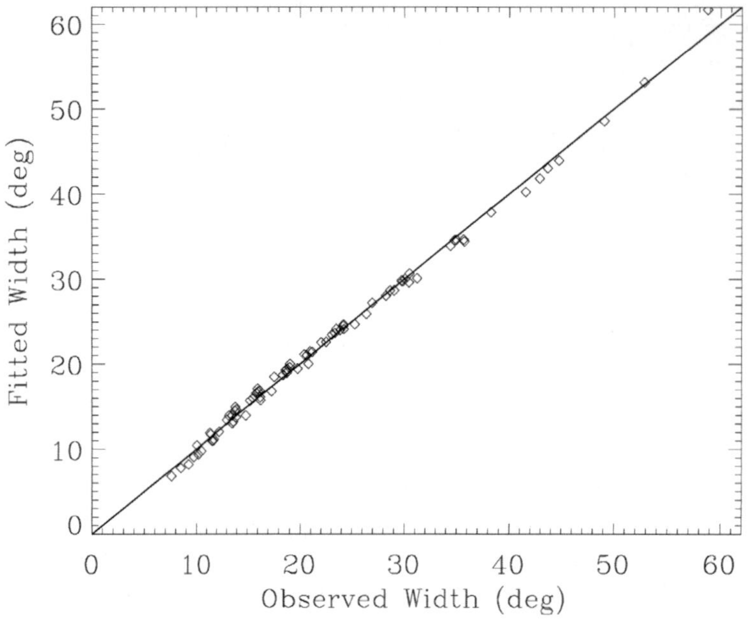

Figure 3. Comparison between the He distribution widths predicted by the power law relation between W, V, T, and D in equation (1); and the observed widths based on numerical models, like those in Figure 2.

therefore would expect that the IBEX data could also be reasonably well fit by a flow with parameters (D, T, V)=(0.94, 21000, 9), with a temperature that is more plausible for the outer heliosheath than the $T = 9500$ K measurement of [9].

Obviously, an actual fit to the IBEX data needs to be performed to verify that such a flow can reproduce the observations. Nevertheless, this exercise illustrates how the He secondaries may provide a way to actually detect and measure the divergence of the ISM flow around the heliopause. This point becomes even more valid if it can be demonstrated that a $D > 0$ flow demonstrably fits the IBEX data better than a $D = 0$ flow.

In future modeling of IBEX data using a parametrized divergent flow, the model will have to be fully 3-D, and it will be necessary to address the question of how to deal with asymmetries in the flow caused by the ISM magnetic field [7, 12]. One approach is to simply allow the flow axis direction to differ from the ISM flow direction, as [9] have done. The apparent deflection of the secondary He flow from the primary flow in a manner consistent with the ISM field orientation is in fact one of the biggest lines of argument in favor of the "Warm Breeze" neutrals originating in the outer heliosheath [9].

However, it is worth noting that the term "deflection" is not the most physically accurate description of what is going on, neither for the secondary He neutrals, nor for the secondary H neutrals studied by [14]. There are, after all, no neutrals that are actually being physically deflected at the heliopause. Instead, what is happening is an asymmetry in the divergence of the flow, which yields an average velocity vector for the flow inside the heliopause that is shifted from the original ISM flow direction. With a parametrized divergent flow it should actually be possible to keep the central flow axis fixed to the flow direction of the primary ISM flow component, but to allow the divergence to vary with direction in some smooth fashion, relative to a plane defined by the ISM magnetic field and flow directions. Such an approach could in principle provide a more physically realistic description of the flow without increasing the number of free parameters of the fit.

Regardless of how one implements a parametrized divergent flow model, it is worthwhile to keep in mind that it will still only be an approximation for the actual complex flow pattern. It will always be useful to use sophisticated global heliospheric models [11, 12] to provide direct predictions for the He secondary flow properties observed by IBEX. Such models provide the most realistic descriptions of the flow pattern in the outer heliosheath. However, a parametrized flow description is necessary to perform model-independent fits to IBEX data, and in such fitting the methodology proposed here should be preferable to the "laminar flow from infinity" approximation.

Acknowledgments
Support for this project was provided by NASA award NNH16AC40I to the Naval Research Laboratory.

References
[1] McComas D J *et al* 2009 *Space Sci. Rev.* **146** 11
[2] Witte M 2004 *A&A* **426** 835
[3] Bzowski M *et al* 2012 *ApJS* **198** 12
[4] Bzowski M, Kubiak M A, Hłond M, Sokól J M, Banaszkiewicz M, and Witte M 2014 *A&A* **569** A8
[5] McComas D J *et al* 2015 *ApJ* **801** 28
[6] Wood B E, Müller H -R, and Witte M 2015 *ApJ* **801** 62
[7] Kubiak M A *et al* 2014 *ApJS* **213** 29
[8] Wood B E, Müller H. -R., and Witte M 2017 *ApJ* submitted
[9] Kubiak M A *et al* 2016 *ApJS* **223** 25
[10] Zank G P, Heerikhuisen J, Wood B E, Pogorelov N V, Zirnstein E, and McComas D J 2013 *ApJ* **763** 20
[11] Müller H -R, Bzowski M, Möbius E, and Zank G P 2013 *Solar Wind 13* (New York: AIP, Vol 1539) p 348
[12] Izmodenov V V and Alexashov D B 2015 *ApJS* **220** 32
[13] Wood B E, Karovska M, and Raymond J C 2002 *ApJ* **575** 1057
[14] Lallement R, Quémerais E, Bertaux J L, Ferron S, Koutroumpa D, and Pellinen R 2005 *Science* **307** 1447

IOP Conf. Series: Journal of Physics: Conf. Series **900** (2017) 012022 doi:10.1088/1742-6596/900/1/012022

Turbulent Equilibria for Charged Particles in Space

Peter H. Yoon[1,2]

[1] IPST, University of Maryland, College Park, USA
[2] SSR, Kyung Hee University, Korea

E-mail: yoonp@umd.edu

Abstract. It is well known that the solar wind electron distribution function is apparently composed of several components, but the energetic tail population is well fitted with kappa distribution function. It is also well established that the solar wind protons possess quasi power-law tail distribution function that is well fitted with an inverse power law model. In the recent past, the present author developed a theory that describes a system of electrons and Langmuir turbulence that are in dynamical steady-state. In such a model, the kappa distribution function for the electrons emerges as a unique solution of the steady-state weak turbulence plasma kinetic equation. For the proton inverse power-law tail problem, Fisk and Gloeckler's theory of compressional turbulence received much attention in the literature. In the present paper, their model is revisited in the light of plasma kinetic theory that involves low-frequency kinetic Alfvén wave fluctuations. It is shown that the proton kappa distribution function satisfies the steady-state proton particle kinetic equation. The steady-state wave kinetic equation for the kinetic Alfvén wave is also solved. This shows that the proton suprathermal distribution with an inverse power law velocity dependence may indeed result from a steady-state wave-particle interaction of the compressional kinetic Alfvénic fluctuations in the solar wind, thus providing support for, and also providing an alternative view of Fisk and Gloeckler's model. However, in the absence of additional constraint that may arise from the balance of nonlinear wave-particle interaction terms within the wave kinetic equations for kinetic Alfvénic waves, the index of inverse power-law velocity tail distribution is undetermined. This calls for further investigation of nonlinear kinetic Alfvénic turbulence.

1. Introduction

It is well known that in the solar wind there exists a permanent low frequency turbulence [1, 2, 3]. While the source of such a turbulence may lie near the solar source, which may get convected outward into the interplanetary space, the role of self-generated low-frequency fluctuations by means of thermal emission is not really understood nor investigated in depth. In one of the noteworthy discoveries in recent heliospheric research, it was shown by means of various *in situ* spacecraft observations that the solar wind is replete with a ubiquitous distribution of energetic protons with an inverse power law suprathermal velocity distribution function, $f(v) \propto v^{-5}$ [4, 5, 6].

Fisk and Gloeckler [7] proposed the so-called "pump acceleration mechanism" in order to explain such a feature. The pump acceleration model requires the following conditions, the essence of which can be found in Fisk and Gloeckler's paper in this volume [8], and are paraphrased below:

- *The plasma contains an embedded mean magnetic field.* The energy in the particles is greater than that of the magnetic field, which has as its only function to couple the motion of the particles.
- *The plasma can be considered to be in a fixed volume that is thermally isolated*, which means simply that there is no net influx or outflow of particles and energy across the boundaries of the volume,. i.e. no large-scale spatial gradients.
- *The plasma contains meso-scale compression and expansion regions.* These compressions and expansions occur normal to the mean magnetic field, and the regions are of size large compared to the gyro-radii of the particles that are to be accelerated in the plasma, but small compared to the total volume of the plasma.

Fisk and Gloeckler [8] also note that the above conditions "are very common conditions to occur in the solar wind." The above stated conditions for pump acceleration mechanism may be repeated and rephrased as follows:

- The uniform magnetized plasma condition is applicable.
- The protons and compressional fluctuations maintain dynamically steady state interaction.
- Magnetic fluctuations must have a compressional component, whose characteristic wavelength can be as small as but generally longer that thermal gyro-radii of the protons.

Even though the model by Fisk and Gloeckler [7, 8] does not specify the precise plasma physical characteristics and properties of the compressional fluctuations, it is quite natural that the kinetic Alfvén waves thermally emitted by the protons are a prime candidate. Such waves and fluctuations satisfy the above re-stated conditions naturally, especially the last condition. Consequently, in the present paper, we put forth a theory in which the protons and kinetic Alfvénic fluctuations for a coupled steady-state solutions in a uniform plasma (conditions 1 and 2).

An analogous dynamical steady-state solution involving the suprathermal population of solar wind electrons, known as the *superhalo* electron component [9], was recently put forth [10, 11, 12]. The quiet time solar wind electrons featuring inverse power law tail with the index ~ -6 or so, $f(v) \propto v^{-6}$, was recently explained in the light of the steady-state wave-particle interaction paradigm that involves high-frequency Langmuir fluctuations . It is thus quite natural to expect that the suprathermal protons in the solar wind may also be interacting with the low-frequency compressional turbulence of the Alfvénic variety in a dynamical steady-state. The present paper may help identify the detailed properties of such low frequency turbulence. It is in this context that the present work may be of relevance. It should be noted, however, that the present paper is preliminary in that our purpose is only to show that the kinetic Alfvénic fluctuations and suprathermal proton velocity distribution modeled by a kappa distribution may coexist in a steady-state fashion. The present paper does not yet show that the inver power-law index of v^{-5} naturally follows. As it will be argued later, the precise determination of the power-law index requires an additional dynamical constraint, which we do not have at this point.

The organization of the present paper is as follows: In Section 2 we discuss the general formalism of particle kinetic equation that describes the dynamical behavior of protons under the influence of kinetic Alfvénic fluctuations. We also discuss the wave kinetic equation for kinetic Alfvén waves and fluctuations. Then in Section 3 we consider the self-consistent steady-state solution of the coupled particle and wave kinetic equation for kinetic Alfvén waves and fluctuations in the limit of steady state, $\partial/\partial t \to 0$ in the sense of asymptotic situation, $t \to \infty$. There it will be shown that the proton kappa distribution function is a legitimate solution of the coupled equations. We also derive the associated kinetic Alfvénic fluctuation spectra for electric and magnetic fields. Finally, Section 4 summarizes the present paper and discuss the findings.

IOP Conf. Series: Journal of Physics: Conf. Series **900** (2017) 012022 doi:10.1088/1742-6596/900/1/012022

2. General Formalism

2.1. Formal Particle Kinetic Equation

The mathematical details and physical models relevant to the present problem of steady-state coupled solutions between the protons and kinetic Alfvénic fluctuations is reported elsewhere [13]. Here, we reproduce the final simplified form only. The general quasilinear kinetic equation including spontaneous emission and velocity space friction term is approximated by retaining only the Landau resonance terms since the kinetic Alfvén waves and fluctuations are of the low frequency type so that the cyclotron resonance conditions are not easily satisfied. The proton particle kinetic equation is expressed in the form of Fokker-Planck equation, which is written in velocity space cylindrical coordinate system as follows:

$$\begin{aligned}\frac{\partial f_i}{\partial t} &= \frac{\partial}{v_\perp \partial v_\perp}(v_\perp A_\perp f_i) + \frac{\partial}{\partial v_\parallel}(A_\parallel f_i)\\ &+ \frac{\partial}{v_\perp \partial v_\perp}\left[v_\perp\left(D_{\perp\perp}\frac{\partial f_i}{\partial v_\perp} + D_{\perp\parallel}\frac{\partial f_i}{\partial v_\parallel}\right)\right] + \frac{\partial}{\partial v_\parallel}\left(D_{\perp\parallel}\frac{\partial f_i}{\partial v_\perp} + D_{\parallel\parallel}\frac{\partial f_i}{\partial v_\parallel}\right),\end{aligned} \tag{1}$$

where

$$\begin{aligned}\begin{pmatrix} A_\perp \\ A_\parallel \end{pmatrix} &= \frac{e^2}{2\pi^2 m_i v_\perp}\int d\mathbf{k}\int \frac{d\omega}{\omega^2}\begin{pmatrix} \omega - k_\parallel v_\parallel \\ k_\parallel v_\perp \end{pmatrix}\\ &\times \mathrm{Im}\left(\frac{[v_\perp J_1(b)]^2}{\Lambda_{yy}^*} + \frac{\Lambda_{xx}^*[v_\parallel J_0(b)]^2}{[\Lambda_{xx}\Lambda_{zz} - (\Lambda_{xz})^2]^*}\right)\delta(\omega - k_\parallel v_\parallel),\\ \begin{pmatrix} D_{\perp\perp} \\ D_{\perp\parallel} \\ D_{\parallel\parallel} \end{pmatrix} &= \frac{\pi e^2}{m_i^2 v_\perp^2}\int d\mathbf{k}\int \frac{d\omega}{\omega^2}\begin{pmatrix} (\omega - k_\parallel v_\parallel)^2 \\ (\omega - k_\parallel v_\parallel)k_\parallel v_\perp \\ (k_\parallel v_\perp)^2 \end{pmatrix}\\ &\times\left(\left[v_\perp J_0'(b)\right]^2\langle\delta E_y^2\rangle_{\mathbf{k},\omega} + \left[v_\parallel J_0(b)\right]^2\langle\delta E_z^2\rangle_{\mathbf{k},\omega}\right)\delta(\omega - k_\parallel v_\parallel).\end{aligned} \tag{2}$$

In arriving at the above equation we assumed that the ambient magnetic field lies along z axis, $\mathbf{B}_0 = B_0\hat{\mathbf{z}}$, the wave vector lies in xz plane, $\mathbf{k} = k_\perp\hat{\mathbf{x}} + k_\parallel\hat{\mathbf{z}}$. In Eq. (2), e and m_i represent the unit electric charge and proton mass, respectively, $J_n(b)$ is the Bessel function of the first kind of order n, and $b = k_\perp v_\perp/\Omega_i$ is the argument, $\Omega_i = eB_0/m_i c$ being the proton cyclotron frequency. Here, c is the speed of light *in vacuo*.

In Eq. (2) the various quantities Λ_{ij} represent the components of simplified dielectric tensor under the assumption of retaining only the Landau resonance terms. In anticipation of the final results we assume isotropic proton velocity distribution function, $f_i = f_i(v)$, and we ignore the displacement current, since the kinetic Alfvénic waves and fluctuations satisfy the slow mode condition, $\omega \ll ck$. For the zz component we retain the influence of isotropic electrons,

$$\begin{aligned}\Lambda_{xx} &= -\frac{c^2 k_\parallel^2}{\omega^2} + \frac{c^2}{v_A^2}\int d\mathbf{v}\,\frac{1 - J_0^2(b)}{b^2}\,v_\perp\,\frac{\partial f_i}{\partial v_\perp},\\ \Lambda_{yy} &= -\frac{c^2 k^2}{\omega^2} + \frac{c^2}{v_A^2}\int d\mathbf{v}\,\frac{1 - J_0^2(b)}{b^2}\,v_\perp\,\frac{\partial f_i}{\partial v_\perp} - i\pi\,\frac{\omega_{pi}^2}{\omega}\int d\mathbf{v}\,\delta(\omega - k_\parallel v_\parallel)\,J_1^2(b)\,v_\perp\,\frac{\partial f_i}{\partial v_\perp},\\ \Lambda_{xz} &= \frac{c^2 k_\perp k_\parallel}{\omega^2},\\ \Lambda_{zz} &= -\frac{c^2 k_\perp^2}{\omega^2} + \sum_{a=e,i}\frac{\omega_{pa}^2}{\omega}\int d\mathbf{v}\,\frac{v_\parallel^2}{\omega - k_\parallel v_\parallel}\frac{\partial f_a}{v_\perp \partial v_\perp}.\end{aligned} \tag{3}$$

In the above $v_A^2 = B_0^2/(4\pi n_0 m_i)$ is the square of the Alfvén speed and $\omega_{pa}^2 = 4\pi n_0 e_a^2/m_a$ is the square of the plasma oscillation frequency defined for species labeled a.

Of the terms that appear within the definitions for A_i and D_{ij} in (2), those terms associated with Λ_{yy} and $\langle \delta E_y^2 \rangle_{\mathbf{k},\omega}$ pertain to the magnetosonic mode branch, while the other terms that contain Λ_{xx}^* and $\langle \delta E_z^2 \rangle_{\mathbf{k},\omega}$ are related to the kinetic Alfvén mode branch. If we are only concerned with the influence of kinetic Alfvénic fluctuations and the associated velocity space friction (or drag) term, then the particle kinetic equation (1) simplifies,

$$\begin{aligned} \frac{\partial f_i}{\partial t} &= \frac{\pi e^2}{m_i^2} \int d\mathbf{k} \int d\omega \frac{\partial}{\partial v_\parallel} \Bigg[J_0^2(b)\, \delta(\omega - k_\parallel v_\parallel) \\ &\quad \times \left(\frac{m_i}{2\pi^3 \omega} \mathrm{Im} \frac{\Lambda_{xx}^*}{[\Lambda_{xx}\Lambda_{zz} - (\Lambda_{xz})^2]^*} f_i + \langle \delta E_z^2 \rangle_{\mathbf{k},\omega} \frac{\partial f_i}{\partial v_\parallel} \right) \Bigg]. \end{aligned} \tag{4}$$

2.2. Formal Wave Kinetic Equation

In recent series of papers the general problem of spontaneous emission of EM fluctuations in magnetized plasmas has been formulated [14, 15, 16, 17]. In particular, Yoon et al. [17] discussed in detail, the spontaneous emission of Alfvénic fluctuations. The formalism outlined in [17] is particularly relevant for the present discussion. Even though Ref. [17] is exclusively relevant to thermal plasma, we may apply the formalism in Ref. [17] to the present case of non-thermal plasma in a straightforward manner. We start from the electric and magnetic field fluctuation spectra obtained in Ref. [17],

$$\begin{aligned} \langle \delta E^2 \rangle_{\mathbf{k},\omega} &= \frac{S_{yy}}{|\Lambda_{yy}|^2} + \frac{(|\Lambda_{xx}|^2 + |\Lambda_{xz}|^2) S_{zz}}{|\Lambda_{xx}\Lambda_{zz} - \Lambda_{xz}^2|^2}, \\ \langle \delta B^2 \rangle_{\mathbf{k},\omega} &= \frac{c^2}{\omega^2} \frac{(|k_\perp \Lambda_{xx}|^2 + |k_\parallel \Lambda_{xz}|^2) S_{zz}}{|\Lambda_{xx}\Lambda_{zz} - \Lambda_{xz}^2|^2} + \frac{c^2 k^2}{\omega^2} \frac{S_{yy}}{|\Lambda_{yy}|^2}, \end{aligned} \tag{5}$$

where

$$\begin{aligned} S_{yy} &= \frac{2e^2 n_0}{\pi \omega^2} \int d\mathbf{v}\, [v_\perp J_1(b)]^2\, \delta(\omega - k_\parallel v_\parallel)\, f_i, \\ S_{zz} &= \frac{2e^2 n_0}{\pi \omega^2} \int d\mathbf{v}\, [v_\parallel J_0(b)]^2\, \delta(\omega - k_\parallel v_\parallel)\, f_i. \end{aligned} \tag{6}$$

If we are only interested in contributions from the kinetic Alfvén wave branch then we may ignore S_{yy} terms in Eq. (5). Consider the electric field spectrum for the kinetic Alfvén wave branch, which can be rewritten as

$$\begin{aligned} (\Lambda_{xx}\Lambda_{zz} - \Lambda_{xz}^2) \langle \delta E^2 \rangle_{\mathbf{k},\omega} &= \frac{1}{(\Lambda_{xx}\Lambda_{zz} - \Lambda_{xz}^2)^*} \frac{2e^2 n_0}{\pi \omega^2} \int d\mathbf{v} \\ &\quad \times (|\Lambda_{xx}|^2 + |\Lambda_{xz}|^2)\, [v_\parallel J_0(b)]^2\, \delta(\omega - k_\parallel v_\parallel) f_i. \end{aligned} \tag{7}$$

Upon introducing the slow time derivative associated with the angular frequency $\omega \to \omega + i\partial/\partial t$, which acts upon both $\delta E_{\mathbf{k},\omega}$ and $\delta E_{\mathbf{k},\omega}^*$ within the ensemble average, $\langle \delta E^2 \rangle_{\mathbf{k},\omega} = \langle \delta E_{\mathbf{k},\omega}\, \delta E_{\mathbf{k},\omega}^* \rangle$, we obtain

$$\begin{aligned} &(\Lambda_{xx}\Lambda_{zz} - \Lambda_{xz}^2) \langle \delta E^2 \rangle_{\mathbf{k},\omega} + \frac{i}{2} \frac{\partial (\Lambda_{xx}\Lambda_{zz} - \Lambda_{xz}^2)}{\partial \omega} \frac{\partial \langle \delta E^2 \rangle_{\mathbf{k},\omega}}{\partial t} \\ &= \frac{1}{(\Lambda_{xx}\Lambda_{zz} - \Lambda_{xz}^2)^*} \frac{2e^2 n_0}{\pi \omega^2} \int d\mathbf{v}\, (|\Lambda_{xx}|^2 + |\Lambda_{xz}|^2)\, [v_\parallel J_0(b)]^2\, \delta(\omega - k_\parallel v_\parallel) f_i. \end{aligned} \tag{8}$$

This is the formal wave kinetic equation for the kinetic Alfvén wave mode.

2.3. Kinetic Alfvén Dispersion Relation

To discuss the dispersion relation, let us assume that the bulk of proton velocity space distribution can be approximated by the quasi-Maxwell form so that the relevant component of the dielectric tensor defined in Eq. (3) can be written as

$$\begin{aligned}\Lambda_{xx} &= -\frac{c^2k_\parallel^2}{\omega^2}+\frac{c^2}{v_A^2}\frac{1-I_0(\lambda)e^{-\lambda}}{\lambda}, \qquad \Lambda_{xz}=\frac{c^2k_\perp k_\parallel}{\omega^2},\\ \Lambda_{zz} &= -\frac{c^2k_\perp^2}{\omega^2}-\frac{\omega_{pe}^2}{k_\parallel^2v_e^2}Z'(\xi_e)-i\pi\frac{\omega_{pi}^2}{\omega}\int d\mathbf{v}\,\delta(\omega-k_\parallel v_\parallel)\frac{v_\parallel^2}{v_\perp}\frac{\partial f_i}{\partial v_\perp},\end{aligned} \tag{9}$$

where $\lambda=k_\perp^2T_i/(m_i\Omega_i^2)$, $v_e^2=2T_e/m_e$ is the square of the electron thermal speed, $\xi_e=\omega/(k_\parallel v_e)$, and $Z(\xi_e)$ is the plasma dispersion function, the prime representing the derivative with respect to the argument.

The dispersion relation is given by [18]

$$\begin{pmatrix}\Lambda_{xx} & \Lambda_{xz}\\ \Lambda_{xz} & \Lambda_{zz}\end{pmatrix}\begin{pmatrix}\delta E^x_{\mathbf{k},\omega}\\ \delta E^z_{\mathbf{k},\omega}\end{pmatrix}=0, \tag{10}$$

which leads to $\Lambda_{xx}\Lambda_{zz}-\Lambda_{xz}^2=0$. Let us approximate $[1-I_0(\lambda)e^{-\lambda}]/\lambda\approx 1-3\lambda/4$. We also approximate $1+\xi_eZ(\xi_e)\sim 1$, which is equivalent to warm electrons. Under such approximations and following Ref. [18], we may obtain

$$\begin{aligned}0 &= \Lambda_{xx}\Lambda_{zz}-\Lambda_{xz}^2\\ &= \frac{c^4}{v_A^4}\frac{2T_i}{T_e}\frac{\Omega_i^2}{v_i^2}\frac{v_A^2}{\omega^2}\left(1-\frac{3}{8}\frac{k_\perp^2v_i^2}{\Omega_i^2}\right)\left[\frac{\omega^2}{k_\parallel^2v_A^2}-1-\left(\frac{3}{8}+\frac{T_e}{2T_i}\right)\frac{k_\perp^2v_i^2}{\Omega_i^2}\right]\\ &\quad +i\pi\frac{c^4}{v_A^4}\frac{\Omega_i^2}{\omega}\left(\frac{k_\parallel^2v_A^2}{\omega^2}-1+\frac{3}{8}\frac{k_\perp^2v_i^2}{\Omega_i^2}\right)\int d\mathbf{v}\,\delta(\omega-k_\parallel v_\parallel)\frac{v_\parallel^2}{v_\perp}\frac{\partial f_i}{\partial v_\perp}.\end{aligned} \tag{11}$$

Setting the real part equal to zero we obtain the desired kinetic Alfvén wave dispersion relation,

$$\omega^2=\omega_{\mathbf{k}}^2\equiv k_\parallel^2v_A^2\left[1+\left(\frac{3}{4}+\frac{T_e}{T_i}\right)\frac{k_\perp^2v_i^2}{2\Omega_i^2}\right]. \tag{12}$$

Shown in Figure 1 is the dispersion "surface" corresponding to the kinetic Alfvén wave. The normalized real frequency ω/Ω_i is plotted in a surface plot format versus normalized parallel and perpendicular wave numbers, $k_\parallel v_A/\Omega_i$ and $k_\perp v_T/\Omega_i$, respectively, for $T_e/T_i=0.1$.

2.4. Wave Kinetic Equation for Kinetic Alfvén Waves

Returning to the formal wave kinetic equation (8), taking the imaginary part, we have

$$\begin{aligned}&\frac{\partial\mathrm{Re}\left(\Lambda_{xx}\Lambda_{zz}-\Lambda_{xz}^2\right)}{\partial\omega}\frac{\partial\left\langle\delta E^2\right\rangle_{\mathbf{k},\omega}}{\partial t}+2\,\mathrm{Im}\left(\Lambda_{xx}\Lambda_{zz}-\Lambda_{xz}^2\right)\left\langle\delta E^2\right\rangle_{\mathbf{k},\omega}\\ &=\mathrm{Im}\frac{1}{(\Lambda_{xx}\Lambda_{zz}-\Lambda_{xz}^2)^*}\frac{4e^2n_0}{\pi\omega^2}\int d\mathbf{v}\left(|\Lambda_{xx}|^2+|\Lambda_{xz}|^2\right)[v_\parallel J_0(b)]^2\,\delta(\omega-k_\parallel v_\parallel)f_i.\end{aligned} \tag{13}$$

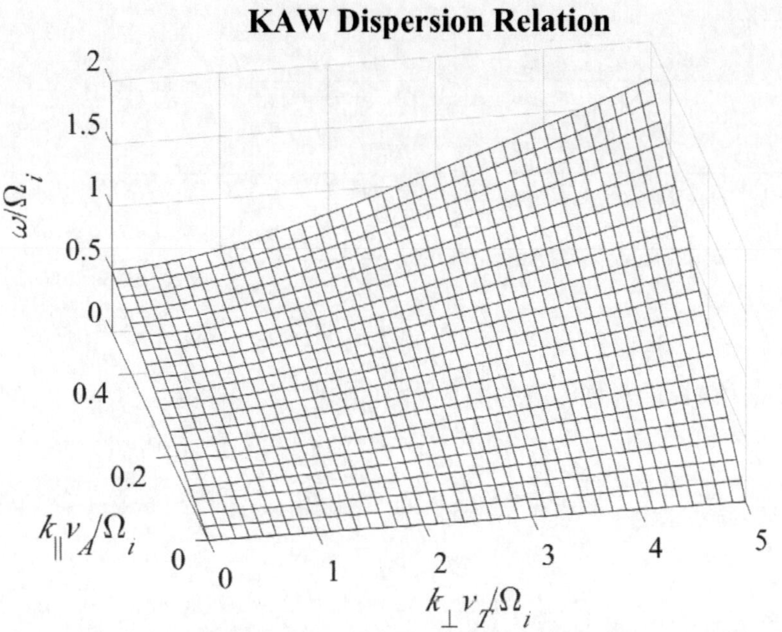

Figure 1. Kinetic Alfvén wave (KAW) dispersion relation (12) for $T_e/T_i = 0.1$.

In what follows, we are interested in ω that satisfies the dispersion relation (12). Taking the residue contribution from the denominator, we have

$$\begin{aligned} \mathrm{Im}\,\frac{1}{(\Lambda_{xx}\Lambda_{zz}-\Lambda_{xz}^2)^*} &= \mathrm{Im}\,\frac{1}{(\omega-\omega_{\mathbf{k}}-i0)[\partial(\Lambda_{xx}\Lambda_{zz}-\Lambda_{xz}^2)/\partial\omega]_{\omega=\omega_{\mathbf{k}}}} \\ &= \frac{\pi\delta(\omega-\omega_{\mathbf{k}})}{\partial\mathrm{Re}(\Lambda_{xx}\Lambda_{zz}-\Lambda_{xz}^2)/\partial\omega]_{\omega=\omega_{\mathbf{k}}}}. \end{aligned} \tag{14}$$

We may also write the electric field spectrum by considering only the eigenmode contribution among all possible ω's,

$$\left\langle \delta E^2 \right\rangle_{\mathbf{k},\omega} = I_{\mathbf{k}}\,\delta(\omega-\omega_{\mathbf{k}}). \tag{15}$$

This leads to

$$\begin{aligned} &\frac{\partial\mathrm{Re}\left(\Lambda_{xx}\Lambda_{zz}-\Lambda_{xz}^2\right)}{\partial\omega_{\mathbf{k}}}\frac{\partial I_{\mathbf{k}}}{\partial t} + 2\,\mathrm{Im}\left(\Lambda_{xx}\Lambda_{zz}-\Lambda_{xz}^2\right) I_{\mathbf{k}} \\ &= \frac{4e^2 n_0}{\omega_{\mathbf{k}}^2[\partial\mathrm{Re}(\Lambda_{xx}\Lambda_{zz}-\Lambda_{xz}^2)/\partial\omega_{\mathbf{k}}]}\left(|\Lambda_{xx}|^2+|\Lambda_{xz}|^2\right)\int d\mathbf{v}[v_\parallel J_0(b)]^2\,\delta(\omega_{\mathbf{k}}-k_\parallel v_\parallel) f_i. \end{aligned} \tag{16}$$

In the above the various dielectric tensor elements are defined by Eq. (9).

We evaluate the following various terms within the context of the dispersion relation, $\omega=\omega_{\mathbf{k}}$, defined in Eq. (12):

$$\begin{aligned}
\frac{\partial \mathrm{Re}\left(\Lambda_{xx}\Lambda_{zz}-\Lambda_{xz}^2\right)}{\partial\omega_{\mathbf{k}}} &= \frac{c^4}{v_A^4}\frac{4T_i}{T_e}\frac{\Omega_i^2}{k_\parallel^2 v_i^2}\frac{1}{\omega_{\mathbf{k}}}\left(1-\frac{3}{8}\frac{k_\perp^2 v_i^2}{\Omega_i^2}\right), \\
\frac{1}{\partial \mathrm{Re}\left(\Lambda_{xx}\Lambda_{zz}-\Lambda_{xz}^2\right)/\partial\omega_{\mathbf{k}}} &= \frac{v_A^4}{c^4}\frac{T_e}{4T_i}\frac{k_\parallel^2 v_i^2}{\Omega_i^2}\frac{\omega_{\mathbf{k}}}{1-3k_\perp^2 v_i^2/(8\Omega_i^2)}, \\
2\,\mathrm{Im}\left(\Lambda_{xx}\Lambda_{zz}-\Lambda_{xz}^2\right) &= 2\pi\frac{c^4}{v_A^4}\frac{\Omega_i^2}{\omega_{\mathbf{k}}}\left(\frac{k_\parallel^2 v_A^2}{\omega_{\mathbf{k}}^2}-1+\frac{3}{8}\frac{k_\perp^2 v_i^2}{\Omega_i^2}\right) \\
&\quad\times\int d\mathbf{v}\, v_\parallel^2\,\delta(\omega_{\mathbf{k}}-k_\parallel v_\parallel)\frac{\partial f_i}{v_\perp\partial v_\perp}.
\end{aligned} \tag{17}$$

Inserting the above to Eq. (16), approximating $J_0(b)\sim 1$, and making use of

$$|\Lambda_{xx}|^2+|\Lambda_{xz}|^2=\frac{c^4}{v_A^4}\left(\frac{k_\parallel^2 v_A^2}{\omega_{\mathbf{k}}^2}-1+\frac{3}{8}\frac{k_\perp^2 v_i^2}{\Omega_i^2}\right)^2+\frac{c^4}{v_A^4}\frac{k_\perp^2 k_\parallel^2 v_A^4}{\omega_{\mathbf{k}}^4}, \tag{18}$$

we have

$$\begin{aligned}
\left(1-\frac{3}{8}\frac{k_\perp^2 v_i^2}{\Omega_i^2}\right)\frac{\partial I_{\mathbf{k}}}{\partial t} &= -\frac{\pi}{2}\frac{T_e}{T_i}\omega_{\mathbf{k}}^2 v_i^2\left(\frac{k_\parallel^2 v_A^2}{\omega_{\mathbf{k}}^2}-1+\frac{3}{8}\frac{k_\perp^2 v_i^2}{\Omega_i^2}\right)\int d\mathbf{v}\,\delta(\omega_{\mathbf{k}}-k_\parallel v_\parallel)\frac{\partial f_a}{v_\perp\partial v_\perp}I_{\mathbf{k}} \\
&\quad+\frac{v_A^4}{c^4}\left(\frac{T_e}{4T_i}\right)^2\frac{k_\parallel^2 v_i^4}{\Omega_i^4}\frac{\omega_{\mathbf{k}}^2}{1-3k_\perp^2 v_i^2/(8\Omega_i^2)}\left[\left(\frac{k_\parallel^2 v_A^2}{\omega_{\mathbf{k}}^2}-1+\frac{3}{8}\frac{k_\perp^2 v_i^2}{\Omega_i^2}\right)^2\right. \\
&\quad\left.+\frac{k_\perp^2 k_\parallel^2 v_A^4}{\omega_{\mathbf{k}}^4}\right]4e^2 n_0\int d\mathbf{v}\,\delta(\omega_{\mathbf{k}}-k_\parallel v_\parallel)f_i.
\end{aligned} \tag{19}$$

Since thermal gyro-radius correction is considered small, we may further approximate the above by retaining only the leading terms in gyro-radius correction. Making use of the dispersion relation (12), it is possible to show that the final simplified form of the wave kinetic equation for kinetic Alfvén mode branch is given by

$$\begin{aligned}
\frac{\partial I_{\mathbf{k}}}{\partial t} &= \pi\left(\frac{T_e}{2T_i}\right)^2\frac{v_A^4}{c^2}\frac{k_\perp^2 k_\parallel^2 v_i^4}{\Omega_i^2}\left(\int d\mathbf{v}\,\delta(\omega_{\mathbf{k}}-k_\parallel v_\parallel)\frac{\partial f_i}{v_\perp\partial v_\perp}\frac{c^2}{v_A^2}I_{\mathbf{k}}\right. \\
&\quad\left.+\frac{m_i\omega_{\mathbf{k}}^2}{4\pi^2 k_\parallel^2 v_A^2}\int d\mathbf{v}\,\delta(\omega_{\mathbf{k}}-k_\parallel v_\parallel)f_i\right).
\end{aligned} \tag{20}$$

2.5. Particle Kinetic Equation for Protons Under the Influence of Kinetic Alfvénic Fluctuations

Returning to the formal particle kinetic equation (4), we make further simplification by approximating $J_0(b)\sim 1$, and making use of Eqs. (14), (15), and (17). We also note from Eq. (10), that

$$\delta E_{\mathbf{k},\omega}^x=-\frac{\Lambda_{xz}}{\Lambda_{xx}}\delta E_{\mathbf{k},\omega}^z, \tag{21}$$

or

$$\begin{aligned}
\left\langle\delta E_x^2\right\rangle_{\mathbf{k},\omega} &= \left\langle\delta E^2\right\rangle_{\mathbf{k},\omega}\frac{|\Lambda_{xz}|^2}{|\Lambda_{xx}|^2+|\Lambda_{xz}|^2}\approx\left\langle\delta E^2\right\rangle_{\mathbf{k},\omega}, \\
\left\langle\delta E_z^2\right\rangle_{\mathbf{k},\omega} &= \left\langle\delta E^2\right\rangle_{\mathbf{k},\omega}\frac{|\Lambda_{xx}|^2}{|\Lambda_{xx}|^2+|\Lambda_{xz}|^2}\approx\left\langle\delta E^2\right\rangle_{\mathbf{k},\omega}\frac{T_e^2}{4T_i^2}\frac{k_\perp^2 k_\parallel^2 v_i^4}{\Omega_i^4}.
\end{aligned} \tag{22}$$

From this we arrive at

$$\begin{aligned} \langle \delta E_x^2 \rangle_{\mathbf{k},\omega} &= I_{\mathbf{k}}\,\delta(\omega - \omega_{\mathbf{k}}), \\ \langle \delta E_z^2 \rangle_{\mathbf{k},\omega} &= \frac{T_e^2}{4T_i^2}\frac{k_\perp^2 k_\parallel^2 v_i^4}{\Omega_i^4}\, I_{\mathbf{k}}\,\delta(\omega - \omega_{\mathbf{k}}). \end{aligned} \tag{23}$$

These lead to the simplification of the formal particle kinetic equation (4),

$$\frac{\partial f_i}{\partial t} = \frac{\pi e^2}{m_i^2}\frac{v_A^2}{c^2}\left(\frac{T_e}{2T_i}\right)^2 \int d\mathbf{k}\, k_\parallel \frac{\partial}{\partial v_\parallel}\left(\frac{m_i \omega_{\mathbf{k}}}{4\pi^2} f_i + \frac{c^2}{v_A^2} I_{\mathbf{k}}\, k_\parallel \frac{\partial f_i}{\partial v_\parallel}\right)\frac{k_\perp^2 v_i^4}{\Omega_i^4}\,\delta(\omega_{\mathbf{k}} - k_\parallel v_\parallel). \tag{24}$$

The coupled set of equations, (20) and (24), can now be considered together in order to discuss the self-consistent steady-state solution for the proton velocity distribution function and a spectrum of kinetic Alfvénic fluctuations.

3. Asymptotic Steady State Between Kinetic Alfvén Fluctuations and Protons

Let us consider the wave and particle kinetic equations, (20) and (24),

$$\begin{aligned} 0 &= \int d\mathbf{v}\left(\frac{\partial f_i}{v_\perp \partial v_\perp}\frac{c^2}{v_A^2} I_{\mathbf{k}} + \frac{m_i \omega_{\mathbf{k}}^2}{4\pi^2 k_\parallel^2 v_A^2} f_i\right)\delta(\omega_{\mathbf{k}} - k_\parallel v_\parallel), \\ 0 &= \int d\mathbf{k}\, k_\parallel \frac{\partial}{\partial v_\parallel}\left(\frac{m_i \omega_{\mathbf{k}}}{4\pi^2} f_i + \frac{c^2}{v_A^2} I_{\mathbf{k}}\, k_\parallel \frac{\partial f_i}{\partial v_\parallel}\right)\frac{k_\perp^2 v_i^4}{\Omega_i^4}\,\delta(\omega_{\mathbf{k}} - k_\parallel v_\parallel). \end{aligned} \tag{25}$$

Our interest as noted in the Introduction, is to consider the possibility of protons and kinetic Alfvénic fluctuations constantly undergoing wave-particle interaction while maintaining a dynamical steady state, so as to form a suprathermal population. In the literature, the kappa velocity distribution function is one of the most successful model to describe a population of particles with suprathermal tail population. Consequently, let us suppose that f_i is given by an isotropic kappa distribution,

$$f_i(v) = \frac{1}{\pi^{3/2} v_T^3}\frac{\Gamma(\kappa+1)}{\left(\kappa - \frac{3}{2}\right)^{\frac{3}{2}}\Gamma\left(\kappa - \frac{1}{2}\right)}\left(1 + \frac{1}{\kappa - \frac{3}{2}}\frac{v^2}{v_T^2}\right)^{-\kappa-1}, \tag{26}$$

where $v_T^2 = 2T_i/m_i$ is square of the thermal speed. Our limited aim is to show that indeed the above kappa distribution satisfies Eq. (25) in a self consistent manner. However, it is important to note that our formalism as it stands, namely Eq. (25), is not capable of determining the actual value of kappa index. From analogy with the present author's theory of suprathermal electrons in the solar wind, the computation of kappa index requires an additional constraint. In the case of suprathermal electrons and steady-state Langmuir turbulence, the balance of nonlinear wave-particle interaction terms in the wave kinetic equation provides such an addition constraint, from which the kappa index for the electrons were determined to be $\kappa = 2.25$ [10, 11]. In the case of solar wind protons, the kappa index must be close to $\kappa = 1.5$ (the exact value of $\kappa = 1.5$ is, of course, not permitted within the context of Eq. (26), as such a value causes divergence). The additional dynamical constraint that will force the specific choice of kappa value will likely come from the nonlinear wave-particle interaction terms in the wave kinetic equation for the kinetic Alfvén wave turbulence, but such terms have not been derived in the literature. With this caveat in mind, let us proceed.

Making use of

$$2\pi\int_0^\infty dv_\perp v_\perp \frac{\partial f_i}{v_\perp \partial v_\perp} = -\frac{2}{\pi^{1/2}v_T^3}\frac{\Gamma(\kappa+1)}{\left(\kappa-\frac{3}{2}\right)^{\frac{3}{2}}\Gamma\left(\kappa-\frac{1}{2}\right)}\left(1+\frac{1}{\kappa-\frac{3}{2}}\frac{v_\parallel^2}{v_T^2}\right)^{-\kappa-1},$$

$$2\pi\int_0^\infty dv_\perp v_\perp \frac{\partial f_i}{\partial v_\parallel} = -\frac{2}{\pi^{1/2}v_T^3}\frac{\Gamma(\kappa+1)}{\left(\kappa-\frac{3}{2}\right)^{\frac{3}{2}}\Gamma\left(\kappa-\frac{1}{2}\right)}v_\parallel\left(1+\frac{1}{\kappa-\frac{3}{2}}\frac{v_\parallel^2}{v_T^2}\right)^{-\kappa-1},$$

$$2\pi\int_0^\infty dv_\perp v_\perp f_i = \frac{1}{\pi^{1/2}v_T}\frac{\Gamma(\kappa+1)}{\left(\kappa-\frac{3}{2}\right)^{\frac{1}{2}}\kappa\Gamma\left(\kappa-\frac{1}{2}\right)}\left(1+\frac{1}{\kappa-\frac{3}{2}}\frac{v_\parallel^2}{v_T^2}\right)^{-\kappa}, \tag{27}$$

one may show that the two equations (25) reduce to the following forms, if we integrate the second equation, that is, the steady-state wave kinetic equation, over perpendicular velocity,

$$\begin{aligned}
0 &= \int_{-\infty}^\infty dv_\parallel\left[-\frac{c^2}{v_A^2}I_\mathbf{k}+\frac{m_i\omega_\mathbf{k}^2}{4\pi^2k_\parallel^2v_A^2}\frac{v_T^2}{2}\frac{\kappa-\frac{3}{2}}{\kappa}\left(1+\frac{1}{\kappa-\frac{3}{2}}\frac{\omega_\mathbf{k}^2}{k_\parallel^2v_T^2}\right)\right]\\
&\quad\times\left(1+\frac{1}{\kappa-\frac{3}{2}}\frac{\omega_\mathbf{k}^2}{k_\parallel^2v_T^2}\right)^{-\kappa-1}\delta(\omega_\mathbf{k}-k_\parallel v_\parallel),\\
0 &= \int d\mathbf{k}\,k_\parallel\frac{\partial}{\partial v_\parallel}\left[\frac{m_i}{4\pi^2}\frac{v_T^2}{2}\frac{\kappa-\frac{3}{2}}{\kappa}\left(1+\frac{1}{\kappa-\frac{3}{2}}\frac{\omega_\mathbf{k}^2}{k_\parallel^2v_T^2}\right)-\frac{c^2}{v_A^2}I_\mathbf{k}\right]\\
&\quad\times\left(1+\frac{1}{\kappa-\frac{3}{2}}\frac{\omega_\mathbf{k}^2}{k_\parallel^2v_T^2}\right)^{-\kappa-1}\frac{k_\perp^2v_i^4}{\Omega_i^4}\delta(\omega_\mathbf{k}-k_\parallel v_\parallel).
\end{aligned} \tag{28}$$

Making use of the dispersion relation and retaining the leading order terms in the small thermal gyro-radius expansion, we obtain from the necessary conditions for the two equalities in Eq. (28), the following form for the kinetic Alfvén wave electric field fluctuation spectrum:

$$\langle\delta E^2\rangle_\mathbf{k}\equiv I_\mathbf{k}=\frac{v_A^2}{c^2}\left\{\frac{\kappa-\frac{3}{2}}{\kappa}\frac{T_i}{4\pi^2}+\frac{1}{\kappa}\frac{B_0^2}{32\pi^3n_0}\left[1+\left(\frac{3}{4}+\frac{T_e}{T_i}\right)\frac{k_\perp^2\rho_i^2}{2}\right]\right\}, \tag{29}$$

where $\rho_i^2=v_T^2/\Omega_i^2$ is the thermal proton gyro-radius. The associated magnetic field fluctuation intensity may be obtained directly from the second equation in (5), by reconsidering all the analyses we have carried out thus far, or more simply, by considering from Faraday's law that $\delta B_{\mathbf{k},\omega}=(c/\omega_\mathbf{k})(k_\parallel\delta E_{\mathbf{k},\omega}^x-k_\perp\delta E_{\mathbf{k},\omega}^z)$. Upon making use of Eq. (23), it follows that the magnetic field fluctuation spectrum associated with the kinetic Alfvén wave mode is given by

$$\langle\delta B^2\rangle_\mathbf{k}\equiv M_\mathbf{k}=\frac{T_i}{4\pi^2}\frac{\dfrac{\kappa-\frac{3}{2}}{\kappa}+\dfrac{1}{\kappa\beta_i}\left[1+\left(\dfrac{3}{4}+\dfrac{T_e}{T_i}\right)\dfrac{k_\perp^2\rho_i^2}{2}\right]}{1+\left(\dfrac{3}{4}+\dfrac{T_e}{T_i}\right)\dfrac{k_\perp^2\rho_i^2}{2}}. \tag{30}$$

Note that the above magnetic field intensity spectrum reduces to the recently discussed solution in the limit of Maxwellian proton distribution ($\kappa\to\infty$) [17].

4. Conclusion and Discussions

To conclude, the assumed proton kappa distribution function (26) and the kinetic Alfvénic fluctuation spectra (29) and (30) satisfy the coupled particle and wave kinetic equations in a

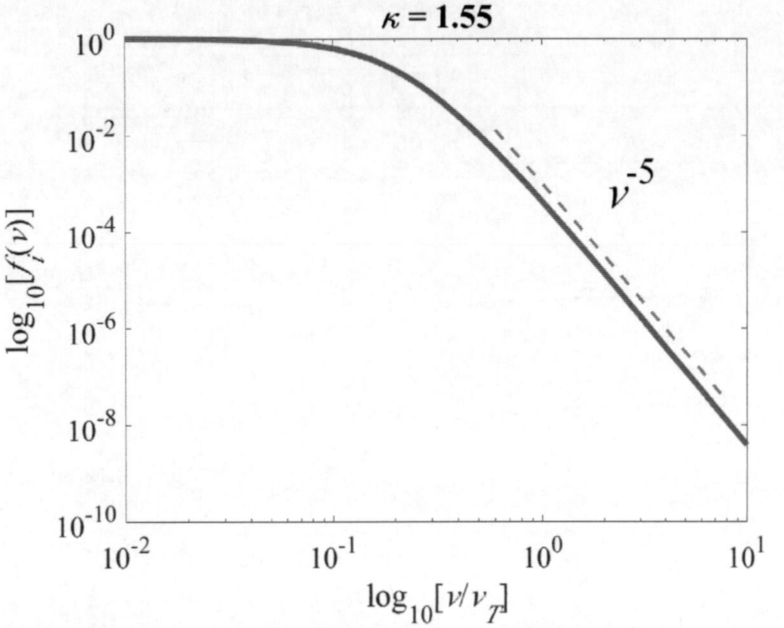

Figure 2. Plot of the proton kappa velocity distribution function (26), in logarithmic vertical scale, versus logarithmic dimensionless speed, for $\kappa = 1.55$. The asymptotic behavior of the solution is $f_i \propto v^{-5}$ in an approximate sense, which is consistent with observation [4, 5, 6].

self-consistent manner, and for a steady-state situation. We have thus proved that the kappa velocity distribution for the protons subject to dynamically steady-state wave-particle interaction with kinetic Alfvénic fluctuations, which are compressive fluctuations as envisioned by Fisk and Gloeckler [8, 7], is indeed a possible solution. What we have not yet proven is that the kappa index close to 1.5 is the unique choice. This requires an additional constraint and physical considerations such as nonlinear wave-particle interaction terms in the wave kinetic equation for kinetic Alfvénic turbulence. We reiterate that the purpose of the present theoretical analysis had been limited in the sense that, our aim was to rephrase Fisk and Gloeckler's *pump acceleration mechanism* in an alternative way, with the emphasis placed on describing the self-consistent wave dynamics. However, at present, the precise value of κ is a free parameter. According to the similar and analogous theory of solar wind electrons and Langmuir fluctuations, the kappa index must be determined from the balance of nonlinear wave-particle interaction term in the Langmuir wave kinetic equation. In the present analysis, we do not yet have nonlinear term in the kinetic Alfvén wave equation. Consequently, the parameter κ must be treated as a free parameter for now.

Thus, in the present analysis, let us choose κ value in an arbitrary manner. According to observation [4, 5, 6] the observed suprathermal proton tail distribution $f_i \propto v^{-5}$ corresponds to $\kappa = 1.5 = 3/2$, but the value of $\kappa = 3/2$ is the minimum critical value for which the model distribution function is not defined. We thus choose κ close to 1.5 but slightly higher than the exact value of 1.5 or equivalently 3/2, namely $\kappa = 1.55$ when plotting the proton distribution function. Figure 2 shows the solution (26) in log-log plot format. Note that the proton kappa distribution asymptotically behaves as $f_i \propto v^{-2\kappa-2}$, which in the case of $\kappa = 1.55$ is $f_i \propto v^{-5}$, in an approximate sense, which is consistent with observation [4, 5, 6].

The electric and magnetic field spectral intensities that form the other part of the self-consistent solution are given by (29) and (30). Recently, Yoon et al. [17] discussed the magnetic

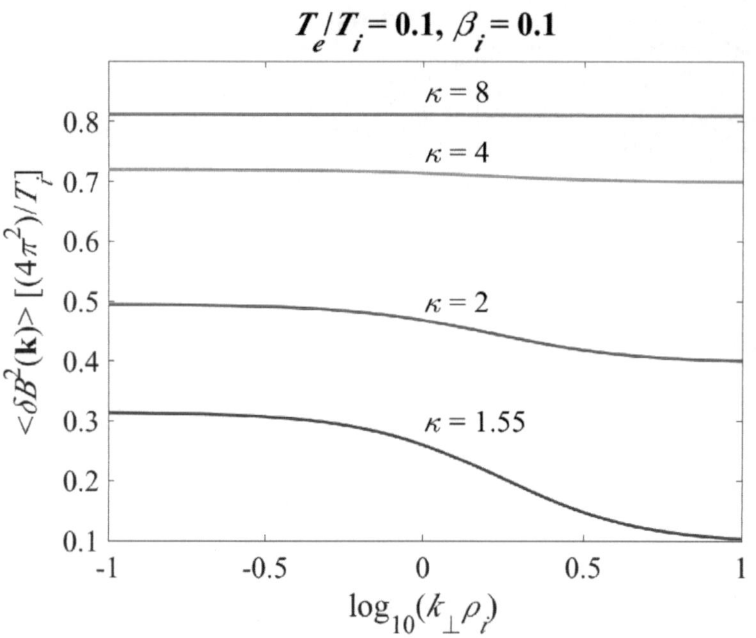

Figure 3. Magnetic form factor $f(k_\perp)$ [Eq. (31)] plotted against normalized perpendicular wave number $k_\perp \rho_i$ for various values of kappa index.

field intensity spectrum for kinetic Alfvén wave when the proton distribution is given by thermal Maxwellian form. The solution (30) is a direct generalization for kappa distribution. It is noteworthy that the kinetic Alfvénic fluctuations are characterized by the magnetic form factor,

$$f(k_\perp) \equiv \frac{4\pi^2 \langle \delta B^2 \rangle_{\mathbf{k}}}{T_i} = \frac{\dfrac{\kappa - \frac{3}{2}}{\kappa} + \dfrac{1}{\kappa \beta_i}\left[1 + \left(\dfrac{3}{4} + \dfrac{T_e}{T_i}\right)\dfrac{k_\perp^2 \rho_i^2}{2}\right]}{1 + \left(\dfrac{3}{4} + \dfrac{T_e}{T_i}\right)\dfrac{k_\perp^2 \rho_i^2}{2}}. \tag{31}$$

Shown in Figure 3 is the above form factor versus normalized perpendicular wave number $k_\perp \rho_i$ for various values of kappa index, $\kappa = 8$ (which is close to the Maxwellian limit), $\kappa = 4$, 2, and $\kappa = 1.55$.

Acknowledgments
The author acknowledges NSF grant AGS1550566 to the University of Maryland, and the BK21 plus program from the National Research Foundation (NRF), Korea, to Kyung Hee University. He also acknowledges the Science Award Grant from the GFT, Inc., to the University of Maryland. Part of this work was carried out while the author was visiting Ruhr University Bochum, Germany, which was made possible by the support from the Ruhr University Research School PLUS, funded by Germany's Excellence Initiative (DFG GSC 98/3), and by a Mercator fellowship awarded by the Deutsche Forschungsgemeinschaft through the grant Schl 201/31-1.

References

[1] C. Briand, Nonlin. Proc. Geophys. **16**, 319 (2009).
[2] Alexandrova O, Chen C H, Sorriso-Valvo L, Horbury T S and Bale S D 2013 *Space Sci. Rev.* **178** 101
[3] Bruno R and Carbone V 2013 *Liv. Rev. Solar Phys.* **10** 2

[4] Gloeckler G, Fisk L A, Zurbuchen T H and Schwadron N A 2000 in *AIP Conf. Proc.* **528**, ed. R. A. Mewaldt et al. (Melville, AIP), p. 221.
[5] Simunac K D C and Armstrong T P 2004, *J. Geophys. Res.* **109**, A10101
[6] Decker R B, Krimigis S M, Roelof E C, Hill M E, Armstrong T P, GLoeckler G, Hamilton D C and Lanzarotti L J 2005 *Science* **309** 2020
[7] Fisk L A and Gloeckler G 2008 *Astrophys. J.* **686** 1466
[8] Fisk L A and Gloeckler G 2017, submitted to *J. Physics: Conf. Ser. (JPCS)*, Proceedings of the 16th Annual International Astrophysics Conference held in Santa Fe, NM, 2017.
[9] Wang L, Lin R P, Salem C, Pulupa M, Larson D E, Yoon P H and Luhmann J G 2012 *Astrophys. J. Lett.* **753** L23
[10] Yoon P H 2012 *Phys. Plasmas* **19** 052301
[11] Yoon P H 2014 *J. Geophys. Res.* **119** 7074
[12] Kim S, Yoon P H, Choe G S and Moon Y-J 2016, *Astrophys. J.* **828** 60
[13] Yoon P H, Schlickeiser R, Kim S and López R A 2017, *A&A* submitted
[14] Yoon P H and López R A 2017a *Phys. Plasmas* **24** 022117
[15] Yoon P H and López R A 2017a *Phys. Plasmas* **24** 049902
[16] López R A and Yoon P H 2017, Simulation of electromagnetic fluctuations in thermal magnetized plasma, *Plasma Phys. Contr. Fusion*, submitted
[17] Yoon P H, López R A, Vafin S, Kim S, and Schlickeiser R 2017, Spontaneous emission of Alfvénic fluctuations, *Plasma Phys. Contr. Fusion*, accepted
[18] Lysak R L and Lotko W 1996 *J. Geophys. Res.* **101** 5085

IOP Conf. Series: Journal of Physics: Conf. Series **900** (2017) 012023 doi:10.1088/1742-6596/900/1/012023

The Theory of Nearly Incompressible Magnetohydrodynamic Turbulence: Homogeneous Description

G P Zank[1,2], L Adhikari[1], P Hunana[1], D Shiota[3], R Bruno[4], D Telloni[5], and K. Avinash[1,6]

[1] Center for Space Plasma and Aeronomic Research (CSPAR), University of Alabama in Huntsville, Huntsville, AL 35805, USA

[2] Department of Space Science, University of Alabama in Huntsville, Huntsville, AL 35805, USA

[3] Institute for Space-Earth Environmental Research, Nagoya University, Nagoya, Aichi 464-8601, Japan

[4] INAF-IAPS Istituto di Astrofisica e Planetologia Spaziali, Via del Fosso del Cavaliere 100, 00133 Roma, Italy

[5] INAF - Astrophysical Observatory of Torino, Via Osservatorio 20, 10025 Pino Torinese, Italy

[6] Department of Physics and Astrophysics, University of Delhi, Delhi, 110007, India

E-mail: garyp.zank@gmail.com

Abstract. The theory of nearly incompressible magnetohydrodynamics (NI MHD) was developed to understand the apparent incompressibility of the solar wind and other plasma environments, particularly the relationship of density fluctuations to incompressible manifestations of turbulence in the solar wind and interstellar medium. Of interest was the identification of distinct leading-order incompressible descriptions for plasma beta $\beta \gg 1$ and $\beta \sim 1$ or $\ll 1$ environments. In the first case, the "dimensionality" of the MHD description is 3D whereas for the latter two, there is a collapse of dimensionality in that the leading-order incompressible MHD description is 2D in a plane orthogonal to the large-scale or mean magnetic field. Despite the success of NI MHD in describing fluctuations in a low-frequency plasma environment such as the solar wind, a basic turbulence description has not been developed. Here, we rewrite the NI MHD system in terms of Elsässer variables. We discuss the distinction that emerges between the three cases. However, we focus on the $\beta \sim 1$ or $\ll 1$ regimes since these are appropriate to the solar wind and solar corona. In both cases, the leading-order turbulence model describes 2D turbulence and the higher-order description corresponds to slab turbulence, which forms a minority component. The Elsässer $\beta \sim 1$ or $\ll 1$ formulation exhibits the nonlinear couplings between 2D and slab components very clearly, and shows that slab fluctuations respond in a passive scalar sense to the turbulently evolving majority 2D component fluctuations. The coupling of 2D and slab fluctuations through the $\beta \sim 1$ or $\ll 1$ NI MHD description leads to a very natural emergence of the "Goldreich-Sridhar" critical balance scaling parameter, although now

IOP Conf. Series: Journal of Physics: Conf. Series **900** (2017) 012023 doi:10.1088/1742-6596/900/1/012023

with a different interpretation. Specifically, the critical balance parameter shows that the energy flux in wave number space is a consequence of the intensity of Alfvén wave sweeping versus passive scalar convection by leading-order 2D Elsässer fluctuations, with critical balance being achieved when Alfvén wave sweeping balances passive scalar convection by leading-order 2D Elsässer fluctuations. Besides yielding predictions of 2D and slab spectra for Elsässer fluctuations, NI MHD shows that density fluctuations are advected by the majority or dominant incompressible velocity fluctuations. In the case of $\beta \sim 1$ or $\ll 1$, the density spectrum is Kolmogorov in the perpendicular wave number, thus providing a possible explanation for the observed extended Kolmogorov-like power law spectrum for electron density fluctuations in the interstellar medium.

1. Introduction

Developed in the early 1990s [1–13] for homogenous flows, and subsequently extended to large-scale inhomogeneous flows [14–17], NI MHD is a formulation of the MHD equations in a weakly compressible or nearly incompressible regime. The early work focused on the apparent incompressibility of the solar wind and other plasma environments, and the relationship of density fluctuations to incompressible manifestations of turbulence in the solar wind and interstellar medium, such a $k^{-5/3}$ Kolmogorov-like density spectrum [18–21]. An important prediction of NI MHD was that solar wind turbulence, for which the plasma beta is $O(1)$ or $\ll 1$, is a superposition of dominant 2D and minority slab components [10; 11].

Density fluctuations observed in the supersonic solar wind have typical amplitudes that deviate no more than 10% from the mean solar wind density [8; 22; 23]. Magnetic field and velocity fluctuations are frequently well described by incompressible turbulence [23], exhibiting power-law distributions in energy, density, temperature, magnetic field with a Kolmogorov exponent $(-5/3)$ or occasionally an Iroshnikov-Kraichnan exponent $(-3/2)$. Solar wind observations suggest that compressible MHD often converges to an incompressible state, passing through a nearly incompressible regime in which weakly compressible corrections are related to the final incompressible state. The inverse square of the "turbulent Mach number," i.e., a Mach number defined by the fluctuating velocity rather than the bulk flow velocity and mean sound speed, is an important scaling parameter in compressible MHD since it is typically very small. The turbulent Mach number therefore introduces a highly singular term into the MHD equations that leads to the generation of high-frequency (magneto)acoustic fluctuations. However, because of the three fundamental wave speeds in MHD, it is necessary to consider three plasma beta β regimes ($\beta = P/(B^2/2\mu_0)$, where P is the thermal pressure, $B = |\mathbf{B}|$, $\mathbf{B}$ the magnetic field, and μ_0 the magnetic permeability). Zank and Matthaeus [11] (hereafter ZM93) considered $\beta \ll 1$, $\beta \sim 1$, and $\beta \gg 1$ and showed that the description of solar wind turbulence is very different in the $\beta \gg 1$ and the $\beta \sim 1$ or $\ll 1$ regimes.

2. The equations of NI MHD

We consider here the NI MHD equations for a homogeneous plasma only. The inhomogeneous development was done by Hunana and Zank [17] and subsequently extended in [24; 25]. We present the equations for the $\beta \gg 1$ and $\beta \sim 1$ cases (the

$\beta \ll 1$ equations are essentially the same as the $\beta \sim 1$ case), where the latter assumes the presence of a large constant magnetic field $\mathbf{B}_0$. A related analysis has been given in [26] for the $\beta \gg 1$ case. The NI MHD description comprises a core set of incompressible equations given by the normalized equations (11) - (13) and (57) - (59) in ZM93 for the $\beta \gg 1$ and $\beta \sim 1$ regimes respectively. These equations are derived by requiring that all fast-scale magnetoacoustic variation to vanish. The elimination of fast-time scale variation is accomplished through the principle of bounded derivatives [27]. The bounding of the time derivatives yields the incompressible MHD equations as secular conditions that ensure the elimination of all fast-scale variation [5; 7]. The $\beta \gg 1$ regime yields a leading-order incompressible MHD description that is fully 3D, whereas the leading-order description for the $\beta \ll 1$, $\beta \sim 1$ regimes is reduced to two dimensions in the plane perpendicular to the mean magnetic field. Higher-order nearly incompressible fluctuations for all β regimes comprise 3D propagating magnetosonic and sound waves, and Alfvén waves propagating parallel to the magnetic field. When $\beta \ll 1$ or ~ 1, nearly incompressible transverse fluctuations propagate parallel to the magnetic field. The weakly compressible equations are given by the normalized equations (69) - (72) in ZM93. The higher-order corrections can be both compressible and incompressible. ZM93 predicted that in the $\beta \ll 1$ or ~ 1 regimes, solar wind turbulence is a superposition of 2D and slab turbulence, dominated by the 2D component. Numerical simulations [28] are "tantalizingly consistent" with the predictions of a superposition of dominant 2D and slab turbulence by [11].

We use the notation of ZM93 in using an "∞" superscript to denote MHD variables that satisfy the core incompressible equations (i.e., the sound speed is "infinite"), and the superscript "*" or subscript "1" to indicate higher-order corrections. ZM93 showed that the incompressible pressure and the NI pressure correction enter at the same order in the square of the turbulent sonic or Alfvénic (depending on the plasma beta regime) Mach number, as does the fluctuating density (however see [17] for the inhomogeneous case where the density enters linearly in the turbulent sonic Mach number). The incompressible magnetic field component $\mathbf{B}^\infty$ is of order the turbulent sonic Mach number and the NI correction $\mathbf{B}^*$ enters at the next order. The NI velocity $\mathbf{u}_1$ is also of order the turbulent sonic Mach number. Zank et al. [24] renormalize the $\beta \sim 1$ form of the NI MHD equations derived in ZM93. This renomalization is easily extended to the $\beta \gg 1$ regime. Accordingly, we can express the NI MHD equations in the $\beta \gg 1$ and $\beta \sim 1$ regimes in the following dimensional fashion,

$$\nabla \cdot \mathbf{u}^\infty = 0; \qquad \nabla \cdot \mathbf{B}^\infty = 0; \tag{1}$$

$$\frac{\partial \mathbf{u}^\infty}{\partial t} + \mathbf{u}^\infty \cdot \nabla \mathbf{u}^\infty = -\frac{1}{\rho_0} \nabla \left(P^\infty + \frac{B^{\infty 2}}{2\mu_0} \right) + \frac{1}{\mu_0 \rho_0} \mathbf{B}^\infty \cdot \nabla \mathbf{B}^\infty; \tag{2}$$

$$\frac{\partial \mathbf{B}^\infty}{\partial t} + \mathbf{u}^\infty \cdot \nabla \mathbf{B}^\infty = \mathbf{B}^\infty \cdot \nabla \mathbf{u}^\infty, \tag{3}$$

(the $\beta \gg 1$ case), and

$$\nabla_\perp \cdot \mathbf{u}^\infty = 0; \qquad \nabla_\perp \cdot \mathbf{B}^\infty = 0; \tag{4}$$

$$\frac{\partial \mathbf{u}^\infty}{\partial t} + \mathbf{u}^\infty \cdot \nabla_\perp \mathbf{u}^\infty = -\frac{1}{\rho_0}\nabla_\perp\left(P^\infty + \frac{1}{2\mu_0}B^{\infty 2}\right) + \frac{1}{\mu_0\rho_0}\mathbf{B}^\infty \cdot \nabla_\perp \mathbf{B}^\infty; \tag{5}$$

$$\frac{\partial \mathbf{B}^\infty}{\partial t} + \mathbf{u}^\infty \cdot \nabla_\perp \mathbf{B}^\infty = \mathbf{B}^\infty \cdot \nabla_\perp \mathbf{u}^\infty, \tag{6}$$

(the $\beta \sim 1$ case). Notice the important difference between the two leading-order descriptions. The $\beta \gg 1$ regime in the incompressible limit is a fully 3D (spatially) description, and corresponds to the "standard" description of incompressible MHD. By contrast, the leading-order description in the $\beta \sim 1$ NI limit are the 2D MHD equations in a plane perpendicular to the mean magnetic field i.e., $\mathbf{u}^\infty = \mathbf{u}^\infty(x, y)$ and $\mathbf{B}^\infty = \mathbf{B}^\infty(x, y)$ where $\hat{\mathbf{z}}$ is defined by the large-scale mean magnetic field $\mathbf{B}_0 = B_0\hat{\mathbf{z}}$. Note too the absence of propagation effects associated with Alfvén waves. Hence, the leading-order incompressible description for a $\beta \sim O(1)$ (and $\ll 1$) plasma is properly 2D MHD in a plane perpendicular to the mean magnetic field.

The NI corrections to both the $\beta \gg 1$ and $\beta \sim 1$ leading-order incompressible equations are formally the same, given by

$$\begin{aligned}
\frac{\partial \rho^*}{\partial t} + \mathbf{u}^\infty \cdot \nabla\rho^* + \rho_0\nabla\cdot\mathbf{u}_1 &= 0; && (7)\\
\frac{\partial \mathbf{u}_1}{\partial t} + \mathbf{u}^\infty \cdot \nabla\mathbf{u}_1 + \mathbf{u}_1\cdot\nabla\mathbf{u}^\infty &= -\frac{1}{\rho_0}\nabla P^* - \frac{1}{\mu_0\rho_0}\nabla\left(\mathbf{B}_0\cdot\mathbf{B}^* + \mathbf{B}^\infty\cdot\mathbf{B}^*\right)\\
&\quad + \frac{1}{\mu_0\rho_0}\left(\mathbf{B}_0 + \mathbf{B}^\infty\right)\cdot\nabla\mathbf{B}^* + \frac{1}{\mu_0\rho_0}\mathbf{B}^*\cdot\nabla\mathbf{B}^\infty; && (8)\\
\frac{\partial P^*}{\partial t} + \mathbf{u}^\infty\cdot\nabla P^* + \gamma p_0\nabla\cdot\mathbf{u}_1 &= -\frac{\partial P^\infty}{\partial t} - \mathbf{u}^\infty\cdot\nabla P^\infty; && (9)\\
\frac{\partial \mathbf{B}^*}{\partial t} + \mathbf{u}^\infty\cdot\nabla\mathbf{B}^* - \mathbf{B}^*\cdot\nabla\mathbf{u}^\infty &= \left(\mathbf{B}_0 + \mathbf{B}^\infty\right)\cdot\nabla\mathbf{u}_1 - \left(\mathbf{B}_0 + \mathbf{B}^\infty\right)\nabla\cdot\mathbf{u}_1\\
&\quad - \mathbf{u}_1\cdot\nabla\mathbf{B}^\infty. && (10)
\end{aligned}$$

However, the variables $\mathbf{u}^\infty$, $\mathbf{B}^\infty$, etc. are functions of (x, y, z) when $\beta \gg 1$, and functions of (x, y), $B_0\hat{\mathbf{z}}$, when $\beta \sim 1$. Notice that equations (7) - (10) are not linearized but instead describe the driving of the higher-order corrections by the dynamics of the leading-order incompressible flow variables. The incompressible flow $\mathbf{u}^\infty$, in particular, introduces an important passive scalar transport component to the dynamics of the higher-order plasma variables. The variables ρ^* and $\mathbf{u}_1$ possess both an incompressible and a compressible component. Furthermore, as shown by equation (9) for example, the incompressible flow acts as a source of higher-order fluctuations, including compressible fluctuations, and therefore represents a generalization of the Lighthill mechanism for the generation of sound. Consequently, incompressible fluctuations can be a possible source of density fluctuations. Since the NI corrections enter at $O(M^2)$, M being either the turbulent sonic or Alfvénic Mach number, turbulent fluctuations in a $\beta \sim 1$ or $\ll 1$ plasma are primarily 2D rather than slab. In the solar wind, the observed turbulent Mach number ordering implies a crude partitioning of energy between 2D and slab components

IOP Conf. Series: Journal of Physics: Conf. Series **900** (2017) 012023 doi:10.1088/1742-6596/900/1/012023

in the ratio 80 : 20 [10], with some observational confirmation [29]. The partitioning of fluctuation energy into 2D and slab also implies a strong variance anisotropy in the solar wind fluctuations since the parallel variance is greatly reduced compared to the perpendicular variance, consistent with that observed (e.g., [30; 31]).

We may introduce Elsässer variables for the leading-order or majority incompressible and minority NI components through

$$\mathbf{z}^{\infty\pm} = \mathbf{u}^{\infty} \pm \frac{\mathbf{B}^{\infty}}{\sqrt{\mu_0 \rho_0}} \quad \text{and} \quad \mathbf{z}^{*\pm} \equiv \mathbf{u}_1 \pm \mathbf{B}^*/\sqrt{\mu_0 \rho_0}. \tag{11}$$

In so doing, we may rewrite the coupled incompressible core equations (1) - (3) ($\beta \gg 1$) as

$$\frac{\partial \mathbf{z}^{\infty\pm}}{\partial t} + \mathbf{z}^{\infty\mp} \cdot \nabla \mathbf{z}^{\infty\pm} = -\frac{1}{\rho_0} \nabla \left(P^{\infty} + \frac{1}{2\mu_0} B^{\infty 2} \right), \tag{12}$$

and (4) - (6) ($\beta \sim 1$) as

$$\frac{\partial \mathbf{z}^{\infty\pm}}{\partial t} + \mathbf{z}^{\infty\mp} \cdot \nabla_{\perp} \mathbf{z}^{\infty\pm} = -\frac{1}{\rho_0} \nabla_{\perp} \left(P^{\infty} + \frac{1}{2\mu_0} B^{\infty 2} \right), \tag{13}$$

where, as a reminder, $\nabla_{\perp}$ refers to a coordinate system orthogonal to the mean magnetic field that is oriented along the $\hat{\mathbf{z}}$-axis.

For the $\beta \sim 1$ or $\ll 1$ regimes, we assume explicitly a strong mean magnetic field $\mathbf{B}_0$, which introduces a small turbulent Alfvénic Mach number M_{A0}. For $\beta \sim 1$, M_{A0} and the corresponding sonic Mach number based on the sound speed are of the same order. For the $\beta \gg 1$ regime, although not explicitly present, we do not rule out the presence of a large-scale of mean magnetic field $\mathbf{B}_0$, assuming only that the mean Alfvén velocity is of the same order as the fluctuating plasma velocity and the fluctuating magnetic field Alfvén velocity i.e., if we assume there exists $\mathbf{B}_0$ and express $\mathbf{B}^{\infty} = \mathbf{B}_0 + \mathbf{b}^{\infty}$, then we may rewrite (11) as

$$\mathbf{z}^{\infty\pm} = \mathbf{u}^{\infty} \pm \frac{\mathbf{b}^{\infty}}{\sqrt{\mu_0 \rho_0}} \pm \frac{\mathbf{B}_0}{\sqrt{\mu_0 \rho_0}} = \mathbf{u}^{\infty} \pm \mathbf{v}_A \pm \mathbf{V}_{A0} \equiv \mathbf{z}_f^{\infty\pm} \pm \mathbf{V}_{A0}. \tag{14}$$

Subsequently, the $\beta \gg 1$ equations (12) assume, for the homogeneous case, the familiar form

$$\frac{\partial \mathbf{z}_f^{\infty\pm}}{\partial t} \mp \mathbf{V}_{A0} \cdot \nabla \mathbf{z}_f^{\infty\pm} + \mathbf{z}_f^{\infty\mp} \cdot \nabla \mathbf{z}_f^{\infty\pm} = -\frac{1}{\rho_0} \nabla \left(P^{\infty} + \frac{1}{2\mu_0} B^{\infty 2} \right). \tag{15}$$

Equation (15) is the "standard" form of the incompressible MHD equations used typically to study turbulence in a variety of settings, including the solar wind and interstellar medium. However, we reemphasize that (15) is appropriate only under the condition $\beta \gg 1$.

In the presence of a large mean magnetic field ($\beta \sim 1$), the dominant fluctuating component is 2D and propagation effects are completely absent at leading-order, as illustrated in (13). Equation (13) indicates that all core modes have zero frequency and the interactions are purely "nonlinear" i.e., there is no mediation by Alfvénic wave packets [32; 33].

On rewriting equations (7) - (10), we obtain

$$\frac{\partial \mathbf{z}^{*\pm}}{\partial t} \mp \mathbf{V}_{A0} \cdot \nabla \mathbf{z}^{*\pm} + \mathbf{z}^{\infty\mp} \cdot \nabla \mathbf{z}^{*\pm} + \mathbf{z}^{*\mp} \cdot \nabla \mathbf{z}^{\infty\pm}$$
$$= -\frac{1}{\rho_0}\nabla\left(P^* + \frac{1}{\mu_0}\mathbf{B}_0 \cdot \mathbf{B}^* + \frac{1}{\mu_0}\mathbf{B}^\infty \cdot \mathbf{B}^*\right) \mp \left(\mathbf{V}_{A0} + \frac{\mathbf{B}^\infty}{\sqrt{\mu_0 \rho_0}}\right)\nabla \cdot \mathbf{u}_1. \quad (16)$$

Note that the last term on the right-hand-side of (16) vanishes if we restrict our attention to incompressible (transverse) NI modes.

Evidently, the NI Elsässer modes propagate at the Alfvén speed $\pm V_{A0}$ along the mean magnetic field. A very important difference between the $\beta \gg 1$ and $\beta \sim 1$ or $\ll 1$ NI MHD descriptions is that the dominant nonlinear interaction for the $\beta \sim 1$ or $\ll 1$ minority slab (see below) fluctuations is due to the coupling of minority Alfvénic fluctuations $\mathbf{z}^{*\pm}$ to the dominant convected 2D fluctuating component $\mathbf{z}^{\infty\pm}$ i.e., the $\beta \sim 1$ or $\ll 1$ NI Elsässer variables respond to the core $\mathbf{z}^{\infty\pm}$ fluctuations in a passive sense. This is quite unlike the nature of nonlinear interactions in the $\beta \gg 1$ regime since equations (15) show that both nonlinear $\mathbf{z}_f^{\infty\mp} \cdot \nabla \mathbf{z}_f^{\infty\pm}$ and propagation $\mp \mathbf{V}_{A0} \cdot \nabla \mathbf{z}_f^{\infty\pm}$ effects are present at the leading order. For the $\beta \sim 1$ or $\ll 1$ minority slab component, nonlinear couplings $\mathbf{z}^{*\mp} \cdot \nabla \mathbf{z}^{*\pm}$ and $\mathbf{z}^{*\pm} \cdot \nabla \mathbf{z}^{*\mp}$ enter only at the next higher order. Thus, the nonlinear cascade of energy in the low-frequency inertial range for slab fluctuations in a $\beta \sim 1$ or $\ll 1$ plasma is governed primarily by the dynamical time scale of the dominant advected 2D component. We show below that this conclusion is analogous in a sense to the conclusions of Shebalin et al. [34] who argued that the nonlinear cascade proceeded via the three-wave coupling of counter-propagating Alfvén waves and a zero frequency mode. However, the Shebalin et al. analysis [34] is appropriate only to the $\beta \gg 1$ regime, is based on a weak turbulence model, and assumed that the mean magnetic field lay within the 2D plane. The clear separation of the 2D and slab components described by equation (16) shows that the 2D core nonlinear time scale rather than the Alfvénic time scale dominates for a $\beta \sim 1$ or $\ll 1$ plasma, and therefore determines the basic spectral characteristics of the minority slab component. Since the spectral time scale is associated with the dominant 2D fluctuating component, it is not a resonant wave coupling that describes how $\mathbf{z}^{*\pm}$ behaves primarily; instead, $\mathbf{z}^{*\pm}$ responds as a passive scalar to the $\mathbf{z}^{\infty\pm}$ fluctuating field.

In the analysis below, we do not neglect the nonlinear terms that enter (16) at the next order. The higher-order nonlinear terms $\mathbf{z}^{*\mp} \cdot \nabla \mathbf{z}^{*\pm}$ enter the NI expansion (7) - (10) with the inclusion of terms such as $\mathbf{u}_1 \cdot \nabla \mathbf{u}_1$, $\mathbf{B}^* \cdot \nabla \mathbf{B}^*$, $\mathbf{u}_1 \cdot \nabla \mathbf{B}^*$, etc. A straightforward extension of ZM93 in which such terms are retained shows that these terms contribute higher-order but equivalent terms as those already contained in (16) and the only structurally new terms introduced are the nonlinear terms $\mathbf{z}^{*\mp} \cdot \nabla \mathbf{z}^{*\pm}$. The "richest" NI evolution equation should therefore include the nonlinear terms in (16) despite it being formally of higher order. Thus, the nonlinear NI transport equation is given by

$$\frac{\partial \mathbf{z}^{*\pm}}{\partial t} \mp \mathbf{V}_{A0} \cdot \nabla \mathbf{z}^{*\pm} + \mathbf{z}^{\infty\mp} \cdot \nabla \mathbf{z}^{*\pm} + \mathbf{z}^{*\mp} \cdot \nabla \mathbf{z}^{\infty\pm} + \mathbf{z}^{*\mp} \cdot \nabla \mathbf{z}^{*\pm}$$
$$= -\frac{1}{\rho_0}\nabla\left(P^* + \frac{1}{\mu_0}\mathbf{B}_0 \cdot \mathbf{B}^* + \frac{1}{\mu_0}\mathbf{B}^\infty \cdot \mathbf{B}^*\right) \mp \left(\mathbf{V}_{A0} + \frac{\mathbf{B}^\infty}{\sqrt{\mu_0 \rho_0}}\right)\nabla \cdot \mathbf{u}_1. \quad (17)$$

IOP Conf. Series: Journal of Physics: Conf. Series **900** (2017) 012023 doi:10.1088/1742-6596/900/1/012023

3. Properties of the $\beta \sim 1$ 2D incompressible MHD equations

We now restrict our attention to the $\beta \sim 1$ regime that is relevant to most of the solar wind. Following Zank et al. [24], we analyze equations (13) using a von Karman-Howarth approach [26; 35–37]. The nonlinear time scales are given by

$$\left(\tau_{\infty}^{\pm}\right)^{-1} \equiv \frac{\left\langle z^{\infty\mp 2}\right\rangle^{1/2}}{\lambda_{\perp}^{\pm}}, \tag{18}$$

[38–42], allowing us to approximate the quadratic nonlinearity in (13) as

$$\mathbf{NL}^{\infty\pm} \equiv \mathbf{z}^{\infty\pm}\frac{\left\langle z^{\infty\mp 2}\right\rangle^{1/2}}{\lambda_{\perp}^{\pm}}. \tag{19}$$

The length scale $\lambda_{\perp}^{\pm}$ is the correlation length in the 2D "forward" and "backward" modes. The total energy in fluctuations is

$$\left\langle z^{\infty 2}\right\rangle \equiv \frac{\left\langle z^{\infty+2}\right\rangle + \left\langle z^{\infty-2}\right\rangle}{2}, \tag{20}$$

where $\langle\bullet\rangle$ is an ensemble-averaging operator. For 2D turbulence we assume that

$$\frac{\left\langle z^{\infty-2}\right\rangle^{1/2}}{\lambda_{\perp}^{+}} = \frac{\left\langle z^{\infty+2}\right\rangle^{1/2}}{\lambda_{\perp}^{-}} \equiv \frac{\left\langle z^{\infty 2}\right\rangle^{1/2}}{2\lambda_{\perp}^{\infty}}, \tag{21}$$

which defines $\lambda_{\perp}^{\infty}$. The ensemble-averaged 1-point energy containing transport equation for the total energy is then

$$\frac{d}{dt}\left\langle z^{\infty 2}\right\rangle = -\frac{\left\langle z^{\infty 2}\right\rangle^{3/2}}{\lambda_{\perp}^{\infty}}. \tag{22}$$

We assume that the steady energy transfer flux $\Pi(\mathbf{k})$ ($\mathbf{k}$ is the wave number vector) is proportional to the time scale for the decay of the transfer function correlations τ_3 (also known as the triple correlation time scale) and that the energy-containing eddies determine the rate of spectral energy transfer in decaying turbulence. We can then adopt the phenomenological steady energy transfer rate model

$$\epsilon = \Pi(\mathbf{k}) = \tau_3\frac{\left\langle z^{\infty 2}\right\rangle}{\tau_{nl}^{2}}, \tag{23}$$

where ϵ is the dissipation rate [43]. The nonlinear dynamical time scale is denoted by τ_{nl} and, from (22) can be expressed as $\lambda_{\perp}^{\infty}/\left\langle z^{\infty 2}\right\rangle^{1/2}$. By assuming isotropy in the 2D plane orthogonal to the mean $\mathbf{B}$ and writing $\left\langle z^{\infty 2}\right\rangle = \int E^{\infty} dk_{\perp}$, it follows from (23) that

$$E^{\infty} = \epsilon_{\infty}^{2/3} k_{\perp}^{-5/3}, \tag{24}$$

where $k_\perp = |\mathbf{k}_\perp| = |(k_x, k_y)|$ is the length of the wave number vector in the 2D plane orthogonal to $B_0\hat{\mathbf{z}}$, and ϵ_∞ is the dissipation rate for 2D incompressible MHD turbulence. Thus, the energy spectrum of the dominant component in the $\beta \sim 1$ or $\ll 1$ NI MHD theory, when expressed in Elsässer variables, is of a Kolmogorov form in the 2D perpendicular wave number $k_\perp$ i.e., $\propto k_\perp^{-5/3}$.

The 2D isotropic spectral result can be generalized by using $\epsilon^\pm = \Pi^\pm(\mathbf{k}) = \tau_3^\pm \left\langle z^{\infty\pm 2} \right\rangle / \tau_{nl}^{\pm 2}$, where $\epsilon^\pm$ refers to the dissipation rates for forward and backward fluctuating Elsässer variables [39; 44]. Use of the nonlinear time scales (18) for $\tau_3^\pm$ then yields the Kolmogorov spectra for the intensities $\left\langle z^{\infty\pm 2} \right\rangle = \int E_k^{\infty\pm} dk_\perp$ as $E^{\infty\pm} = \left(\epsilon_\infty^\pm / \epsilon_\infty^{\mp\,1/2}\right)^{4/3} k_\perp^{-5/3}$ with $E^{\infty+}/E^{\infty-} = (\epsilon_\infty^+/\epsilon_\infty^-)^2$.

The core 2D MHD equations (4) - (6) admit a class of exact nonlinear solutions that are the analogue of hydrodynamic vortices [45–47].

On introducing the vector potential $\mathbf{A}^\infty$ by $\mathbf{B}^\infty \equiv \nabla \times \mathbf{A}^\infty = \nabla \times A^\infty \hat{z}$, the induction equation (6) shows that $\mathbf{A}^\infty$ is a passive scalar in the 2D flow, satisfying

$$\frac{\partial A^\infty}{\partial t} + \mathbf{u}^\infty \cdot \nabla A^\infty = 0. \tag{25}$$

The flow vorticity is expressed as $\xi^\infty \equiv \nabla \times \mathbf{u}^\infty$ and a flux function Ψ^∞ can be introduced through $\mathbf{u}^\infty \equiv \hat{z} \times \nabla \Psi^\infty$. It then follows that

$$\frac{\partial A^\infty}{\partial t} + \{\Psi^\infty, A^\infty\} = 0, \tag{26}$$

where $\{a, b\} \equiv (\partial a/\partial x)(\partial b/\partial y) - (\partial a/\partial y)(\partial b/\partial x)$ denotes a Poisson bracket. Since the current $\mathbf{J}^\infty$ is defined by $\mu_0 \mathbf{J}^\infty = \nabla \times \mathbf{B}^\infty$, for 2D flows we find that $\xi^\infty \parallel \mathbf{J}^\infty$ and

$$\frac{\partial \xi^\infty}{\partial t} + \mathbf{u}^\infty \cdot \nabla_\perp \xi^\infty = \frac{1}{\rho} \mathbf{B}^\infty \cdot \nabla \mathbf{J}^\infty. \tag{27}$$

Since $\xi_z^\infty = \nabla_\perp^2 \Psi^\infty$ and $\mu_0 J_z^\infty = -\nabla_\perp^2 A^\infty$, we can rewrite (27) as

$$\begin{gathered} \frac{\partial \xi_z^\infty}{\partial t} + \{\Psi^\infty, \xi_z^\infty\} = -\frac{1}{\rho}\{A^\infty, J_z^\infty\}; \\ \frac{\partial}{\partial t}\left(\nabla_\perp^2 \Psi^\infty\right) + \left\{\Psi^\infty, \nabla_\perp^2 \Psi^\infty\right\} = \frac{1}{\mu_0 \rho}\left\{A^\infty, \nabla_\perp^2\right\}. \end{gathered} \tag{28}$$

For non-propagating structures, equations (26) and (28) become

$$\begin{gathered} \{\Psi^\infty, A^\infty\} = 0; \\ \left\{\Psi^\infty, \nabla_\perp^2 \Psi^\infty\right\} = \frac{1}{\mu_0 \rho}\left\{A^\infty, \nabla_\perp^2\right\}, \end{gathered} \tag{29}$$

the first condition of which ensures that $A^\infty = f(\Psi^\infty)$ for an arbitrary function f. The second condition of (29) implies that

$$\nabla_\perp^2 \Psi^\infty = -\frac{1}{\rho} f' J_z^\infty + f_1(\Psi^\infty), \tag{30}$$

where f_1 is another arbitrary function. The simplest spherically symmetric solution of (30) is $A^\infty = \Psi^\infty$ (which implies $f' = 1$). For J_z^∞ linear in A^∞, say $\mu_0 J_z^\infty = k^2 A^\infty$ for some constant k, (30), written in polar coordinates, reduces to Bessel's equation of order 0,

$$\frac{d^2 A^\infty}{dx^2} + \frac{1}{x}\frac{dA^\infty}{dx} + A^\infty = 0, \tag{31}$$

after setting $x = \alpha r$, where $\alpha = k/\sqrt{\mu_0 \rho}$. Solutions are therefore $A^\infty(r) = C J_0\left(k/\sqrt{\mu_0 \rho} r\right)$, where J_0 is Bessel's function of order 0. The parameter k is chosen to ensure that $A(r_0) = 0$. This then yields the vortical solution

$$A^\infty(r) = \begin{cases} C J_0\left(\frac{k}{\sqrt{\mu_0 \rho}} r\right) & r < r_0 \\ 0 & r \geq r_0 \end{cases}. \tag{32}$$

Here r is the polar coordinate distance with origin at the center of the vortex structure, r_0 is the edge of the vortex, and C and k are constants. The vorticity $\xi^\infty \equiv \nabla \times \mathbf{u}^\infty$ and current $\mu_0 \mathbf{J}^\infty \equiv \nabla \times \mathbf{B}^\infty$ are aligned and parallel to the large-scale magnetic field $\mathbf{B}_0$. Since the contours of $\mathbf{A}^\infty$ determined by (32) correspond to closed magnetic field lines, the vortex solutions (32) are a subclass of magnetic islands in the 2D plane perpendicular to the mean field $\mathbf{B}_0$.

The existence of 2D magnetic islands or vortex structures is an explicit prediction of $\beta \sim 1$ or $\ll 1$ NI MHD. Observations of magnetic islands on a variety of scales in the supersonic solar wind have been reported, ranging from hourly time scales [48–50] to minute time scales [46] to ion kinetic scales [51]. In the last case, the plasma beta of the solar wind environment in which vortex/magnetic island structures were embedded was $\beta < 1$ ($\sim 0.5 - 0.7$).

4. Properties of the NI transport equation for $\mathbf{z}^*$

Consider now the minority fluctuating component described by the NI corrections (7) - (10). It is illuminating to consider a "linearized" form of equations (7) - (10) in which the nonlinear coupling terms between the core and nearly incompressible corrections, such as $\mathbf{u}^\infty \cdot \nabla \mathbf{u}_1$, etc., are neglected. We neglect also the incompressible source terms. Equations (7) - (10) then reduce to a linear system in the NI variables

$$\frac{\partial \rho^*}{\partial t} + \rho_0 \nabla \cdot \mathbf{u}_1 = 0; \tag{33}$$

$$\frac{\partial \mathbf{u}_1}{\partial t} = -\frac{1}{\rho_0}\nabla P^* - \frac{1}{\mu_0 \rho_0}(\nabla \times \mathbf{B}^*) \times \mathbf{B}_0; \tag{34}$$

$$\frac{\partial \mathbf{B}^*}{\partial t} = \mathbf{B}_0 \nabla \cdot \mathbf{u}_1 - \mathbf{B}_0 \nabla \cdot \mathbf{u}_1 = \nabla \times (\mathbf{u}_1 \times \mathbf{B}_0); \tag{35}$$

$$\frac{\partial P^*}{\partial t} + \gamma p_0 \nabla \cdot \mathbf{u}_1 = 0. \tag{36}$$

Strictly speaking, equations (33) - (36) do not simply represent a linearization of the compressible MHD equations since terms of similar order such as P^∞ and $\mathbf{B}^\infty$ are excluded. On assuming that $\mathbf{B}_0 = B_0\hat{\mathbf{z}}$, we obtain the wave equation [52]

$$\frac{\partial^2 \Delta}{\partial t^2} = C_{s0}^2 \nabla^2 \Delta + V_{A0}^2 \nabla^2 (\Delta - \Gamma), \tag{37}$$

where $\Delta \equiv \nabla \cdot \mathbf{u}_1$ and $\Gamma \equiv \partial u_{1z}/\partial z$. The z-component of the NI vorticity $\nabla \times \mathbf{u}_1$, $\xi \equiv -\partial u_{1x}/\partial y + \partial u_{1y}/\partial x$, satisfies

$$\frac{\partial^2 \xi}{\partial t^2} = V_{A0}^2 \frac{\partial^2 \xi}{\partial z^2}, \tag{38}$$

showing that ξ propagates along the mean magnetic field at the Alfvén speed V_{A0}. By contrast, the gradient in $\mathbf{u}_1$ along the magnetic field direction Γ is coupled to the compressive component of the NI velocity $\mathbf{u}_1$, propagating at the sound speed according to

$$\frac{\partial^2 \Gamma}{\partial t^2} = C_{s0}^2 \frac{\partial^2 \Delta}{\partial z^2}. \tag{39}$$

Equations (37) - (39) completely determine $\mathbf{u}_1$. However, it is more useful to introduce the x- and y-components of the vorticity

$$\eta \equiv \frac{\partial u_{1z}}{\partial y} - \frac{\partial u_{1y}}{\partial z}; \qquad \zeta \equiv \frac{\partial u_{1x}}{\partial z} - \frac{\partial u_{1z}}{\partial x},$$

since

$$\frac{\partial^2 \eta}{\partial t^2} = V_{A0}^2 \left(\frac{\partial^2 \eta}{\partial z^2} - \frac{\partial^2 \Delta}{\partial y \partial z} \right); \qquad \frac{\partial^2 \zeta}{\partial t^2} = V_{A0}^2 \left(\frac{\partial^2 \zeta}{\partial z^2} - \frac{\partial^2 \Delta}{\partial x \partial z} \right). \tag{40}$$

The x- and y-vortical components are compressive and propagate obliquely to the mean magnetic field. The NI fluctuations are therefore a superposition of incompressible and compressible modes. If we restrict our attention to incompressible NI fluctuations satisfying $\Delta = 0$ i.e., $\nabla \cdot \mathbf{u}_1 = 0$, then equations (40) are identical to (38),

$$\frac{\partial^2 \eta}{\partial t^2} = V_{A0}^2 \frac{\partial^2 \eta}{\partial z^2}; \qquad \frac{\partial^2 \zeta}{\partial t^2} = V_{A0}^2 \frac{\partial^2 \zeta}{\partial z^2}. \tag{41}$$

Incompressible NI modes therefore all propagate at the Alfvén speed in the parallel direction. The minority incompressible fluctuating component is therefore an admixture of counter-propagating Alfvén modes i.e., slab turbulence, that are coupled nonlinearly to the zero-frequency, non-propagating 2D fluctuations described by the core equations (4) - (6).

Consider now the higher-order nonlinearity in (17) when the passive scalar terms are neglected. Since $\mathbf{z}^{*\mp} \cdot \nabla \mathbf{z}^{*\pm}$ enters formally at the next order in the NI expansion, we may express $\mathbf{z}^{*\pm} = \mathbf{z}_1^{*\pm} + \mathbf{z}_2^{*\pm}$, where $\mathbf{z}_2^{*\pm}$ is of higher order such that $O(\mathbf{z}_2^{*\pm}) = O(\mathbf{z}_1^{*\mp} \cdot \nabla \mathbf{z}_1^{*\pm})$. Then, neglecting the passive scalar terms and the RHS of (17), we have

$$\frac{\partial \mathbf{z}_1^{*\pm}}{\partial t} \mp \mathbf{V}_{A0} \cdot \nabla \mathbf{z}_1^{*\pm} = 0; \tag{42}$$

$$\frac{\partial \mathbf{z}_2^{*\pm}}{\partial t} \mp \mathbf{V}_{A0} \cdot \nabla \mathbf{z}_2^{*\pm} = -\mathbf{z}_1^{*\mp} \cdot \nabla \mathbf{z}_1^{*\pm}. \tag{43}$$

Equation (42) is a linear wave equation with solutions $\mathbf{z}_1^{*\pm} \propto \exp[i\mathbf{k}\cdot\mathbf{x} - i\omega^{\pm}t]$ provided the frequency ω and wave number $\mathbf{k}$ are related via the dispersion relation $\omega^{\pm} = \mp\mathbf{V}_{A0}\cdot\mathbf{k}$. Equation (43) is a wave equation too with a source term. On expressing

$$\mathbf{z}_1^{*\pm} = \sum_k \mathbf{A}_k^{\pm} e^{i\mathbf{k}\cdot(\mathbf{x}\pm\mathbf{V}_{A0}t)},$$

and substituting in the RHS of (43), secular terms are absent provided the resonance conditions

$$\mathbf{k}''' = \mathbf{k}' + \mathbf{k}''; \qquad \mathbf{k}'''\cdot\mathbf{V}_{A0} = -\mathbf{k}'\cdot\mathbf{V}_{A0} + \mathbf{k}''\cdot\mathbf{V}_{A0}, \tag{44}$$

hold. Condition (44) is identical to the resonance condition found by [34] (see also [53]) from their perturbation analysis of the incompressible 3D MHD ($\beta \gg 1$) equations. The distinction here from [34] is that the resonance condition holds only for the NI Elsässer variables i.e., the slab component, and the perturbation ordering is a natural consequence of the NI expansion. Equation (44) implies of course that

$$\mathbf{k}'\cdot\mathbf{V}_{A0} = 0 \quad \Leftrightarrow \quad \mathbf{k}' \perp \mathbf{B}_0, \tag{45}$$

meaning that the nonlinear cascade of energy for slab turbulence proceeds via the coupling of forward and backward propagating Alfvénic fluctuations with 2D zero frequency modes. This is consistent with the standard picture of the cascade process in MHD turbulence [34; 54]. At the NI level, the nonlinear cascade of energy therefore drives 2D modes, which are then transferred to the dissipation range and heat the plasma. Consequently, the NI MHD 2D core equations acquire a source term corresponding to the nonlinear loss term in the slab NI description, thereby coupling the leading-order 2D and higher-order NI correction descriptions.

Slab turbulence spectra can be derived from (17). The forward and backward intensities $\left\langle z^{*\pm 2}\right\rangle$ satisfy the 1-point transport equations

$$\frac{\partial}{\partial t}\left\langle z^{*\pm 2}\right\rangle \mp \mathbf{V}_{A0}\cdot\nabla\left\langle z^{*\pm 2}\right\rangle \simeq -2\frac{\left\langle z^{*\pm 2}\right\rangle\left\langle z^{\infty\mp 2}\right\rangle^{1/2}}{\lambda_{\perp}^{\pm}} - 2\frac{\left\langle z^{*\mp 2}\right\rangle^{1/2}\left\langle z^{*\pm 2}\right\rangle^{1/2}}{\lambda_{*}^{\pm}}, \tag{46}$$

after invoking orthogonality between the NI and core Elsässer variables i.e., $\langle \mathbf{z}^{*\pm}\cdot\mathbf{z}^{\infty\pm}\rangle \simeq 0$. Consider the total energy

$$\left\langle z^{*2}\right\rangle \equiv \frac{\left\langle z^{*+2}\right\rangle + \left\langle z^{*-2}\right\rangle}{2}.$$

On exploiting the Kolmogorov phenomenology, we have instead of (23),

$$\epsilon_* = \Pi(\mathbf{k}) = \tau_3 \frac{\left\langle z^{*2}\right\rangle}{\tau_{nl}^2}. \tag{47}$$

We now need to include the passive scalar coupling of the 2D core variable $\left\langle z^{\infty 2}\right\rangle$ to $\left\langle z^{*2}\right\rangle$. The nonlinear coupling terms enter at the next higher order, as expressed in equation (17), and this corresponds to τ_{nl} in (47).

We should not assume that $\left\langle z^{*2}\right\rangle$ is isotropic[1], so we define

$$\begin{aligned} \left\langle z^{*2}\right\rangle &= \frac{1}{2\pi}\int E^*(\mathbf{k})d^3\mathbf{k} = \frac{1}{2\pi}\int\int\int E^*(\mathbf{k}_\perp, k_\parallel)d\mathbf{k}_\perp dk_\parallel \\ &= \int\int E^*(k_\perp, k_\parallel)k_\perp dk_\perp dk_\parallel \sim E^*(k_\perp, k_\parallel)k_\perp^2 k_\parallel, \end{aligned} \tag{48}$$

on assuming isotropy in the 2D plane. Consider three cases. (i) If we assume $\tau_3^{-1} = \tau_{nl}^{-1} = \left\langle z^{*2}\right\rangle^{1/2} k_\perp$, then, from (47), we obtain

$$E^*(k_\perp, k_\parallel)k_\perp^2 = \epsilon_*^{2/3} k_\perp^{-2/3} k_\parallel^{-1}, \tag{49}$$

which is a slight generalization of the Kolmogorov spectrum. (ii) If instead we assume that $\tau_3^{-1} = \tau_A^{-1} = V_A k_\parallel$ and $\tau_{nl}^{-1} = \left\langle z^{*2}\right\rangle^{1/2} k_\perp$, we recover the extended Iroshnikov-Kraichnan (IK) spectrum

$$E^*(k_\perp, k_\parallel)k_\perp^2 = (\epsilon_* V_A)^{1/2} k_\perp^{-1} k_\parallel^{-1/2}. \tag{50}$$

(iii) We do however need to incorporate the dominant passive scalar interaction of $\left\langle z^{*2}\right\rangle$ with $\left\langle z^{\infty 2}\right\rangle$. As before,

$$\left\langle z^{\infty 2}\right\rangle = \frac{1}{2\pi}\int \tilde{E}^\infty(\mathbf{k}_\perp)d\mathbf{k}_\perp = \int \tilde{E}^\infty(k_\perp)k_\perp dk_\perp \sim \tilde{E}^\infty(k_\perp)k_\perp^2 \equiv E^\infty(k_\perp)k_\perp, \tag{51}$$

on assuming isotropy of $\left\langle z^{\infty 2}\right\rangle$ in the 2D plane and relating $\tilde{E}^\infty(k_\perp)$ to $E^\infty(k_\perp)$ given by equation (24). Following the suggestion of [43; 55; 56], we introduce the triple correlation time as the sum of the relevant inverse time scales, of which the dominant ones are the passive scalar term $\tau_s^{-1} \equiv \left\langle z^{\infty 2}\right\rangle^{1/2} k_\perp = \epsilon_\infty^{1/3} k_\perp^{-5/6} k_\perp^{3/2}$ and the Alfvén term $\tau_A^{-1} = V_A k_\parallel$, i.e.,

$$\tau_3^{-1} = V_A k_\parallel + \epsilon_\infty^{1/3} k_\perp^{2/3} = V_A k_\parallel \left(1 + \left(\frac{\epsilon_\infty}{V_A^3}\right)^{1/3} k_\perp^{2/3} k_\parallel^{-1}\right). \tag{52}$$

In (52), the term

$$\left(\frac{\epsilon_\infty}{V_A^3}\right)^{1/3} k_\perp^{2/3} k_\parallel^{-1} \equiv k_A^{1/3} k_\perp^{2/3} k_\parallel^{-1}, \tag{53}$$

is the critical balance parameter identified by [57], although here with a quite different physical interpretation. In our case ($\beta \sim 1$ or $\ll 1$), if the term (53) is of order 1 ("critical balance"), then the energy flux in wave number space is a consequence of a

[1] To put the analysis here into perspective, suppose that $\left\langle z^{*2}\right\rangle$ is isotropic. In this case, $\left\langle z^{*2}\right\rangle = (4\pi)^{-1}\int \tilde{E}^*(\mathbf{k})d^3\mathbf{k} = (4\pi)^{-1}\int\int\int \tilde{E}^*(k)k^2 \sin\theta dk d\theta d\phi = \int \tilde{E}^*(k)k^2 dk \equiv \int E^*(k)dk \sim E^*(k)k$, where $E^*(k)$ is the omni-directional or reduced spectrum. We utilize equation (47). (i) Suppose $\tau_3^{-1} = \tau_{nl}^{-1} = \left\langle z^{*2}\right\rangle^{1/2} k$, which implies $\tilde{E}^*(k) = \epsilon_*^{2/3} k^{-11/3}$ or $E^*(k) = \epsilon_*^{2/3} k^{-5/3}$, which is the Kolmogorov spectrum. (ii) If $\tau_3^{-1} = \tau_A^{-1} = V_A k$, then $\tilde{E}(k) = (\epsilon_* V_A)^{1/2} k^{-7/2}$ or $E^*(k) = (\epsilon_* V_A)^{1/2} k^{-3/2}$, i.e., the Iroshnikov-Kraichnan spectrum. However, there is no good reason to suppose that NI turbulence should be isotropic.

IOP Conf. Series: Journal of Physics: Conf. Series **900** (2017) 012023 doi:10.1088/1742-6596/900/1/012023

balance of Alfvén wave sweeping and passive scalar convection by leading-order 2D $\mathbf{z}^\infty$ fluctuations.

The phenomenological steady-state energy transfer rate equation (47), yields the NI or slab spectrum as

$$E^*(k_\perp, k_\parallel)k_\perp^2 = (\epsilon_* V_A)^{1/2} k_\perp^{-1} k_\parallel^{-1/2} \left(1 + \left(\frac{\epsilon_\infty}{V_A^3}\right)^{1/3} k_\perp^{2/3} k_\parallel^{-1}\right)^{1/2}. \tag{54}$$

If $\epsilon_\infty^{1/3} V_A^{-1} k_\perp^{2/3} k_\parallel^{-1} \sim$ const. or $\ll 1$, we recover the IK spectrum (50) from (54). Conversely, if $\epsilon_\infty^{1/3} V_A^{-1} k_\perp^{2/3} k_\parallel^{-1} \gg 1$, then $E^*(k_\perp, k_\parallel)k_\perp^2 = \epsilon_*^{1/2} \epsilon_\infty^{1/6} k_\perp^{-2/3} k_\parallel^{-1}$, and we obtain the Kolmogorov spectrum from (54).

Since NI MHD is a superposition of the core 2D MHD equations and the NI corrections, the total turbulent energy spectrum may be expressed as $\langle z^2 \rangle \equiv \langle z^{\infty 2} \rangle + \langle z^{*2} \rangle$. On using $\langle z^{\infty 2} \rangle = \int \tilde{E}^\infty(\mathbf{k}_\perp)\delta(k_\parallel) d\mathbf{k}_\perp dk_\parallel$, the total spectrum (core plus NI) is given by

$$E(k_\perp, k_\parallel)k_\perp^2 = C_\infty \epsilon_\infty^{2/3} k_\perp^{-2/3} k_\parallel^{-1} + C_* (\epsilon_* V_A)^{1/2} k_\perp^{-1} k_\parallel^{-1/2} \left(1 + \left(\frac{\epsilon_\infty}{V_A^3}\right)^{1/3} k_\perp^{2/3} k_\parallel^{-1}\right)^{1/2}, \tag{55}$$

where C_∞ and C_* are constants reflecting the assumed ratio of 2D to slab turbulence. The majority 2D component always contributes a Kolmogorov power law spectrum $k_\perp^{-5/3}$, whereas the minority component contributes a "broken power law" distribution along both the $k_\perp$ or $k_\parallel$ directions. The break in the spectrum is determined by the parameter k_A and the values of $k_\perp$ or $k_\parallel$ relative to it. The total spectrum, being the sum of the dominant 2D and minority slab components, is therefore not a power law necessarily but can possess a somewhat complex "broken power law" structure.

5. The fluctuating density variance determined from NI MHD

NI MHD allows us describe the basic physics and dynamics of density fluctuations in a turbulent flow, including the determination of the fluctuating density variance spectrum. For incompressible NI fluctuations satisfying $\nabla \cdot \mathbf{u}_1 = 0$, the homogeneous density transport equation (7) describes the passive scalar advection of density fluctuations,

$$\frac{\partial \rho^*}{\partial t} + \mathbf{u}^\infty \cdot \nabla \rho^* = 0. \tag{56}$$

The density fluctuations are purely entropic and the passive scalar behavior is due to convection by the dominant core incompressible velocity fluctuations for all plasma beta regimes. This is in marked contrast to the conjecture by Lithwick and Goldreich [58] that entropic density modes are "passively mixed by the cascade of shear Alfvén waves" - see also the related discussion by [59]. The transport equation for the density fluctuation variance $\langle \rho^{*2} \rangle$ is given by

$$\frac{d}{dt}\left\langle \rho^{*2} \right\rangle = -\left\langle \mathbf{u}^\infty \cdot \nabla \rho^{*2} \right\rangle \simeq -\frac{\langle u^{\infty 2} \rangle^{1/2} \langle \rho^{*2} \rangle}{\ell_u}, \tag{57}$$

where we have introduced a correlation length ℓ_u appropriate to the core 2D incompressible velocity fluctuations $\mathbf{u}^\infty$. Since $\langle u^{\infty 2}\rangle = \left(\left\langle u^{\infty +2}\right\rangle + \left\langle u^{\infty -2}\right\rangle\right)/2$ and the residual energy $E_D^\infty \equiv \langle \mathbf{z}^{\infty +} \cdot \mathbf{z}^{\infty -}\rangle = \langle u^{\infty 2}\rangle - \langle B^{\infty 2}/(\mu_0\rho_0)\rangle$, we have

$$\langle u^{\infty 2}\rangle = \frac{\langle z^{\infty 2}\rangle + E_D^\infty}{2}. \tag{58}$$

Consider three cases. (i) $E_D^\infty = 0$ implies $\langle u^{\infty 2}\rangle = \langle B^{\infty 2}/(\mu_0\rho_0)\rangle$ and hence $\langle u^{\infty 2}\rangle = \langle z^{\infty 2}\rangle/2$. (ii) The turbulent kinetic energy dominates i.e., $E_D^\infty \simeq \langle u^{\infty 2}\rangle$ and $\langle B^{\infty 2}/(\mu_0\rho_0)\rangle = 0$, which implies $\langle u^{\infty 2}\rangle = \langle z^{\infty 2}\rangle/2$. Finally, (iii) if the magnetic energy dominates, then $\langle u^{\infty 2}\rangle \simeq 0$, in which case there is no passive scalar convection of the entropic density fluctuation. This result is interesting because, as shown by Adhikari et al. [60], solar wind turbulence beyond ~ 2 AU becomes increasingly dominated by the magnetic energy, and by ~ 10 AU, the normalized residual energy $\sigma_D = -1$ (see also [37] and references therein). Thus, within ~ 2 AU, density turbulence in the solar wind evolves dynamically as a passive scalar, being mixed by the dominant 2D turbulent velocity component. Beyond 2 AU, the density turbulence "freezes" into a non-evolving statistical state, until about 10 AU when pickup ion driven turbulence [37; 60; 61] begins to generate turbulent velocity fluctuations in the outer heliosphere. In this region, the density turbulence "thaws" and begins to evolve dynamically again. This is discussed further below when we extend these results to the inhomogeneous solar wind.

For either $E_D^\infty = 0$ or kinetic energy dominated MHD turbulence, equation (57) can be approximated as

$$\frac{d}{dt}\left\langle \rho^{*2}\right\rangle \simeq -\frac{\langle z^{\infty 2}\rangle^{1/2}\langle \rho^{*2}\rangle}{2\ell_u} = -\frac{\epsilon_\rho}{2}, \tag{59}$$

where ϵ_ρ is the density dissipation rate. Consider again $\beta \sim 1$ or $\ll 1$ so that the core incompressible majority description is 2D. For 2D isotropic density turbulence, $\langle \rho^{*2}\rangle = \int E_\rho dk_\perp \sim E_\rho(k_\perp)k_\perp$. On using $E^\infty(k_\perp) = \epsilon_\infty^{2/3} k_\perp^{-5/3}$, it follows that

$$E_\rho(k_\perp) = \epsilon_\rho \epsilon_\infty^{-1/3} k_\perp^{-5/3}, \tag{60}$$

i.e., the density variance spectrum is Kolmogorov. Alternatively, if we do not assume isotropy, we may, as before, express

$$\left\langle \rho^{*2}\right\rangle = \int\int E_\rho(k_\perp, k_\parallel)k_\perp dk_\perp dk_\parallel \simeq E_\rho(k_\perp, k_\parallel)k_\perp^2 k_\parallel. \tag{61}$$

On using $\langle z^{\infty 2}\rangle \simeq \tilde{E}^\infty(k_\perp)k_\perp^2 = E^\infty(k_\perp)k_\perp$ as above, together with $\epsilon_\rho = \langle z^{\infty 2}\rangle^{1/2} k_\perp E_\rho(k_\perp, k_\parallel)k_\perp^2 k_\parallel$, we obtain

$$E_\rho(k_\perp, k_\parallel)k_\perp^2 = \epsilon_\rho \epsilon_\infty^{-1/3} k_\perp^{-2/3} k_\parallel^{-1}. \tag{62}$$

The results (60) and (62) show that the density variance spectrum is of a Kolmogorov form for plasma beta regimes ~ 1 or $\ll 1$, providing a possible explanation for the observed interstellar electron density spectrum [18–21].

IOP Conf. Series: Journal of Physics: Conf. Series **900** (2017) 012023 doi:10.1088/1742-6596/900/1/012023

6. Conclusions

Zank et al. [24] presented a detailed development of the theory of nearly incompressible MHD turbulence for homogeneous and inhomogeneous flows. Here we summarize the [24] results for homogeneous NI MHD, extending the results to the $\beta \gg 1$ regime and contrasting these with the $\beta \sim 1$ and $\ll 1$ results. We re-expressed the original dimensionless form of the NI MHD equations in dimensional units. For $\beta \gg 1$, NI MHD describes a majority incompressible component that is fully 3D and can admit low-frequency Alfvén waves together with a NI component that is essentially slab in the incompressible limit. By contrast, in the case of plasma beta values $\beta \sim O(1)$ or $\ll 1$, we find that NI MHD describes a superposition of majority 2D and minority slab fluctuations. This regime appears to be appropriate to the solar wind, unlike the $\beta \gg 1$ case. We therefore focused more attention on the $\beta \sim 1$ regime. Our conclusions are enumerated as follows.

(i) For $\beta \gg 1$, the dominant turbulence component is described by the "standard" 3D incompressible MHD equations. If a mean magnetic field $\mathbf{B}_0$ is present, the majority component of the Elsässer form of the 3D incompressible MHD equations includes the usual Alfvén velocity propagation term. This system is very well studied, although it is frequently forgotten that it is valid only in the $\beta \gg 1$ regime.

(ii) For $\beta \sim 1$ or $\ll 1$, the majority fluctuating component is fully 2D in planes perpendicular to the large-scale mean magnetic field, being advected by the large-scale background flow. The Alfvén velocity is absent, ensuring that the majority component interaction occurs only between 2D modes, ensuring that the spectrum for the inertial range of the energy density for Elsässer modes is Kolmogorov in the perpendicular wave number $k_\perp = |\mathbf{k}_\perp|$, i.e., $\propto k_\perp^{-5/3}$.

(iii) For $\beta \sim 1$ or $\ll 1$, 2D magnetic islands or vortex structures are a nonlinear component of the dominant core 2D incompressible MHD fluctuations.

(iv) For all plasma beta regimes, the homogeneous NI transport equations show that the component of vorticity parallel to the mean magnetic field satisfies the 1D Alfvén wave equation. The remaining NI modes are magnetosonic. By considering only the incompressible NI corrections, we show that all three components of the vorticity satisfy a 1D Alfvén wave equation, illustrating that the incompressible NI corrections correspond to a minority slab component. NI MHD in the $\beta \sim 1$ or $\ll 1$ limits is therefore a superposition of 2D plus slab fluctuations.

(v) For $\beta \sim 1$ or $\ll 1$, the NI Elsässer formulation shows that the slab components are mixed passively by the coupling of the higher-order NI corrections to the leading order 2D incompressible fluctuations, this being the dominant nonlinear interaction. The nonlinear interaction of counter-propagating Alfvén wave packets, which are included in the higher-order NI transport equation, is via the generation of zero frequency 2D modes [34]. This nonlinear interaction is a source of 2D fluctuations in the leading-order 2D incompressible description, serving to couple the NI description back to the majority component for the $\beta \sim 1$ or $\ll 1$ regimes.

(vi) In considering the spectral characteristics of the slab component for $\beta \sim 1$ or $\ll 1$, we identified the inverse triple correlation time as the sum of the inverse passive scalar time scale associated with the leading-order 2D Elsässer variables and the inverse Alfvén time scale. The "critical balance" parameter emerges from the triple correlation time as the term that determines whether the slab energy flux in wave number space is dominated by either passive scalar convection by leading-order 2D Elsässer fluctuations or Alfvén wave sweeping. This is a quite different physical interpretation of the critical balance parameter than that of Goldreich and Sridhar [57] and does not emerge as a conjecture.

(vii) Use of the triple correlation time scale and the leading order 2D spectrum allows us to derive the anisotropic slab energy spectrum $E^*(k_\perp, k_\parallel)k_\perp^2$ (equation (54)), in which the critical balance parameter enters by demarcating the wave number regime that has a Kolmogorov spectrum from the region with an IK spectrum. The complete wave number spectrum can be expressed as a superposition of the majority 2D and minority slab spectra (equation (55)). The composite Elsässer energy spectrum is anisotropic and does not possess a simple power law in either $k_\perp$ or $k_\parallel$, sometimes exhibiting a somewhat complex "broken power law" structure.

(viii) Although NI density fluctuations enter at different orders in the turbulent Mach number for the homogeneous and inhomogeneous formulations ($O(M^2)$ and $O(M)$ respectively), NI MHD shows that density fluctuations behave as passive scalars in response to advection by the majority incompressible velocity fluctuations. This is in contrast to the conjecture [58] that ascribes the mixing of density fluctuations to shear Alfvén waves. For $\beta \sim 1$ or $\ll 1$, the predicted isotropic and anisotropic wave number spectra for the fluctuating density variance are $\propto k_\perp^{-5/3}$ (equation (60)) or $\propto k_\perp^{-2/3} k_\parallel^{-1}$ (equation (62)) respectively. If, at some point, the majority 2D component of the magnetic energy dominates the kinetic energy, i.e., $\sigma_D^\infty \simeq -1$, the density turbulence "freezes" into a non-evolving statistical state. The predicted density spectra are a possible explanation for the observed Kolmogorov-like interstellar density spectrum.

Acknowledgments
We acknowledge the partial support of NASA grants NNX08AJ33G, Subaward 37102-2, NNX14AC08G, NNX14AJ53G, A99132BT, RR185-447/4944336 and NNX12AB30G. GPZ thanks R. Bruno for his kind hospitality while visiting the INAF-IAPS. This work is partly supported by the International Space Science Institute (ISSI) in the framework of International Team 504 entitled Current Sheets, Turbulence, Structures and Particle Acceleration in the Heliosphere.

References
[1] Klainerman S and Majda A 1981 *Communications in Pure Applied Mathematics* 34 481–524
[2] Klainerman S and Majda A 1982 *Communications in Pure Applied Mathematics* 35 629-651
[3] Montgomery D, Brown M R and Matthaeus W H 1987 *J. Geophys. Res.* 92 282–284
[4] Matthaeus W H and Brown M R 1988 *Physics of Fluids* 31 3634–3644
[5] Zank G P and Matthaeus W H 1990 *Physical Review Letters* 64 1243–1246
[6] Zank G P, Matthaeus W H and Klein L W 1990 *Geophys. Res. Lett.* 17 1239–1242
[7] Zank G P and Matthaeus W H 1991 *Physics of Fluids* 3 69–82
[8] Matthaeus W H, Klein L W, Ghosh S and Brown M R 1991 *J. Geophys. Res.* 96 5421–5435
[9] Zank G P and Matthaeus W H 1992 *J. Plasma Phys.* 48 85
[10] Zank G P and Matthaeus W H 1992 *J. Geophys. Res.* 97 17189

[11] Zank G P and Matthaeus W H 1993 *Phys. Fluids* 5 257–273
[12] Bayly B J, Levermore C D and Passot T 1992 *Physics of Fluids* 4 945–954
[13] Ghosh S and Matthaeus W H 1992 *Physics of Fluids* 4 148–164
[14] Bhattacharjee A, Ng C S and Spangler S R 1998 *ApJ* 494 409
[15] Hunana P, Zank G P and Shaikh D 2006 *Phys. Rev. E* 74 026302
[16] Hunana P, Zank G P, Heerikhuisen J and Shaikh D 2008 *J. Geophys. Res.* 113 A11105
[17] Hunana P and Zank G P 2010 *ApJ* 718 148–167
[18] Armstrong J W, Cordes J M and Rickett B J 1981 *Nature* 291 561–564
[19] Armstrong J and Woo R 1980 *Rep. IOM* 3331-80-070
[20] Spangler S R and Armstrong J W 1990 *Low Frequency Astrophysics from Space (Lecture Notes in Physics, Berlin Springer Verlag vol 362)* ed Kassim N E and Weiler K W pp 155–164
[21] Armstrong J W, Rickett B J and Spangler S R 1995 ApJ 443 209–221
[22] Bavassano B and Bruno R 1995 *J. Geophys. Res.* 100 9475–9480
[23] Bruno R and Carbone V 2013 *Living Reviews in Solar Physics* 10
[24] Zank G P, Adhikari L, Hunana P, Shiota D, Bruno R and Telloni D 2017 *ApJ* 835 147
[25] Adhikari L, Zank G P, Hunana P, Shiota D, Bruno R, Hu Q and Telloni D 2017 *ApJ* 841-85
[26] Zank G P, Dosch A, Hunana P, Florinski V, Matthaeus W H and Webb G M 2012 *ApJ* 745 35
[27] Kreiss H O 1980 *Communications on Pure and Applied Mathematics* 33 399–439
[28] Ghosh S and Parashar T N 2015 *Physics of Plasmas* 22 042302
[29] Bieber J W, Wanner W and Matthaeus W H 1996 *J. Geophys. Res.* 101 2511–2522
[30] Belcher J W and Davis Jr L 1971 *J. Geophys. Res.* 76 3534–3563
[31] Parashar T, Oughton S, Matthaeus W and Wan M 2016 ApJ 600 763–775
[32] Fyfe D and Montgomery D 1976 *Journal of Plasma Physics* 16 181–191
[33] Fyfe D, Montgomery D and Joyce G 1977 *Journal of Plasma Physics* 17 369–398
[34] Shebalin J V, Matthaeus W H and Montgomery D 1983 *Journal of Plasma Physics* 29 525–547
[35] von Karman T and Howarth L 1938 *Proc. R. Soc. London, Ser. A* 164 192–215
[36] Matthaeus W H, Oughton S, Pontius Jr D H and Zhou Y 1994 *J. Geophys. Res.* 99 19267
[37] Zank G P, Matthaeus W H and Smith C W 1996 *J. Geophys. Res.* 101 17093–17108
[38] Pouquet A, Frisch U and Leorat J 1976 *J. Fluid Mech.* 77 321–354
[39] Dobrowolny M, Mangeney A and Veltri P 1980 *Physical Review Letters* 45 144–147
[40] Grappin R, Frisch U, Pouquet A and Leorat J 1982 *Astron. Astroph.* 105 6–14
[41] Grappin R, Leorat J and Pouquet A 1983 *Astron. Astroph.* 126 51–58
[42] Matthaeus W H and Zhou Y 1989 *Phys. Fluids B* 1 1929–1931
[43] Zhou Y, Matthaeus W H and Dmitruk P 2004 *Reviews of Modern Physics* 76 1015–1035
[44] Marsch E 1990 *Turbulence in the solar wind (Springer Verlag Berlin)* p 43 Reviews of Modern Astronomy
[45] Kadomtsev B B and Pogutse O P 1973 *Zhurnal Eksperimentalnoi i* Teoreticheskoi Fiziki 65 575–589
[46] Verkhoglyadova O P, Dasgupta B and Tsurutani B T 2003 *Nonlinear Processes in Geophysics* 10 335–343
[47] Alexandrova O 2008 *Nonlinear Processes in Geophysics* 15 95–108
[48] Khabarova O, Zank G P, Li G, le Roux J A, Webb G M, Dosch A and Malandraki O E 2015 *ApJ* 808 181 (Preprint 1504.06616)
[49] Khabarova O V, Zank G P, Li G, le Roux J A, Webb G M, Malandraki O E and Zharkova V V 2015 *Journal of Physics Conference Series* 642 012033
[50] Khabarova O V, Zank G P, Li G, Malandraki O E, le Roux J A and Webb G M 2016 *ApJ* 827 122
[51] Lion S, Alexandrova O and Zaslavsky A 2016 *ApJ* 824 47 (Preprint 1602.07213)
[52] Lighthill M J 1960 *Philosophical Transactions of the Royal Society of London Series A* 252 397–430
[53] Montgomery D 1989 *in Lecture Notes on Turbulence: Magnetohydrodynamic Turbulence (World Scientific)*
[54] Montgomery D and Matthaeus W H 1995 *ApJ* 447 706
[55] Matthaeus W H and Zhou Y 1989 *Physics of Fluids B* 1 1929–1931
[56] Zhou Y and Matthaeus W H 1990 *J. Geophys. Res.* 95 14881–14892
[57] Goldreich P and Sridhar S 1995 *ApJ* 438 763–775

[58] Lithwick Y and Goldreich P 2001 *ApJ* 562 279-296 (Preprint astro-ph/0106425)
[59] Chandran B D G, Quataert E, Howes G G, Xia Q and Pongkitiwanichakul P 2009 *ApJ* 707 1668–1675 (Preprint 0908.0757)
[60] Adhikari L, Zank G P, Bruno R, Telloni D, Hunana P, Dosch A, Marino R and Hu Q 2015 *ApJ* 805 63
[61] Adhikari L, Zank G P, Hu Q and Dosch A 2014 *ApJ* 793 52

IOP Conf. Series: Journal of Physics: Conf. Series **900** (2017) 012024 doi:10.1088/1742-6596/900/1/012024

Automated Detection of Small-scale Magnetic Flux Ropes and Their Association with Shocks

Jinlei Zheng[1], Qiang Hu[1,2], Yu Chen[1], Jakobus le Roux[1,2]
[1]Department of Space Science, University of Alabama in Huntsville, USA
[2]Center for Space Plasma and Aeronomic Research, University of Alabama in Huntsville, USA
E-mail: jz0006@uah.edu, qh0001@uah.edu

Abstract. We have quantitatively examined one type of fundamental space plasma structures in the solar wind, the magnetic flux ropes, especially those of relatively small scales. They usually are of durations ranging from a few minutes to a few hours. The main objectives are to reveal the existence in terms of their occurrence and distributions in the solar wind, to quantitatively examine their configurations and properties, and to relate to other relevant processes, involving particle energization and intermittent structures in the solar wind. The technical approach is a combination of time-series analysis methods with the Grad-Shafranov reconstruction technique. This modeling method is capable of characterizing two and a half dimensional cross section of space plasma structures, based on in-situ spacecraft measurements along a single path across. We present the automated detection of flux ropes, construction of an online magnetic flux rope database, and detailed case studies of such structures identified downstream of interplanetary shocks.

1. Introduction

Small-scale magnetic flux ropes in the solar wind, i.e., those possessing the signatures of flux rope configuration from in-situ spacecraft measurements, were usually of relatively short durations, ranging from minutes to a few hours, compared with their large-scale counterparts, i.e., magnetic clouds (MCs) [1]. They were initially identified from the in-situ spacecraft measurements [2] at a much later time than the discovery of MCs. The same set of observations and analysis tools has been utilized in analyzing both sets of events. For example, these include the earlier identifying and categorizing effort of such events [3, 4], and associated modeling by the simple linear force-free cylindrical flux rope model. These events of durations $\sim$ half an hour were also examined by more sophisticated Grad-Shafranov (GS) reconstruction technique in Hu and Sonnerup [5] which revealed the cylindrical helical field configuration of two small-scale flux rope events. Further detailed statistical analysis and associated analysis of additional observation signatures were carried out recently by Yu et al.[6]. However, up to date, the questions regarding their origination and their role in local particle energization remain.

Figure 1 (a) shows the rare observation of direct white-light imaging of solar atmosphere (corona) surrounding the solar disk during a total solar eclipse in 2008. The same observations but with much higher resolution are expected for the upcoming total solar eclipse over north America in August 2017. This black white picture reveals the plasma outflow as bright strands,

IOP Conf. Series: Journal of Physics: Conf. Series **900** (2017) 012024 doi:10.1088/1742-6596/900/1/012024

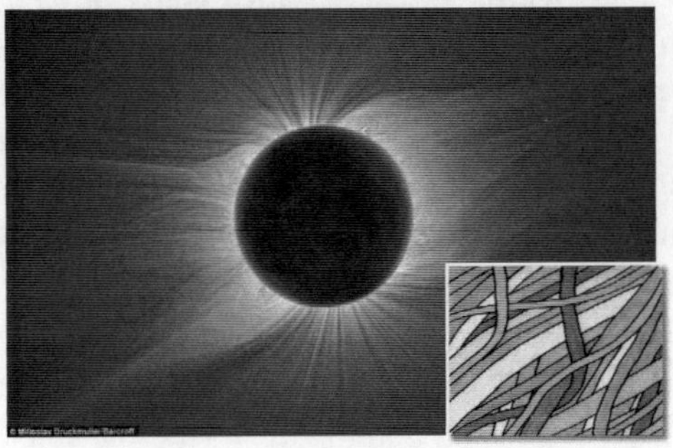

(a)

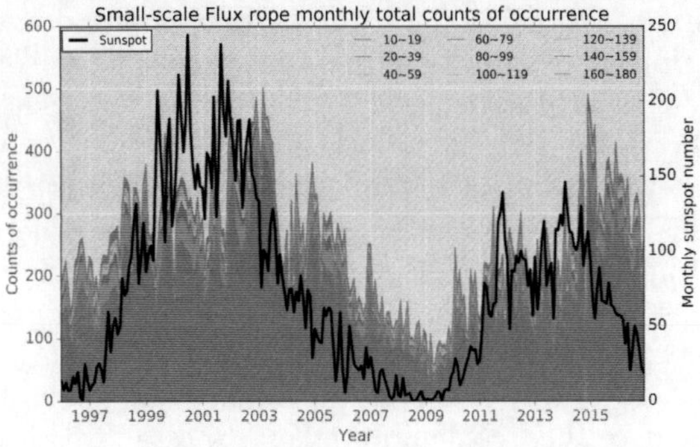

(b)

Figure 1. (a) The structured solar corona during a total eclipse. The bright streamline-like features represent plasma and magnetic structures. The inset depicts artistically one hypothetical scenario of "flux ropes" permeating the interplanetary space [7], originating from those filamentary structures near the Sun. (b) The monthly counts of flux rope events occurrence and the sunspot numbers (right axis) during 1996-2016.

channeled by the underlying magnetic field structures, otherwise invisible under normal conditions. It exhibits filamentary and fine-scale structures that constantly stretch outwards. At larger radial distances from the Sun, the outflows (i.e., solar wind) are further accelerated and become supersonic due to the interplay with the magnetic field. They carry along the embedded magnetic field structures and reach the distance of 1 AU at high but fairly constant speeds of hundreds of kilometers per second. A space probe, sitting in-between Sun and Earth, will sample these flows when they pass by, yielding time-series data equivalent to a set of spatial samples across the structure along a single path. The inset shows the artistic depiction of one hypothesis of the manifestation of such filamentary structures permeating the Sun-Earth space, where a spacecraft would cross such a "sea" of "spaghetti"-like strands. Each strand, or a tube in 3D is distinct from one another, as indicated by different colors, separated by well-defined boundaries. The internal structure of each tube can be characterized by a cylindrical flux rope configuration as to be demonstrated. So a spacecraft would cross multiple flux rope structures and associated boundaries along a single path at 1 AU.

Figure 1 (b) shows our preliminary result in identifying small-scale flux rope structures in the solar wind, with durations ranging from 10 to 180 minutes. We analyzed Wind spacecraft measurements from 1996 to 2016, covering nearly two solar cycles. We identified these small-scale structures via a novel approach to be described in the next section. Here we show the histogram of monthly event counts color-coded by event durations. The events of smaller durations generally have greater rates of occurrence. Clearly the total counts including all events of variable durations follow the monthly sunspot numbers, hinting at solar-cycle dependency of these events.

In contrast to their large-scale counterparts in the solar wind as direct consequences of energetic solar eruptions, these small to medium sized flux ropes are much less studied. The trend of their occurrence and properties varying with solar cycles as revealed from our preliminary analysis has yet to be firmly established. Lately, research interest in these structures and related processes has gained significant momentum [8, 9, 10, 11, 12, 13, 14, 15, 16, 17, 18]. It has been shown that these structures play critical roles in energizing suprathermal particles, an

IOP Conf. Series: Journal of Physics: Conf. Series **900** (2017) 012024 doi:10.1088/1742-6596/900/1/012024

important component of the particle population in the solar wind. They can also contribute to the presence of intermittent structures in the solar wind, an intrinsic ingredient in solar wind turbulence [19, 20, 21, 22, 23, 14, 24, 25]. Such intermittency, fundamental to the dynamic processes in solar wind plasmas, can be naturally explained by the presence of these flux rope structures, as depicted in Figure 1 (a), where the crossing of the flux ropes and their boundaries leads to discrete and abrupt changes in magnetic field and plasma properties, as manifested in time-series data. However such a hypothesis has yet to be verified. In addition, theoretical investigations [10, 26, 27, 28] envisaged the indispensable role of the existence of and interaction among these structures, responsible for trapping and energization of charged particles in space. Therefore it is of significant intellectual merit to quantitatively examine these structures, to elucidate their configuration and to establish their correspondence to closely related processes, from direct spacecraft measurements.

2. Small-scale Magnetic Flux Rope Database

To uncover the intuitive image of small-scale flux ropes in the solar wind, we decided to build a comprehensive database via extensively event searching. Until now, we have built a small-scale magnetic flux rope database based on the observational data from Wind spacecraft, and we will extend the spacecraft data coverage to include ACE and Ulysses data in near future. This database covers the time period from 1996 to 2016, including 63,460 flux rope records in total.

The distinct feature of the database is that it is based on unique Grad-Shafranov reconstruction technique. The Grad-Shafranov method is a true two-dimensional (2D) method that can derive a 2D cross section of magnetic and plasma configurations in space from 1D spacecraft data. It has been developed and applied for a number of years [5, 29, 30, 31, 32, 33, 34, 35, 36]. We adopt the quasi-static magnetic flux rope configuration described by the Grad-Shafranov equation, and developed an detecting algorithm to scan in-situ spacecraft observational data to look for flux rope structures. The data used in flux rope detecting process include time-series magnetic field vectors, plasma bulk properties such as flow velocity, plasma temperature and density, including electrons when available. To guarantee the quality of the flux rope database, we set a series of criteria to rule out the non-flux-rope structures which may survive in the flux rope searching process. For example, to exclude Alfvénic structures, we applied Walen test to each flux rope candidate, and removed the records with slopes below -0.3 or above 0.3. Since the average magnetic field magnitude in the solar wind is about 5 nT, we removed the records with average magnetic field less than 5 nT. In this paper, we put emphasis on the statistical results derived from the flux rope database, leaving the detailed elaboration on the flux rope detecting method to a separate paper.

Figure 2 (a) is flux rope duration histogram. The y-axis represents the monthly counts of flux ropes in logarithmic scale. This histogram hints at two slopes separated by the 60 minutes tick, which means the occurrence frequency of flux ropes exponentially decreases as the increase of duration, and the power indices are different between the flux ropes with durations less than 60 minutes and those longer than 60 minutes. Note that the histogram has an elevated tail distribution (160$\sim$180 minutes bin). This high tail does not follow the frequency distribution trend of the flux ropes with durations longer than 60 minutes. Because of the different axial orientation and moving speed of each flux rope, the flux rope duration cannot accurately reflect the flux rope scale size. If a spacecraft passes through the flux rope cross section with a trajectory perpendicular to flux rope axial direction, it will surely take less time than the situation in which the spacecraft obliquely passes through the flux rope at equal speed. To check the frequency distribution of flux rope scale sizes, we derive the the flux rope size from its duration, orientation,

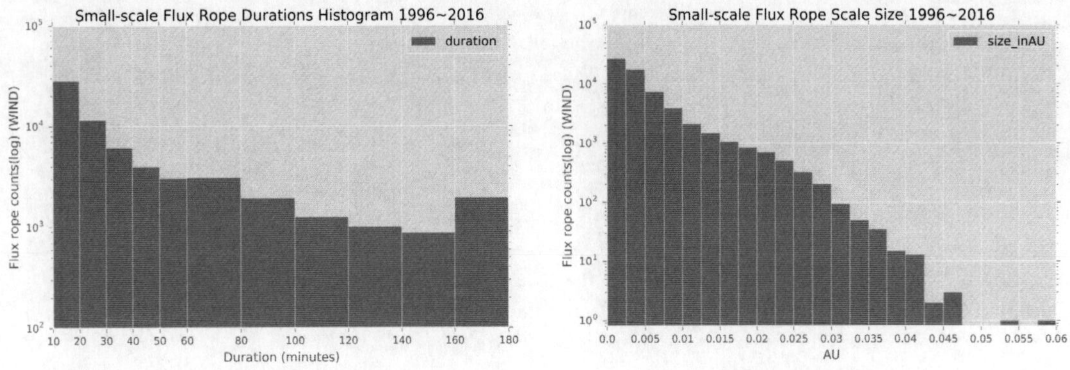

(a) (b)

Figure 2. (a) Flux rope duration histogram. In flux rope detecting process, we split the entire task into 11 iterations with different window sizes. The window sizes are 10~20 minutes, 20~30 minutes, 30~40 minutes, 40~50 minutes, 50~60 minutes, 60~80 minutes, 80~100 minutes, 100~120 minutes, 120~140 minutes, 140~160 minutes, and 160~180 minutes. The bar plot represents the counts (log scale) of detected flux rope in each bin. (b) Flux rope scale size histogram. Derived from flux rope duration, orientation and the velocity of co-moving frame.

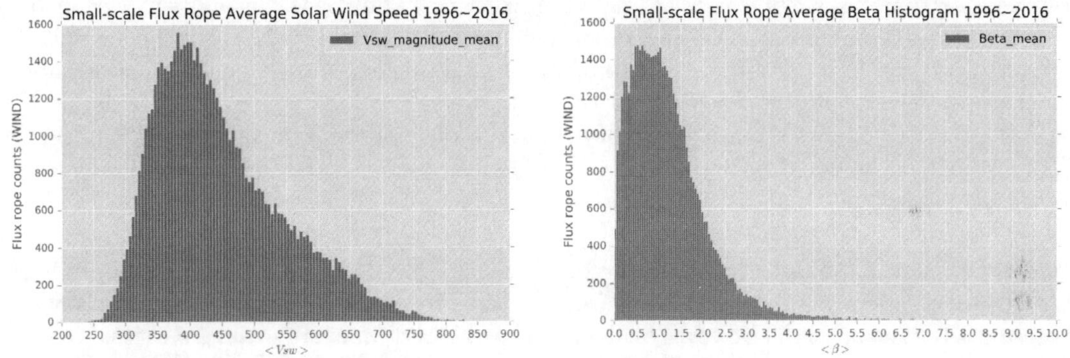

(a) (b)

Figure 3. (a) Flux rope solar wind speed histogram. For each flux rope case, we calculated the average solar wind speed between the flux rope start and end time. (b) Flux rope plasma β histogram. The plasma β is calculated from proton and electron number density (assuming $N_e = N_p = N$), proton and electron temperature, and magnetic field. The electron temperature data has a gap from 2001/05/31 23:59:57 to 2002/08/16 00:00:05. The plasma β in this gap is not counted in the histogram.

and the co-moving frame velocity. Figure 2 (b) shows the flux rope scale size histogram. In this distribution, the relation between log(counts) and scale sizes seems to be bi-linear of similar slopes, with an "ankle"-like transition in between. There is no high tail as that in duration histogram. This indicates that effect on scale sizes by oblique passages is more pronounced for long-duration events. Figure 3 (a) is solar wind histogram corresponding to flux rope events. It indicates that the small scale flux ropes are more likely to occur in low-speed streams, with

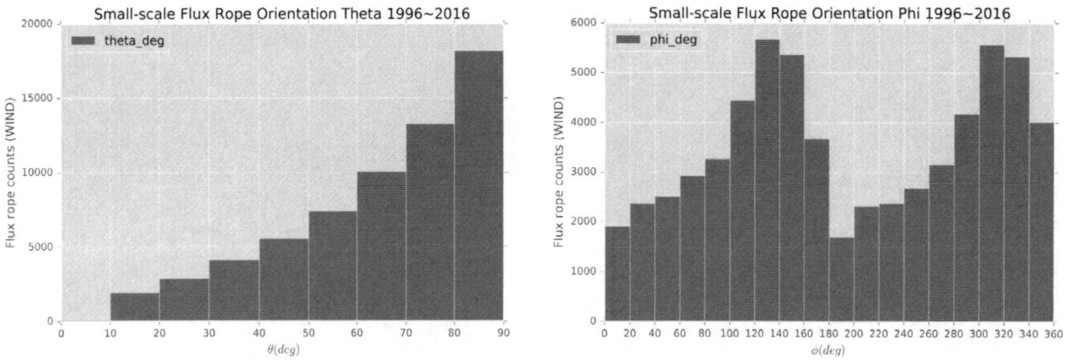

(a) (b)

Figure 4. (a) Flux rope polar angle θ histogram. The actual polar angle range is from $0°$ to $180°$. To cut down the time consuming in flux rope detecting process, we restrict end points of the direction vectors in upper hemisphere, and do not distinguish the vectors with opposite directions with each other. (b) Flux rope azimuthal angle ϕ histogram. This angle is the projection of flux rope axial directions onto ecliptic plane.

peek speed around 400 km/s. Figure 3 (b) is the histogram of average plasma beta within each flux rope interval. There is a clear and broad peak around value 1, indicating the significance of plasma pressure. Figure 4 (a) is the histogram of polar angle of flux rope axis, which implies that the projection of flux rope axis tend to lay between negative X-axis and Y-axis in GSE frame. Figure 4 (b) shows the histogram of azimuthal angle of the flux rope axis orientation in GSE coordinate. The unflat pattern implies a preference of small scale flux rope axis orientation is in the ecliptic plane.

3. Case Studies Associated with Shocks

Interplanetary shocks are known sources contributing to energetic particle acceleration and transport in the heliosphere. However evidences exist for certain energy-range particles to show behaviors departing from expected shock related acceleration processes. The association between structures (mainly discontinuities) other than shocks and local energetic particle flux enhancement was also reported by a number of studies [15, 21, 22, 9, 37]. Theoretical studies suggested that the combined diffusive shock acceleration and downstream multiple magnetic island acceleration would result in the energization of charged particles [27, 10]. We found in a previous study presented in the last year's conference [8] the existence of flux rope structure downstream of shocks and hinted at the correspondence to local particle energization. In the present study, we follow up with further analysis of the occurrence and properties of flux ropes downstream of shocks and the associated energetic particle signatures. In this section, we are going to present two sets of case studies in which the enhanced ions flux was observed within the flux rope region. We will use the combined observations of ACE and Wind spacecraft to illustrate the flux rope morphological change leading to energizing particles.

Our analysis employs existing online resources such as databases containing identified shock waves and other types of discontinuities or structures. For example, they include, but are not limited to, the online shock database from the ACE spacecraft at Earth, and the similar but more comprehensive Finnish Interplanetary Shock Database (*http://ipshocks.fi/*) that lists identified shock waves from multiple spacecraft in the solar wind. The associated shock parameters such as the geometry of the shock, the shock speed, and compression ratios etc. are also given in these databases, which will be utilized by our analysis.

We checked the conditions near two interplanetary shocks. One shock was observed by ACE on March 10th, 1999, at 0:40 UT, and observed again by Wind after about 53 minutes. The other shock was observed by ACE on December 7th, 2003, at 13:41 UT, but not observed by Wind. We examine the magnetic field, solar wind plasma, and energetic ions data beginning from 1 hour upstream the shock and ending at 9 hours downstream the shock. In the 10 hours interval, a few flux ropes were detected, and the combined observations of ACE and Wind suggested the flux rope merging behavior in the region of enhanced energetic ion flux.

3.1. Case study: March 9th, 1999, 23:40 UT to March 10th, 1999, 09:40 UT

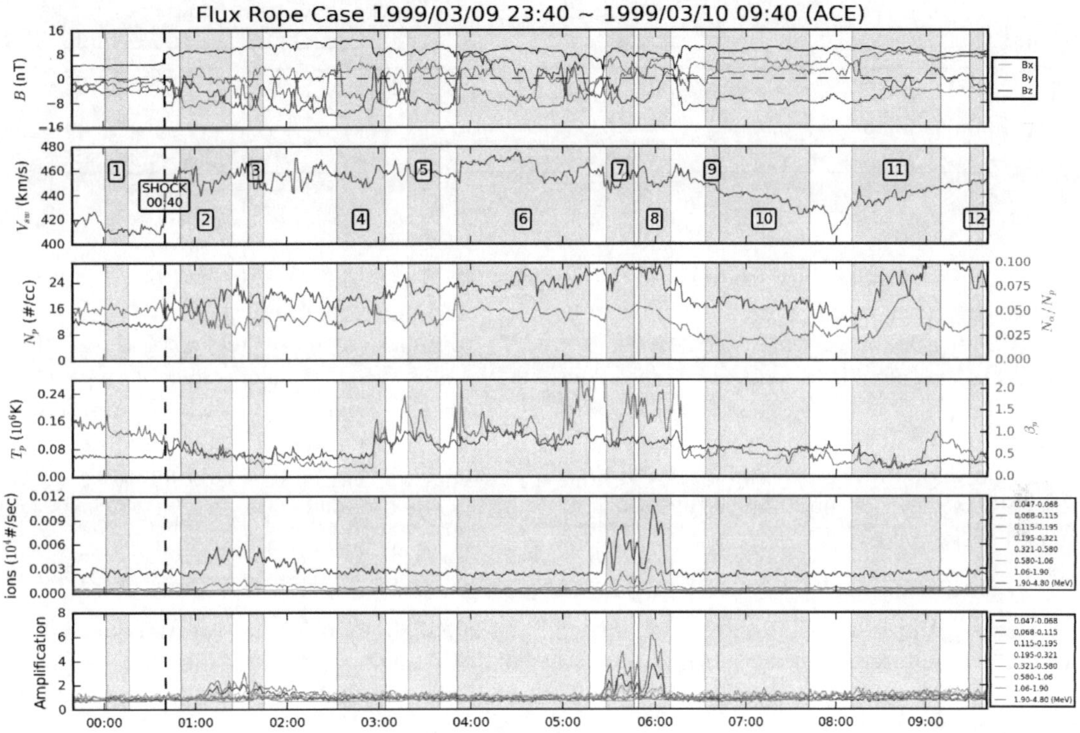

Figure 5. Time series data from March 9th, 1999, 23:40 UT to March 10th, 1999, 09:40 UT. All data are from ACE spacecraft. This time interval contains one interplanetary shock and twelve flux rope structures which are labeled from 1 to 12 and shaded by yellow color. The first panel is the magnetic field data. The black solid line denotes the total magnetic field magnitude. The second panel is the solar wind bulk speed. In the third panel, the proton number density is shown in blue and the alpha/proton ratio is shown in green. The fourth panel are the proton temperature (in blue line) and proton plasma beta (in green line). The fifth panel is ions flux, and the sixth panel is the same ions flux, but normalized by the value at shock time. The energy range of each curve is provided in the box to the right. The shock is denoted by a vertical black dashed line across six panels.

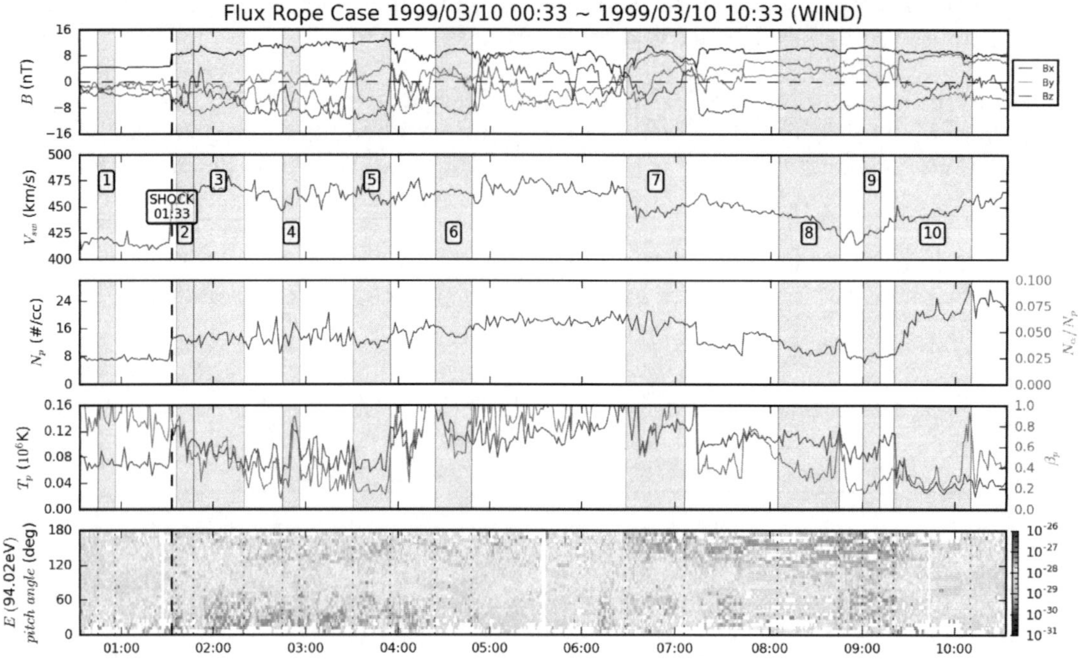

Figure 6. Time series data from March 10th, 1999, 00:33 UT to March 10th, 1999, 10:33 UT. All data are from Wind spacecraft. This time interval contains one interplanetary shock and ten flux rope structures which are labeled from 1 to 10 and shaded by yellow color. The first panel is the magnetic field data. The black solid line denotes the total magnetic field magnitude. The second panel is the solar wind bulk speed. In the third panel, the proton number density is shown in blue line and the alpha/proton ratio is shown in green line. The fourth panel shows the proton temperature (in blue line) and proton plasma beta (in green line). The fifth panel is suprathermal electron pitch angle distribution. The shock is denoted by a vertical black dashed line across five panels.

Figures 5 and 6 are time series data around the same shock observed by ACE and Wind successively. Figure 5 shows a shock observed at 00:40 UT, while in Figure 6, the shock time is 01:33 UT. The two observations have 53 minutes time shift. In Figure 5, we note that during the intervals of flux ropes No.7 and No.8, the ions flux is enhanced. Within the interval of flux rope No.7, the ions flux is enhanced by 5 times in 0.068 - 0.115 MeV channel and by 3 times in 0.047 - 0.068 MeV channel. In each case, the enhancement peak is near the center of the flux rope interval. Panels five and six show that after the shock time, there is a weak ion flux enhancement from 1:00 UT to 2:00 UT, and after 2:00 UT, the ions flux level is steady. Thereafter, starting at about 5:30 UT, the ion flux reaches two consecutive peaks. We believe that these two ion flux peaks are less likely caused by direct shock acceleration. Instead, they may be caused by other co-located processes, such as flux rope merging or contraction. Noting that these two flux ropes are next to each other, which makes the interacting with each other possible. If they are merging, we should be able to see one single flux rope after some time. In order to check their temporal evolution, we examined the flux rope events observed by Wind which are radially aligned with ACE at a distance of 247 Earth radii toward Earth. We found the correspondence to flux rope No.7 and No.8. In Figure 6, we can see that the Wind flux rope No. 7 has similar magnetic filed structure to the combined intervals of ACE flux rope No.7 and No.8 in Figure 5.

IOP Conf. Series: Journal of Physics: Conf. Series **900** (2017) 012024 doi:10.1088/1742-6596/900/1/012024

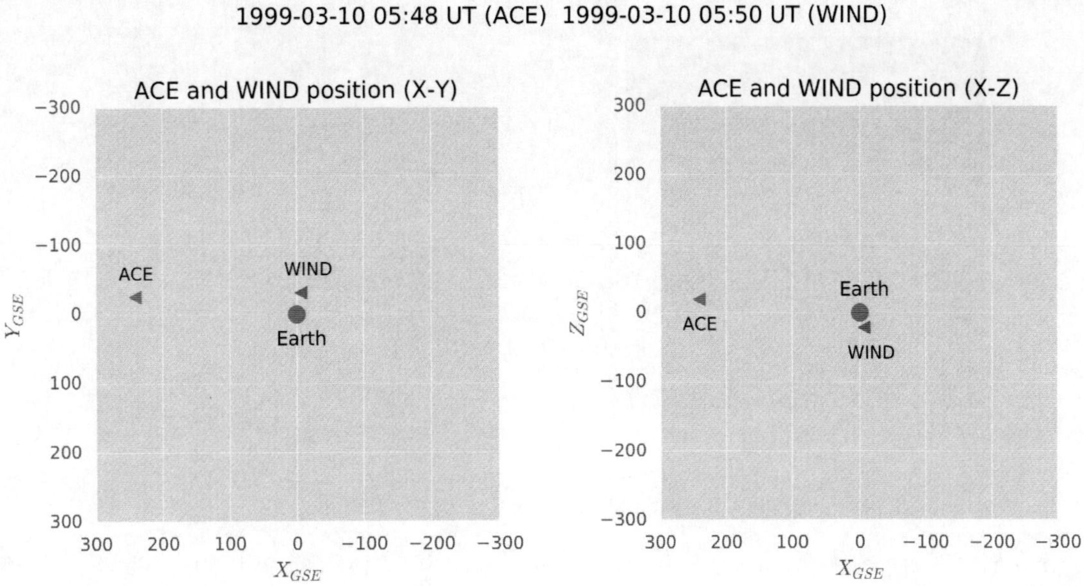

Figure 7. The positions of ACE and Wind spacecraft in GSE coordinate. The left panel is the projection on XY-plane, and the right panel is the projection on XZ-plane. The time stamp of ACE is March 10th, 1999, 5:48 UT, and the time stamp of Wind is March 10th, 1999, 5:50 UT. The unit of all axes is Earth radius.

Also, their surrounding magnetic filed structures are highly similar. We believe the Wind flux rope No. 7 is the correspondence of ACE flux rope No.7 and No.8. To verify our speculation, we check the time shift of ACE and Wind flux ropes, and compare it with propagating time of solar wind. Figure 7 shows the positions of ACE (red triangle) spacecraft, Wind (green triangle) spacecraft, and the Earth (blue dot) in GSE coordinate when ACE is in the center of flux rope No.7 in Figure 5 (at about 5:48 UT). Since the flux rope propagating time is less than 2 hours, in such a short time, we can assume both spacecraft are stationary. We can see that the ACE and Wind are aligned along X_{GSE} axis, and ACE is in front of Wind in the direction toward the Sun. The GSE coordinate of ACE is $(X, Y, Z) = (1.53359 \times 10^6, -1.63047 \times 10^5, 1.25898 \times 10^5)$ km, and the GSE coordinate of Wind is $(X, Y, Z) = (-4.0486 \times 10^4, -2.0099 \times 10^5, -1.3872 \times 10^5)$ km. From their coordinates, we can calculate that the separation of ACE and Wind in X_{GSE} direction is 1.57408×10^6 km. From the second panels of Figure 5 and Figure 6, we can see that the average solar wind speed during the flux rope intervals under checking (No.7 and No.8 in Figure 5, and No.7 in Figure 6) is about 450 km/s. Therefore, the propagating time for a flux rope from ACE to Wind is about 58 minutes. Then we are going to check the time shift of flux rope No.7 and No.8 in Figure 5, and the flux rope No.7 in Figure 6. The center time of flux fope No.7 and No.8 in Figure 5 is 5:49 UT (the center between start time of flux rope No.7 and end time of flux rope No.8), while the center time of flux rope No.7 in Figure 6 is 6:47 UT. The time shift is 58 minutes, with is the same as the calculated propagating time. Now we are confident that the flux rope observed by Wind is the same ones but at later evolution phase of the flux rope No.7 and No.8 observed by ACE.

In order to show the flux rope evolution from ACE to Wind, we reconstructed the flux rope No.7 and No.8 in Figure 5, and the flux rope No.7 in Figure 6. Figure 8 (a) shows the recon-

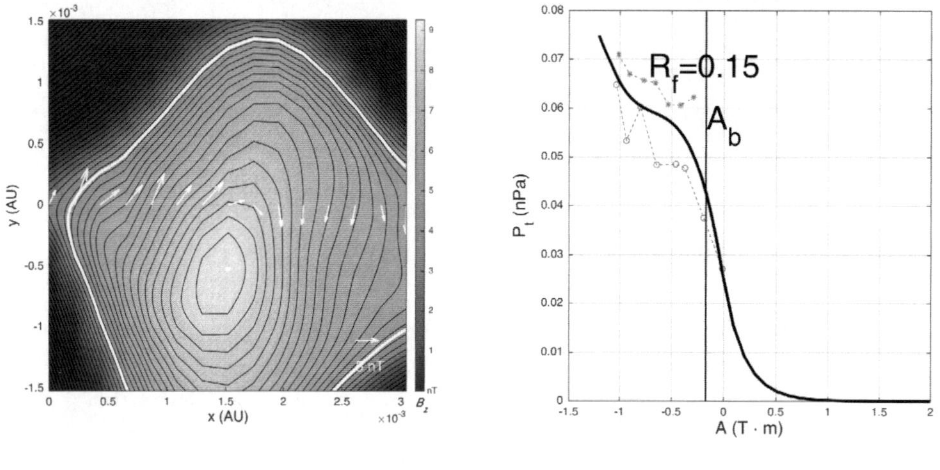

(a) $\hat{\mathbf{z}} = [0.3830, -0.3214, -0.8660]$ (b)

Figure 8. GS reconstruction result for flux rope record No.7 (March 9th, 1999, 05:28 - 05:46 UT) in Figure 5. (a) The cross-section magnetic field configuration of the flux rope. The unit vector $\hat{z}$ represents the z axis orientation of the flux rope. The filled color contour represents the axial magnetic field B_z and the line contour overlapped in black represents the transverse magnetic field. The white arrows across the map horizontally are projected transverse magnetic field vectors along the spacecraft path. The white contour line encloses the region $A < A_b$ (or in some case, $A > A_b$), denoted in (b). The reconstruction map within this line is reconstructed by double-folded satellite data, and the region beyond this line is reconstructed based on the extrapolated data. (b) The fitting curve (bold solid line) of $P_t(A)$ for this flux rope case. The fitting is based on the data points denoted by blue circles and red stars, where the circles represent the data as satellite was crossing the first half of flux rope structure and the stars represent those of second half. The thin vertical solid line denoted by A_b marks the boundary corresponding to the white line in (a). A fitting residue R_f is denoted.

structed cross section of flux rope No.7 in Figure 5. The white arrows are the transverse field vectors along the projection of spacecraft trajectory on xy-plane. The color represents the B_z field of this flux rope. The arrow directions indicate that this flux rope has the left-handed chirality. The maximum B_z field is about 9 nT in the flux rope center. Figure 8 (b) is the fitting curve of $P_t(A)$, in which P_t is the transverse plasma pressure, the sum of plasma and axial magnetic pressure, and A is the magnetic vector potential. Though the fitting residue R_f is acceptable, the two branches of measured data points (in red and blue, respectively) still have notable deviations, which indicates that the current axial orientation we found may not be the optimal one. We will discuss it later when we analyze the axial orientation. Figure 9 (a) is the reconstructed cross section of flux rope No.8 in Figure 5, which is next to flux rope No.7. It's maximum B_z field is about 8.5 nT, which is relatively weaker than its neighbor. No.8 has the same left-handed chirality as No.7. Their $\hat{\mathbf{z}}$ axis orientations differ by 67°. Two neighboring flux ropes with the same chirality and interacting with an oblique angle between their cylindrical axes may merge duo to magnetic reconnection [38].

Since No.7 and No.8 are close to each other, we tried to put these two flux rope together to see if we can find the combined structure. Figure 10 (a) is the reconstructed cross section when

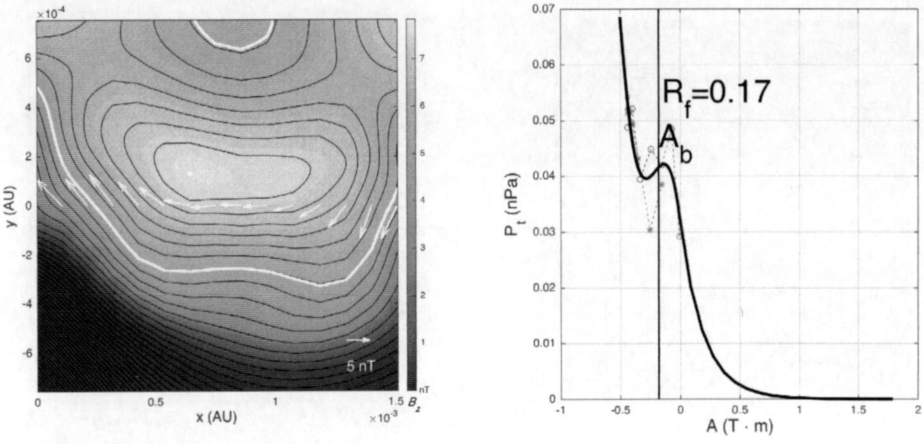

(a) $\hat{\mathbf{z}} = [0.9254, 0.3368, -0.1736]$ (b)

Figure 9. GS reconstruction result for flux rope record No.8 (March 9th, 1999, 05:49 - 06:10 UT) in Figure 5. The format is the same as Figure 8.

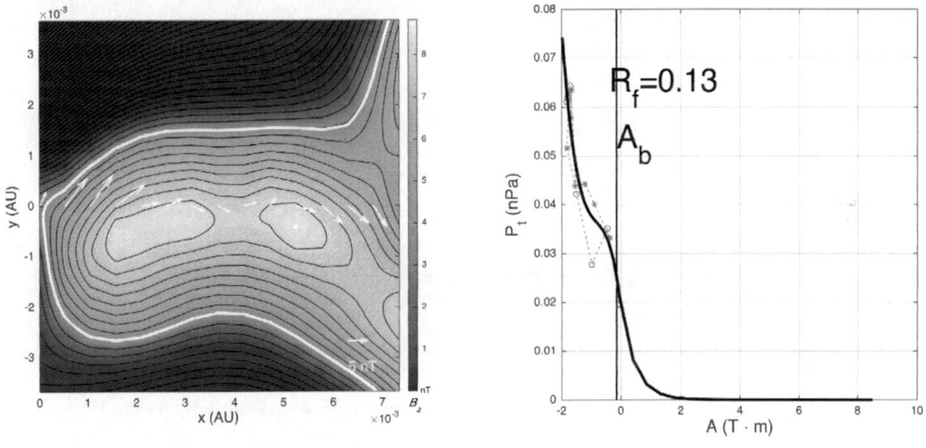

(a) $\hat{\mathbf{z}} = [-0.2179, 0.2872, -0.9328]$ (b)

Figure 10. GS reconstruction result for the combination of flux rope No.7 and No.8 (March 9th, 1999, 05:28 - 06:10 UT) in Figure 5. The format is the same as Figure 8.

we put No.7 and No.8 together. A twin flux rope shape cross section is obtained. The fitting residue shown in Figure 10 (b) is smaller than both of the individual ones. When checking the axial orientation of each individual flux rope and comparing them with the combined flux rope, we found that they have three different axial orientations. This means the structures of flux rope No.7 and No.8 are not strictly 2-D structures. This is also partially the reason why the two individually reconstructed flux ropes do not have very small fitting residues.

Figure 11 (a) is the reconstructed cross section of flux rope No.7 in Figure 6. This flux

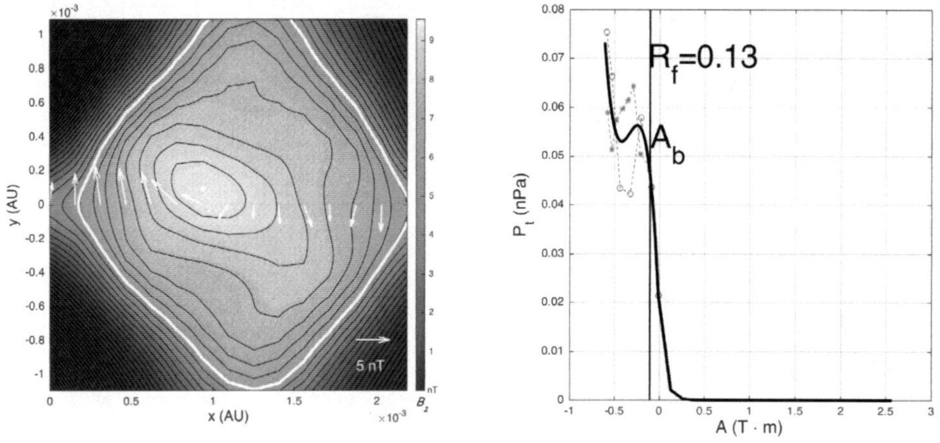

(a) $\hat{\mathbf{z}} = [0.9397, 0.0000, -0.34207]$ (b)

Figure 11. GS reconstruction result for flux rope record No.7 (March 9th, 1999, 05:49 - 06:10 UT) in Figure 6. The format is the same as Figure 8.

rope is observed by Wind and is the correspondence of the combined flux rope No.7 and No.8 observed by ACE in Figure 5. It has similar core field strength but smaller size compared with the combined flux rope observed by ACE. From Figure 10 (a) to Figure 11 (a), we see the twin flux rope structure merges into a single one. The irregular transverse field lines in Figure 11 (a) implies that this flux rope just underwent a non-steady evolution process. It is interesting to note that the axial orientation of the Wind flux rope is also different from the combined flux rope from ACE. The angle between their axial orientations is 83°. This shows that the flux rope orientation is subject to change in the merging or propagating process, either due to the overall rotation or the more complicated 3-D structures formed in merging process.

3.2. Case study: December 7th, 2003, 12:41 UT to 2003 December 7th, 22:41 UT

In this case study, another ions flux enhancement event during flux rope interval was observed, and the temporal evolution of the corresponding flux rope is analyzed. Figure 12 shows ACE time series data from December 7th, 2003, 12:41 UT to 22:41 UT. Figure 13 shows Wind time series data from December 7th, 2003, 12:00 UT to 21:00 UT. Figure 12 shows a shock observed at 13:41 UT, however, this shock was not observed by Wind in Figure 13. The 5th and 6th panels in Figure 12 show that there are two major flux rope enhancement events during the plotted time interval. The first event is right after the shock, and the other ions flux enhancement is from 15:29 UT to 16:41 UT, during which, a flux rope structure is detected and labeled as No.3. The first enhancement event is probably due to direct shock acceleration, while the second enhancement event may be caused by flux rope related acceleration [28, 27, 26]. Next, we will check the temporal evolution of flux rope No.3 in Figure 12. If this flux rope shows contracting feature when propagating, it could has the ability to boost the ions flux. In Figure 13, we found a possible correspondence to flux rope No.3 in Figure 12, whose time interval is from 17:17 to 18:08 and labeled as No.5. Same as the method we used in last case study, we are going to verify the correspondence by checking the ACE and Wind positions and the flux rope propagating speed.

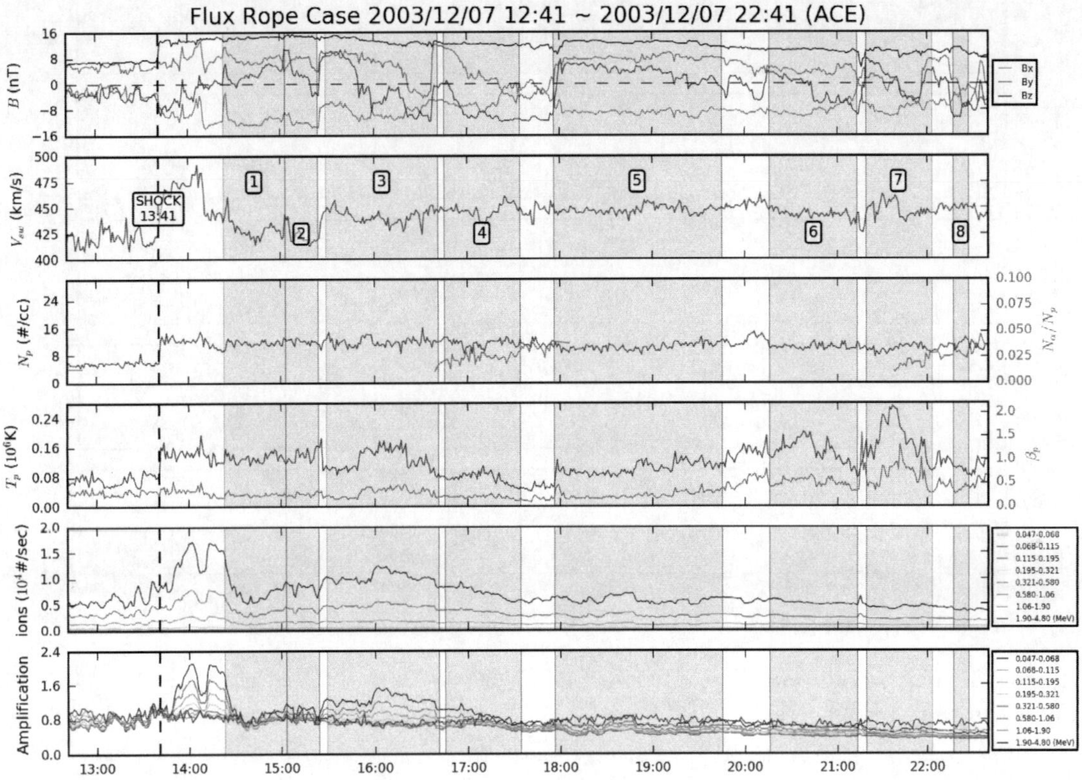

Figure 12. Time series data (ACE) from Dec. 7, 2003, 12:41 UT to Dec. 7, 2003, 22:41 UT. The format is the same as Figure 5.

Figure 14 shows that ACE and Wind are aligned along X_{GSE} axis on the two sides of the Earth. The GSE coordinate of ACE is $(X, Y, Z) = (1.54422 \times 10^6, 2.68164 \times 10^4, 2.28218 \times 10^4)$ km, and the GSE coordinate of Wind is $(X, Y, Z) = (-1.4851 \times 10^6, -1.3105 \times 10^5, -1.1180 \times 10^5)$ km. The separation of ACE and Wind in X_{GSE} direction is 3.02932×10^6 km. The second panels of Figure 12 and Figure 13 show that the average solar wind speed during the interval of flux rope No.3 in Figure 12 and flux rope No.5 in Figure 13 is about 440 km/s. So the propagating time for the flux rope from ACE to Wind is about 115 minutes. The center time of flux rope No.3 in Figure 12 is 16:05 UT, and the center time of flux rope No.5 in Figure 13 is 17:43 UT. The center time shift is 98 minutes, which is close to the estimated propagating time, 115 minutes. Now we confirm that the flux rope No.3 from ACE corresponds to the flux rope No.5 from Wind.

Figure 15 (a) is the reconstructed cross section of flux rope No.3 in Figure 12. Its core B_z field is about 13.5 nT, which is very strong for a small-scale flux rope. The field configuration indicates that this flux rope has right-handed chirality. This flux rope is distorted along x direction. It looks like the remains of two merged flux ropes. Figure 16 (a) is the reconstructed cross section of flux rope No.5 in Figure 13. The core B_z field of flux rope No.5 is about 10 nT, which is relatively weaker compared with its earlier phase. Flux rope No.5 also has right-handed chirality. This is a necessary condition to assure the correspondence between flux rope No.3 from ACE and flux rope No.5 from Wind. Figure 16 (a) shows a flux rope

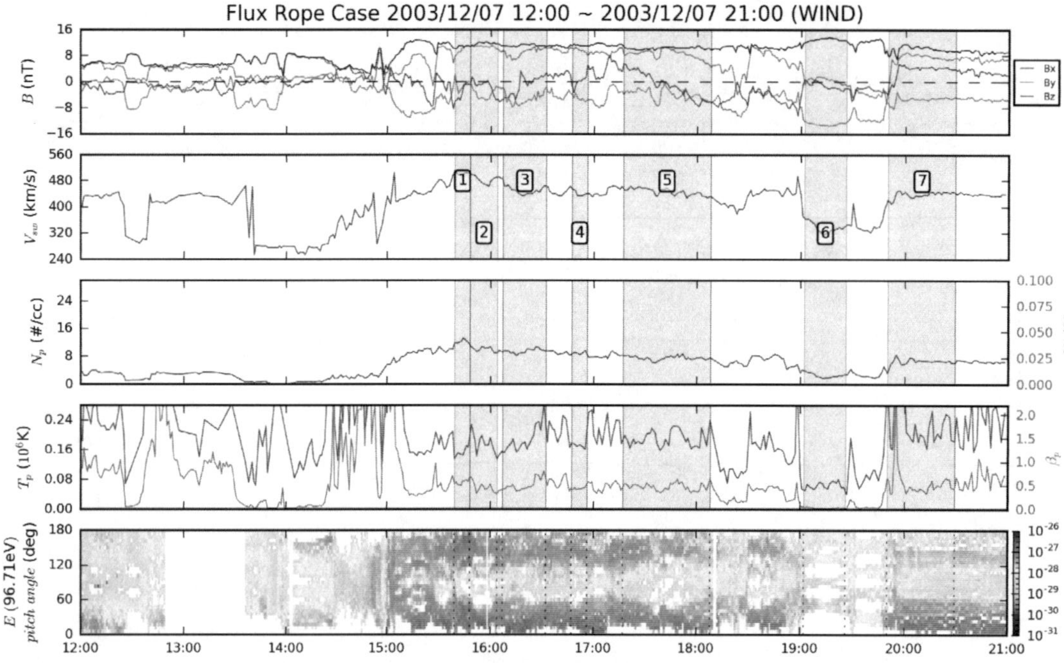

Figure 13. Time series data (Wind) from Dec. 7, 2003, 12:00 UT to Dec. 7, 2003, 21:00 UT. The format is the same as Figure 6.

pinched in the center. This feature can be considered to be evolved from the distorted feature in Figure 15 (a). By checking the axial orientations of flux rope No.3 and flux rope No.5, we found that the angle between them is 45°, indicating the axial orientation change as evolution. The scale sizes of these two flux ropes are similar, which are about 8×10^{-3} AU, i.e., 188 Earth radii. The overall B_z field of flux rope No.5 is lower than that of No.3, which is a reasonable phenomenon in plasma relaxation process. From Figure 15 (a) to Figure 16 (a), we can see a flux rope undergoing the end phase of a merging process, in which its chirality and scale size were kept, but the axial orientation changed by 45° and the overall magnetic field decreased.

The 5th panel in Figure 13 shows that the strong counterstreaming suprathermal electron flux appeared during the interval of flux rope No.5. The counterstreaming suprathermal electrons did not show up in the last case studies (see the 5th panel of Figure 6, in the interval of flux rope No.7). Also we note that the average proton β of flux rope No.3 in Figure 12 and flux rope No.5 in Figure 13 is about 0.5, whereas, in the last case study, this value is higher than 1.0. The strong counterstreaming suprathermal electron flux and low proton β suggest that flux rope No.3 and No.5 may originate from the Sun, instead of being formed in the solar wind. The ions flux within flux rope interval in this case is much stronger than that in the last case (See the 5th panel of Figure 5 and the 5th panel of Figure 12).

4. Conclusions and Discussion

In this paper, we present the preliminary statistical results from our newly-built small-scale flux rope database. The results showed that the occurrence rate of small-scale flux ropes has strong solar cycle dependency. These flux ropes are more likely to appear in slow solar wind streams

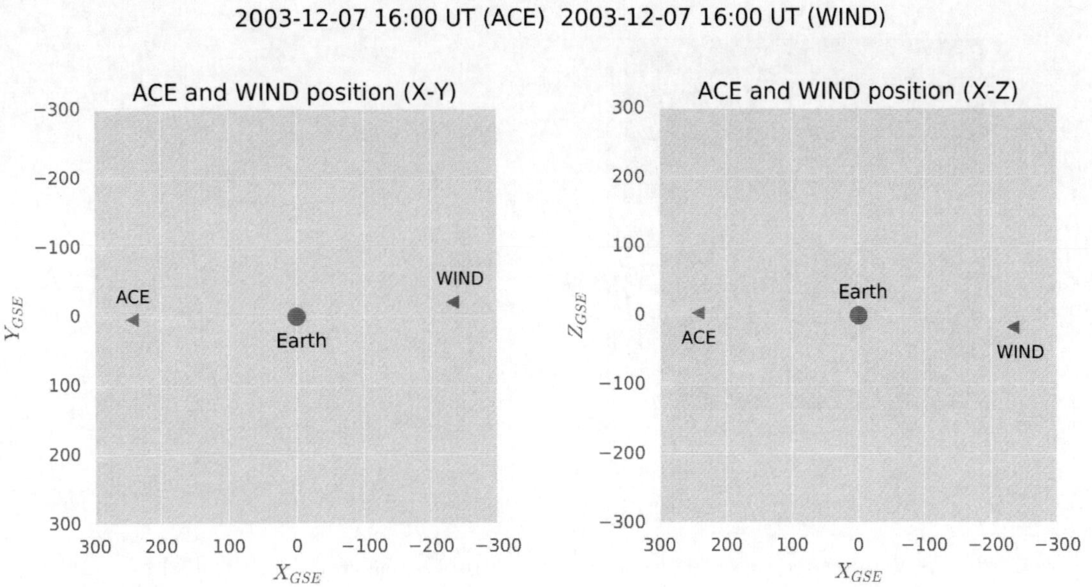

Figure 14. The position of ACE and Wind spacecraft in GSE coordinate. The left panel is the projection on XY-plane, and the right panel is the projection on XZ-plane. Both time stamps of ACE and Wind are July 12th, 2003, 16:00 UT. The unit of all axes is Earth radius.

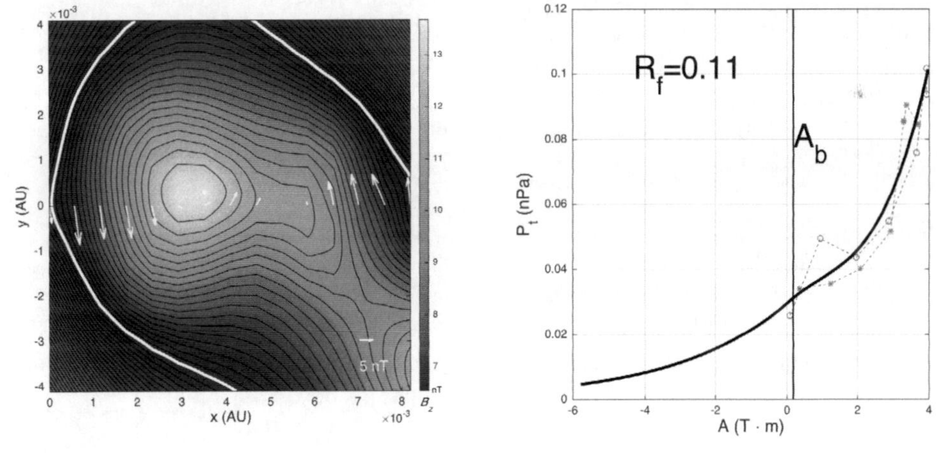

(a) $\hat{\mathbf{z}} = [-0.8586, 0.4957, -0.1305]$ (b)

Figure 15. GS reconstruction result for flux rope record No.3 (2003 Dec. 7, 15:29 - 16:41 UT). The format is the same as Figure 8.

(speeds about 400 km/s), and possess relatively modest to hight plasma β (distribution peaks around 1). Most of the small-scale flux ropes lay on ecliptic plane with the axial directions along parker spirals.

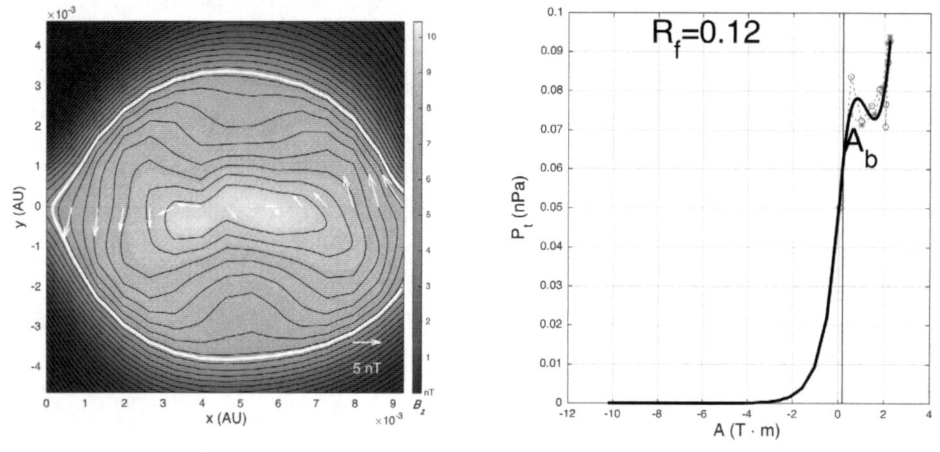

(a) $\hat{\mathbf{z}} = [-0.2566, 0.9577, -0.1305]$ (b)

Figure 16. GS reconstruction result for flux rope record No.5 (2003 Dec. 7, 17:18 - 18:08 UT) in Figure 13. The format is the same as Figure 8.

With the combined observation of ACE and Wind spacecraft, we analyzed the temporal evolution of flux ropes and the effect to particle energization. Two case studies are presented, in which the flux rope merging process and the particle energization were observed. In the first case study, we see that a twin flux rope structure was observed by ACE which then became one single flux rope when it was observed by Wind 58 minutes later. In the second case study, we showed the end phase of a flux rope merging process. In both cases, the energetic ions flux was enhanced during the flux rope merging process. These two cases suggest that the flux rope dynamics may play an important role in local particle energization.

Acknowledgments
We would like to thank NASA's Space Physics Data Facility for the Wind and ACE spacecraft data, and University of Helsinki for the Heliospheric Shock Database. We acknowledge NASA grants NNX12AH50G, NNX14AF41G, and NRL contract N00173-14-1-G006 (NNH13ZDA001N) for support. We acknowledge useful discussion with and initial suggestion by Dr. Pete Riley and colleagues on the automated detection of flux ropes by the GS method.

References

[1] Burlaga L F 1995 *Interplanetary magnetohydrodynamics, by L. F. Burlag. International Series in Astronomy and Astrophysics, Vol. 3, Oxford University Press. 1995. 272 pages; ISBN13: 978-0-19-508472-6* **3**
[2] Moldwin M B, Ford S, Lepping R, Slavin J and Szabo A 2000 *Journal of Geophysical Research: Space Physics* **27** 57–60
[3] Cartwright M L and Moldwin M B 2010 *Journal of Geophysical Research (Space Physics)* **115** A08102
[4] Feng H Q, Wu D J, Lin C C, Chao J K, Lee L C and Lyu L H 2008 *Journal of Geophysical Research (Space Physics)* **113** A12105
[5] Hu Q and Sonnerup B U Ö 2001 *Geophysical Research Letters* **28** 467–470
[6] Yu W, Farrugia C J, Lugaz N, Galvin A B, Kilpua E K, Kucharek H, Moestl C, Leitner M, Torbert R B, Simunac K, Luhmann J G, Szabo A, Ogilvie K W and Sauvaud J 2013 *AGU Fall Meeting Abstracts*
[7] Borovsky J E 2008 *Journal of Geophysical Research (Space Physics)* **113** A08110

[8] Zheng J and Hu Q 2016 *15th AIAC/The Science of Ed Stone: Celebrating his 80th Birthday* IOP Conf. Series ed Zank G P
[9] Khabarova O V, Zank G P, Li G, Malandraki O E, le Roux J A and Webb G M 2016 *The Astrophysical Journal* **827** 122 URL `http://stacks.iop.org/0004-637X/827/i=2/a=122`
[10] le Roux J A, Zank G P, Webb G M and Khabarova O V 2016 *The Astrophysical Journal* **827** 47 URL `http://stacks.iop.org/0004-637X/827/i=1/a=47`
[11] Matsumoto Y, Amano T, Kato T N and Hoshino M 2015 *Science* **347** 974–978
[12] Khabarova O, Zank G P, Li G, le Roux J A, Webb G M, Dosch A and Malandraki O E 2015 *The Astrophysical Journal* **808** 181 (*Preprint* `1504.06616`)
[13] Karimabadi H, Roytershteyn V, Vu H X, Omelchenko Y A, Scudder J, Daughton W, Dimmock A, Nykyri K, Wan M, Sibeck D, Tatineni M, Majumdar A, Loring B and Geveci B 2014 *Physics of Plasmas* **21** 062308
[14] Trenchi L, Bruno R, Telloni D, D'amicis R, Marcucci M F, Zurbuchen T H and Weberg M 2013 *The Astrophysical Journal* **770** 11
[15] Tessein J A, Matthaeus W H, Wan M, Osman K T, Ruffolo D and Giacalone J 2013 *The Astrophysical Journall* **776** L8
[16] Retinò A, Sundkvist D, Vaivads A, Mozer F, André M and Owen C J 2007 *Nature Physics* **3** 236–238
[17] Chen L J, Bhattacharjee A, Puhl-Quinn P A, Yang H, Bessho N, Imada S, Mühlbachler S, Daly P W, Lefebvre B, Khotyaintsev Y, Vaivads A, Fazakerley A and Georgescu E 2008 *Nature Physics* **4** 19–23
[18] Drake J F, Swisdak M, Che H and Shay M A 2006 *Nature* **443** 553–556
[19] Zank G P, Adhikari L, Hunana P, Shiota D, Bruno R and Telloni D 2017 *The Astrophysical Journal* **835** 147 URL `http://stacks.iop.org/0004-637X/835/i=2/a=147`
[20] Tessein J A, Ruffolo D, Matthaeus W H and Wan M 2016 *Geophysical Research Letters* **43** 3620–3627
[21] Tessein J A, Ruffolo D, Matthaeus W H, Wan M, Giacalone J and Neugebauer M 2015 *The Astrophysical Journal* **812** 68
[22] Neugebauer M and Giacalone J 2015 *Journal of Geophysical Research (Space Physics)* **120** 8281–8287
[23] Osman K T, Matthaeus W H, Gosling J T, Greco A, Servidio S, Hnat B, Chapman S C and Phan T D 2014 *Physical Review Letters* **112** 215002 (*Preprint* `1403.4590`)
[24] Karimabadi H, Roytershteyn V, Wan M, Matthaeus W H, Daughton W, Wu P, Shay M, Loring B, Borovsky J, Leonardis E, Chapman S C and Nakamura T K M 2013 *Physics of Plasmas* **20** 012303
[25] Greco A, Matthaeus W H, Servidio S, Chuychai P and Dmitruk P 2009 *The Astrophysical Journall* **691** L111–L114
[26] le Roux J A, Zank G P, Webb G M and Khabarova O 2015 *The Astrophysical Journal* **801** 112 URL `http://stacks.iop.org/0004-637X/801/i=2/a=112`
[27] Zank G P, Hunana P, Mostafavi P, Le Roux J A, Li G, Webb G M, Khabarova O, Cummings A, Stone E and Decker R 2015 *The Astrophysical Journal* **814** 137
[28] Zank G P, le Roux J A, Webb G M, Dosch A and Khabarova O 2014 *The Astrophysical Journal* **797** 28 URL `http://stacks.iop.org/0004-637X/797/i=1/a=28`
[29] Hu Q and Sonnerup B U O 2002 *Journal of Geophysical Research: Space Physics* **107** SSH 10–1–SSH 10–15 ISSN 2156-2202 URL `http://dx.doi.org/10.1029/2001JA000293`
[30] Hu Q, Smith C W, Ness N F and Skoug R M 2003 *Geophysical Research Letters* **30** 1385
[31] Hu Q, Smith C W, Ness N F and Skoug R M 2005 *Journal of Geophysical Research (Space Physics)* **110** A09S03
[32] Hu Q, Farrugia C J, Osherovich V A, Möstl C, Szabo A, Ogilvie K W and Lepping R P 2013 *Solar Physics* **284** 275–291
[33] Hu Q, Qiu J, Dasgupta B, Khare A and Webb G M 2014 *The Astrophysical Journal* **793** 53 (*Preprint* `1408.1470`)
[34] Hu Q, Qiu J and Krucker S 2015 *Journal of Geophysical Research: Space Physics* **120** 1
[35] Hu Q 2016 *International Solar Wind 14 Conference* (*AIP Conf. Series* vol 1720) ed Wang L, Bruno R, Möbius E, Vourlidas A and Zank G p 040005
[36] Hu Q 2016 *Solar Physics* Submitted
[37] Khabarova O V, Zank G P, Malandraki O E, Li G, le Roux J A and Webb G M 2017 *Sun and Geosphere* **12** 23–30
[38] Linton M G 2006 *Journal of Geophysical Research: Space Physics* **111** n/a–n/a ISSN 2156-2202 a12S09 URL `http://dx.doi.org/10.1029/2006JA011891`